Vassilis G. Kaburlasos and Gerhard X. Ritter (Eds.)

Computational Intelligence Based on Lattice Theory

Studies in Computational Intelligence, Volume 67

Editor-in-chief
Prof. Janusz Kacprzyk
Systems Research Institute
Polish Academy of Sciences
ul. Newelska 6
01-447 Warsaw
Poland
E-mail: kacprzyk@ibspan.waw.pl

Vassilis G. Kaburlasos
Gerhard X. Ritter
(Eds.)

Computational Intelligence Based on Lattice Theory

With 101 Figures, 19 in Color and 52 Tables

 Springer

Professor Dr. Vassilis G. Kaburlasos
Department of Industrial Informatics
Division of Computing Systems
Technological Educational
  Institution of Kavala
65404 Kavala
Greece
E-mail:- vgkabs@teikav.edu.gr

Professor Dr. Gerhard X. Ritter
Department of Computer and Information
  Science and Engineering, CSE 364
University of Florida
Gainesville, FL 32611–6120
USA
E-mail:- ritter@cise.ufl.edu

Library of Congress Control Number: 2007927312

ISSN print edition: 1860-949X
ISSN electronic edition: 1860-9503
ISBN 978-3-540-72686-9 Springer Berlin Heidelberg New York

Springer is a part of Springer Science+Business Media

springer.com

© Springer-Verlag Berlin Heidelberg 2007

Cover design: deblik, Berlin
Typesetting by the SPi using a Springer LaTeX macro package
Printed on acid-free paper   SPIN: 11757986   89/SPi   5 4 3 2 1 0

to novelty

Preface

*A number of different instruments for design can be unified
in the context of lattice theory towards cross-fertilization*

By "lattice theory" [1] we mean, equivalently, either a *partial ordering* relation
[2, 3] or a *couple of binary algebraic* operations [3, 4]. There is a growing
interest in computational intelligence based on lattice theory.

A number of researchers are currently active developing lattice theory
based models and techniques in engineering, computer and information sciences, applied mathematics, and other scientific endeavours. Some of these
models and techniques are presented here.

However, currently, lattice theory is not part of the mainstream of computational intelligence. A major reason for this is the "learning curve" associated
with novel notions and tools. Moreover, practitioners of lattice theory, in specific domains of interest, frequently develop their own tools and/or practices
without being aware of valuable contributions made by colleagues. Hence,
(potentially) useful work may be ignored, or duplicated. Yet, other times,
different authors may introduce a conflicting terminology.

The compilation of this book is an initiative towards proliferating established knowledge in the hope to further expand it, soundly.

There was a critical mass of people and ideas engaged to produce this
book. Around two thirds of this book's chapters are substantial enhancements
of preliminary works presented lately in a three-part special session entitled
"Computational Intelligence Based on Lattice Theory" organized in the context of the World Congress in Computational Intelligence (WCCI), FUZZ-
IEEE program, July 16-21, 2006 in Vancouver, BC, Canada. The remaining
book chapters are novel contributions by other researchers.

This book is a balanced synthesis of four parts emphasizing, in turn,
neural computation, mathematical morphology, machine learning, and (fuzzy)
inference/logic.

PART I focuses on neural computation. More specifically, chapter 1 (by Kaburlasos) presents a granular enhancement of two popular neural classifiers, namely fuzzy-ART and SOM, based on lattice (partial) ordering.

Chapter 2 (by Ritter and Urcid) introduces novel algorithms for learning described in terms of lattice algebraic computations in dendritic (neural) structures according to recent discoveries in neuroscience.

Chapter 3 (by Barmpoutis and Ritter) demonstrates, successfully, novel dendritic lattice computing in difficult classification problems.

Chapter 4 (by Healy and Caudell) uses generalized lattices, or equivalently categories, in neural networks in order to model distributed world semantics.

PART II focuses on mathematical morphology applications. In particular, chapter 5 (by Urcid and Ritter) demonstrates the capacity of novel lattice matrix associative memory techniques to recall images degraded by noise.

Chapter 6 (by Graña, Villaverde, Moreno, and Albizuri) introduces a new feature extraction technique, which is employed for visual pattern recognition.

Chapter 7 (by De Witte, Schulte, Nachtegael, Mélange, and Kerre) presents a new approach for extending mathematical morphology from binary- to both greyscale- and color-images including also an image denoising method.

Chapter 8 (by Sussner and Valle) describes the storage and recall phases of morphological and certain fuzzy morphological associative memories including applications to classification and prediction.

PART III focuses on machine learning applications. In particular, chapter 9 (by Athanasiadis) introduces a rule-based perspective for fuzzy lattice reasoning (FLR) including also two environmental data mining applications.

Chapter 10 (by Petridis and Syrris) investigates the application of fuzzy lattice neurocomputing (FLN) in an environmental prediction problem.

Chapter 11 (by Piedra-Fernández, Cantón-Garbín, and Guindos-Rojas) demonstrates applications of a fuzzy lattice classifier to ocean satellite images.

Chapter 12 (by Al-Daraiseh, Kaylani, Georgiopoulos, Mollaghasemi, Wu, and Anagnostopoulos) presents three genetically optimized neural algorithms for classification applicable to a lattice of hyperboxes.

Chapter 13 (by Cripps and Nguyen) uses fuzzy lattice reasoning (FLR) classifier with similarity measures whose effectiveness is shown experimentally.

PART IV focuses on logic and inference. More specifically, chapter 14 (by Munoz-Hernandez and Vaucheret) enhances Fuzzy Prolog by default knowledge in order to represent incomplete information in logic programming.

Chapter 15 (by Knuth) considers lattice fuzzification by valuation functions towards a unification of probability- with information- theory.

Chapter 16 (by Hatzimichailidis and Papadopoulos) studies useful connections between L-fuzzy sets and intuitionistic fuzzy sets.

Chapter 17 (by Kehagias) studies extensions of both t-norms and t-conorms to a superlattice, the latter is a multi-valued analog of a lattice.

Chapter 18 (by Kehagias) describes the construction of fuzzy-valued both t-norms and t-conorms from families of multi-valued t-norms and t-conorms.

Algorithms are presented in several chapters. Often, there are cross references among book chapters. References and (possible) Appendices are shown per chapter. There is a single Index for all chapters at the end of this book.

We thank the authors for their contribution.

Kavala, Greece, EU
Gainesville, Florida, USA

Vassilis G. Kaburlasos
Gerhard X. Ritter

April 2007

References

1. Birkhoff G (1967) Lattice Theory, Colloquium Publications 25. American Mathematical Society, Providence, RI
2. Kaburlasos VG (2006) Towards a Unified Modeling and Knowledge-Representation Based on Lattice Theory — Computational Intelligence and Soft Computing Applications, Studies in Computational Intelligence 27. Springer, Heidelberg, Germany
3. Ritter GX, Gader PD (2006) Fixed points of lattice transforms and lattice associative memories. In: Hawkes P (ed) Advances in Imaging and Electron Physics 144:165–242. Elsevier, Amsterdam, The Netherlands
4. Ritter GX, Wilson JN (2000) Handbook of Computer Vision Algorithms in Image Algebra, 2nd ed. CRC Press, Boca Raton, FL

Contents

Part I Neural Computation

1 Granular Enhancement of Fuzzy-ART/SOM Neural Classifiers Based on Lattice Theory
Vassilis G. Kaburlasos 3

2 Learning in Lattice Neural Networks that Employ Dendritic Computing
Gerhard X. Ritter, Gonzalo Urcid 25

3 Orthonormal Basis Lattice Neural Networks
Angelos Barmpoutis, Gerhard X. Ritter 45

4 Generalized Lattices Express Parallel Distributed Concept Learning
Michael J. Healy, Thomas P. Caudell 59

Part II Mathematical Morphology Applications

5 Noise Masking for Pattern Recall Using a Single Lattice Matrix Associative Memory
Gonzalo Urcid, Gerhard X. Ritter 81

6 Convex Coordinates From Lattice Independent Sets for Visual Pattern Recognition
Manuel Graña, Ivan Villaverde, Ramon Moreno, and Francisco X. Albizuri 101

7 A Lattice-Based Approach to Mathematical Morphology for Greyscale and Colour Images
Valérie De Witte, Stefan Schulte, Mike Nachtegael, Tom Mélange, and Etienne E. Kerre 129

XII   Contents

8 Morphological and Certain Fuzzy Morphological Associative Memories for Classification and Prediction
Peter Sussner, Marcos Eduardo Valle 149

Part III Machine Learning Applications

9 The Fuzzy Lattice Reasoning (FLR) Classifier for Mining Environmental Data
Ioannis N. Athanasiadis 175

10 Machine Learning Techniques for Environmental Data Estimation
Vassilios Petridis, Vassilis Syrris................................195

11 Application of Fuzzy Lattice Neurocomputing (FLN) in Ocean Satellite Images for Pattern Recognition
J.A. Piedra-Fernández, M. Cantón-Garbín, F. Guindos-Rojas 215

12 Genetically Engineered ART Architectures
Ahmad Al-Daraiseh, Assem Kaylani, Michael Georgiopoulos, Mansooreh Mollaghasemi, Annie S. Wu, Georgios Anagnostopoulos 233

13 Fuzzy Lattice Reasoning (FLR) Classification Using Similarity Measures
Al Cripps, Nghiep Nguyen..263

Part IV Logic and Inference

14 Fuzzy Prolog: Default Values to Represent Missing Information
Susana Munoz-Hernandez, Claudio Vaucheret 287

15 Valuations on Lattices: Fuzzification and its Implications
Kevin H. Knuth 309

16 L-fuzzy Sets and Intuitionistic Fuzzy Sets
Anestis G. Hatzimichailidis, Basil K. Papadopoulos 325

17 A Family of Multi-valued t-norms and t-conorms
Athanasios Kehagias 341

18 The Construction of Fuzzy-valued t-norms and t-conorms
Athanasios Kehagias 361

Index 371

List of Contributors

Albizuri, Francisco X.
(chapter 6)
Universidad del País Vasco
Dept Ciencias Comp e Intel Artif
20080 San Sebastian, Spain, EU
ccpgrrom@si.ehu.es

Al-Daraiseh, Ahmad
(chapter 12)
University of Central Florida
School Electr Eng & Comp Science
Orlando, FL 32816, USA
ah1001@hotmail.com

Anagnostopoulos, Georgios
(chapter 12)
Florida Institute of Technology
Dept Electr & Computer Engineering
Melbourne, FL 32796, USA
georgio@fit.edu

Athanasiadis, Ioannis N.
(chapter 9)
Istituto Dalle Molle di Studi
sull'Intelligenza Artificiale
6928 Lugano, Switzerland
ioannis@athanasiadis.info

Barmpoutis, Angelos
(chapter 3)
University of Florida

Dept Comp & Info Science & Eng
Gainesville, FL 32611, USA
abarmpou@cise.ufl.edu

Cantón-Garbín, M.
(chapter 11)
University of Almería
Dept Languages and Computation
04120 Almería, Spain, EU
mcanton@ual.es

Caudell, Thomas P.
(chapter 4)
University of New Mexico
Dept Computer Science and Dept
Electr & Computer Engineering
Albuquerque, NM 87131, USA
tpc@ece.unm.edu

Cripps, Al
(chapter 13)
Middle Tennessee State University
Dept Computer Science
Murfreesboro, TN 37132, USA
acripps@mtsu.edu

De Witte, Valérie
(chapter 7)
Ghent University
Dept Appl Math & Comp Science
9000 Gent, Belgium, EU
Valerie.DeWitte@UGent.be

Georgiopoulos, Michael
(chapter 12)
University of Central Florida
Dept Electr Eng & Comp Science
Orlando, FL 32816, USA
michaelg@mail.ucf.edu

Graña, Manuel
(chapter 6)
Universidad del País Vasco
Dept Ciencias Comp e Intel Artif
20080 San Sebastian, Spain, EU
ccpgrrom@si.ehu.es

Guindos-Rojas, F.
(chapter 11)
University of Almería
Dept Languages and Computation
04120 Almería, Spain, EU
fguindos@ual.es

Hatzimichailidis, Anestis G.
(chapter 16)
Democritus University of Thrace
Dept Civil Engineering
67100 Xanthi, Greece, EU
hatz_ane@hol.gr

Healy, Michael J.
(chapter 4)
University of New Mexico
Dept Electr & Computer Engineering
Albuquerque, NM 87131, USA
mjhealy@ece.unm.edu

Kaburlasos, Vassilis G.
(chapter 1)
Techn Educ Inst of Kavala
Dept Industrial Informatics
65404 Kavala, Greece, EU
vgkabs@teikav.edu.gr

Kaylani, Assem
(chapter 12)
University of Central Florida
School Electr Eng & Comp Science
Orlando, FL 32816, USA
akaylani@productivityapex.com

Kehagias, Athanasios
(chapters 17, 18)
Aristotle University of Thessaloniki
Faculty of Engineering
Dept Math, Phys & Comp Sciences
54006 Thessaloniki, Greece, EU
kehagiat@auth.gr

Kerre, Etienne E.
(chapter 7)
Ghent University
Dept Appl Math & Comp Science
9000 Gent, Belgium, EU
Etienne.Kerre@UGent.be

Knuth, Kevin H.
(chapter 15)
University at Albany (SUNY)
Dept Physics
Albany, NY 12222, USA
kknuth@albany.edu

Mélange, Tom
(chapter 7)
Ghent University
Dept Appl Math & Comp Science
9000 Gent, Belgium, EU
Tom.Melange@UGent.be

Mollaghasemi, Mansooreh
(chapter 12)
University of Central Florida
Dept Industr Eng & Managem Syst
Orlando, FL 32816, USA
mmollagha@mail.ucf.edu

Moreno, Ramon
(chapter 6)
Universidad del País Vasco
Dept Ciencias Comp e Intel Artif
20080 San Sebastian, Spain, EU
ccpgrrom@si.ehu.es

Munoz-Hernandez, Susana
(chapter 14)
Universidad Politécnica de Madrid
Dept Leng, Sist Info e Ing Software
28660 Madrid, Spain, EU
susana@fi.upm.es

Nachtegael, Mike
(chapter 7)
Ghent University
Dept Appl Math & Comp Science
9000 Gent, Belgium, EU
Mike.Nachtegael@UGent.be

Nguyen, Nghiep
(chapter 13)
Middle Tennessee State University
Dept Economics and Finance
Murfreesboro, TN 37132, USA
nguyen@mtsu.edu

Papadopoulos, Basil K.
(chapter 16)
Democritus University of Thrace
Dept Civil Engineering
67100 Xanthi, Greece, EU
papadob@civil.duth.gr

Petridis, Vassilios
(chapter 10)
Aristotle University of Thessaloniki
Dept Electr & Computer Engineering
54006 Thessaloniki, Greece, EU
petridis@eng.auth.gr

Piedra-Fernández, J.A.
(chapter 11)
University of Almería
Dept Languages and Computation
04120 Almería, Spain, EU
jpiedra@ual.es

Ritter, Gerhard X.
(chapters 2, 3, 5)
University of Florida
Dept Comp & Info Science & Eng
Gainesville, FL 32611, USA
ritter@cise.ufl.edu

Schulte, Stefan
(chapter 7)
Ghent University
Dept Appl Math & Comp Science
9000 Gent, Belgium, EU
Stefan.Schulte@UGent.be

Sussner, Peter
(chapter 8)
State University of Campinas
Inst Math Stat & Sci Computing
13081 Campinas, Brazil
sussner@ime.unicamp.br

Syrris, Vassilis
(chapter 10)
Aristotle University of Thessaloniki
Dept Electr & Computer Engineering
54006 Thessaloniki, Greece, EU
vsyrris@auth.gr

Urcid, Gonzalo
(chapters 2, 5)
Nat Inst Astroph Optics & Electron
Dept Optics
Puebla 72000, Mexico
gurcid@inaoep.mx

Valle, Marcos Eduardo
(chapter 8)
State University of Campinas
Inst Math Stat & Sci Computing
13081 Campinas, Brazil
mevalle@ime.unicamp.br

Vaucheret, Claudio
(chapter 14)
Universidad Nacional del Comahue
Dept Ciencias Computación
Buenos Aires 1400, Argentina
vaucheret@gmail.com

Villaverde, Ivan
(chapter 6)
Universidad del País Vasco
Dept Ciencias Comp e Intel Artif
20080 San Sebastian, Spain, EU
ccpgrrom@si.ehu.es

Wu, Annie S.
(chapter 12)
University of Central Florida
School Electr Eng & Comp Science
Orlando, FL 32816, USA
aswu@cs.ucf.edu

Part I

Neural Computation

Granular Enhancement of Fuzzy-ART/SOM Neural Classifiers Based on Lattice Theory

Vassilis G. Kaburlasos

Technological Educational Institution of Kavala, Dept. of Industrial Informatics
65404 Kavala, Greece, EU
vgkabs@teikav.edu.gr

Summary. Fuzzy adaptive resonance theory (fuzzy-ART) and self-organizing map (SOM) are two popular neural paradigms, which compute lattice-ordered granules. Hence, lattice theory emerges as a basis for unified analysis and design. We present both an enhancement of fuzzy-ART, namely fuzzy lattice reasoning (FLR), and an enhancement of SOM, namely granular SOM (grSOM). FLR as well as grSOM can rigorously deal with (fuzzy) numbers as well as with intervals. We introduce inspiring novel interpretations. In particular, the FLR is interpreted as a reasoning scheme, whereas the grSOM is interpreted as an energy function minimizer. Moreover, we can introduce tunable nonlinearities. The interest here is in classification applications. We cite evidence that the proposed techniques can clearly improve performance.

1.1 Introduction

Two novel approaches to neural computing were proposed lately by different authors based on lattice theory [22, 23, 35]. More specifically, one approach is based on the "order" definition for a lattice [22, 23], whereas the other one is based on the "algebra" definition for a lattice [35] as explained later. This chapter builds on the former definition; furthermore, it proposes granular enhancements of two popular neural paradigms, namely adaptive resonance theory (ART) and self-organizing map (SOM).

Stephen Grossberg, the founder of ART, points out occassionally an inherent affinity of the biologically-motivated clustering mechanisms of ART with Kohonen's SOM [26]. This work shows yet another aspect of the aforementioned affinity since both ART and SOM can be studied analytically (and also be further improved) based on mathematical lattice theory. Here we consider the fuzzy enhancement of ART known as fuzzy-ART [5].

The operation of both fuzzy-ART and SOM is based on the computation of clusters in R^N. In particular, a cluster for fuzzy-ART corresponds to a hyperbox, whereas a cluster for SOM corresponds to a Voronoi region in a metric space. Another term for cluster is *granule*. Lately, there is a growing interest in *granular computing* as explained next.

V.G. Kaburlasos: *Granular Enhancement of Fuzzy-ART/SOM Neural Classifiers Based on Lattice Theory*, Studies in Computational Intelligence (SCI) **67**, 3–23 (2007)
www.springerlink.com   © Springer-Verlag Berlin Heidelberg 2007

Granular computing is a category of theories, methodologies, techniques and tools that make use of *information granules* in the process of problem solving. Where, an (information) granule can be conceived as a collection of entities grouped together by similarity, functional adjacency, indistinguishability, coherency, etc. The basic notions and principles of granular computing have appeared under different names in many related fields such as information hiding in programming, granularity in artificial intelligence, divide and conquer in theoretical computer science, interval computing, cluster analysis, fuzzy and rough set theories, and many other. Granular computing is an emerging computational paradigm [27, 30].

Granules in R^N are partially-ordered, in particular they are lattice-ordered. Hence, lattice theory emerges as a basis for analysis and design in granular computing. This work shows granular enhancements of the popular fuzzy-ART and SOM classifiers. Moreover, novel interpretations are introduced. In particular, fuzzy-ART is interpreted as an interactive reasoning scheme, whereas SOM is interpreted as an energy minimizer. Note also that the techniques presented here could be useful elsewhere in granular computing.

This chapter is organized as follows. Section 1.2 summarizes, in context, the learning mechanisms of fuzzy-ART and SOM. Section 1.3 covers the mathematics required for describing the enhancements proposed later. Section 1.4 describes enhancements of both fuzzy-ART and SOM. Section 1.5 summarizes, in perspective, the contribution of this work.

1.2 Fuzzy-ART and SOM

This section summarizes the operation of both *fuzzy Adaptive Resonance Theory (fuzzy-ART)* and *Self-Organizing Map (SOM)* for unsupervised learning, i.e. clustering. It also presents interesting extensions by different authors.

1.2.1 Fuzzy-ART Operation

The original fuzzy-ART neural network regards a two-layer architecture [5]. Layer F1 of fuzzy-ART fans out an input vector to the fully-interconnected, competitive neurons in layer F2. A layer F2 neuron filters an input vector $\mathbf{x}$ by computing vector $\mathbf{x} \wedge \mathbf{w}$, where $\mathbf{w}$ is the *code* (vector) stored on interlayer links. More specifically, an entry of vector $\mathbf{x} \wedge \mathbf{w}$ equals the minimum of the corresponding (positive real number) entries of vectors $\mathbf{x}$ and $\mathbf{w}$. Algorithm fuzzy-ART for training (learning) by clustering is briefly described next.

Algorithm fuzzy-ART for training

ART-1: Do while there are more inputs.

Apply the *complement coding* technique in order to represent input $[x_{i,1}, ..., x_{i,N}] \in [0, 1]^N$ by $\mathbf{x}_i = [x_{i,1}, ..., x_{i,N}, 1 - x_{i,1}, ..., 1 - x_{i,N}] \in$

R^{2N}, $i = 1, ..., n$. Then, present $\mathbf{x}_i$ to the (initially) "set" neurons in layer F2.

ART-2: Each layer F2 neuron with code $\mathbf{w}_j \in \mathsf{R}^{2N}$ computes its *choice* (*Weber*) function $T_j = |\mathbf{x}_i \wedge \mathbf{w}_j|/(\alpha + |\mathbf{w}_j|)$.

ART-3: If there are no "set" neurons in layer F2 then memorize input $\mathbf{x}_i$. Else, competition among the "set" neurons in layer F2: Winner is neuron J such that $T_J \doteq \underset{j}{arg\mathsf{max}}T_j$.

ART-4: *Similarity Test*: $(|\mathbf{x}_i \wedge \mathbf{w}_J|/|\mathbf{x}_i|) \geq \rho$, where $|\mathbf{x}_i \wedge \mathbf{w}_J|/|\mathbf{x}_i|$ is the *match* function and $\rho \in (0, 1]$ is the user-defined *vigilance parameter*.

ART-5: If the *Similarity Test* is not satisfied then "reset" the winner neuron; goto step ART-3 to search for another winner. Else, replace the winner neuron code $\mathbf{w}_J$ by $\mathbf{x}_i \wedge \mathbf{w}_J$; goto step ART-1.

We remark that $|\mathbf{x}|$ above equals, by definition, the sum of vector $\mathbf{x}$ (positive) entries. Parameter "α" in the choice (Weber) function T_j is a very small positive number whose role is to break ties in case of multiple winners [16].

As soon as training (learning) completes, each neuron defines a cluster by a hyperbox. It follows algorithm fuzzy-ART for testing (generalization).

Algorithm fuzzy-ART for testing

art-1: Feed an input vector $\mathbf{x}_0 = [x_{0,1}, ..., x_{0,N}, 1 - x_{0,1}, ..., 1 - x_{0,N}] \in \mathsf{R}^{2N}$.

art-2: A layer F2 neuron with code $\mathbf{w}_j \in \mathsf{R}^{2N}$ computes the choice (Weber) function $|\mathbf{x}_0 \wedge \mathbf{w}_j|/(\alpha + |\mathbf{w}_j|)$.

art-3: Competition among the neurons in layer F2: Winner is neuron J such that $T_J \doteq \underset{j}{arg\mathsf{max}}T_j$. Assign input $\mathbf{x}_0$ to the cluster represented by neuron J.

1.2.2 SOM Operation

Kohonen's self-organizing map (SOM) architecture [26] includes a two dimensional $L \times L$ *grid* (or, *map*) of *neurons* (or, *cells*). Each cell $C_{i,j}$ stores a vector $\mathbf{m}_{i,j} = [m_{i,j,1}, ..., m_{i,j,N}]^T \in \mathsf{R}^N$, $i = 1, ..., L$, $j = 1, ..., L$. Vectors $\mathbf{m}_{i,j}$ are called *code vectors* and they are initialized randomly. A version of algorithm SOM for training (learning) by clustering is briefly described next.

Algorithm SOM for training

SOM-1: Memorize the first input datum $\mathbf{x}_1 \in \mathsf{R}^N$ by committing, randomly, a neuron on the $L \times L$ grid. Repeat the following steps a user-defined number N_{epochs} of epochs, $t = 1, ..., N_{epochs}$.

SOM-2: For each training datum $\mathbf{x}_k \in \mathsf{R}^N$, $k = 1, ..., n$ carry out the following computations.

SOM-3: Calculate the Euclidean distance $d(\mathbf{m}_{i,j}, \mathbf{x}_k)$, $i, j \in \{1, ..., L\}$.

6   V.G. Kaburlasos

SOM-4: Competition among the neurons on the $L \times L$ grid: Winner is neuron $(I, J) \doteq arg \min\limits_{i,j \in \{1,...,L\}} d_1(\mathbf{m}_{i,j}, \mathbf{x}_k)$.

SOM-5: *Assimilation Condition*: Vector $\mathbf{m}_{i,j}$ is in the neighborhood of vector $\mathbf{m}_{I,J}$ on the $L \times L$ grid.

SOM-6: If the *Assimilation Condition* is satisfied then compute a new value $\mathbf{m}'_{i,j}$:

$$\mathbf{m}'_{i,j} = \mathbf{m}_{i,j} + a(t)(\mathbf{x}_k - \mathbf{m}_{i,j}) = [1 - a(t)]\mathbf{m}_{i,j} + a(t)\mathbf{x}_k, \qquad (1.1)$$

where $a(t) \in (0, 1)$ is a decreasing function in time (t).

As soon as training (learning) completes, each cell $C_{i,j}$ defines a cluster by a code vector $\mathbf{m}_{i,j}$. It follows algorithm SOM for testing (generalization).

Algorithm SOM for testing

som-1: Present an input $\mathbf{x}_0 \in \mathsf{R}^N$, $k = 1, ..., n$ to the neurons of the $L \times L$ grid.

som-2: Calculate the Euclidean distance $d(\mathbf{m}_{i,j}, \mathbf{x}_0)$, $i, j \in \{1, ..., L\}$.

som-3: Competition among the neurons on the $L \times L$ grid: Winner is neuron $(I, J) \doteq arg \min\limits_{i,j \in \{1,...,L\}} d_1(\mathbf{m}_{i,j}, \mathbf{x}_k)$. Assign input $\mathbf{x}_0$ to the cluster represented by neuron J.

Note that the set of clusters computed during training by both fuzzy-ART [10, 16] and SOM depends on the order of data presentation.

1.2.3 Extensions by Different Authors in Context

Both fuzzy-ART and SOM compute information granules. In particular, fuzzy-ART computes fuzzy-sets with hyperbox cores, whereas SOM partitions its data domain in Voronoi-regions.

Both fuzzy-ART and SOM for clustering have been extended by a number of authors. More specifically, on the one hand, fuzzy-ARTMAP has been proposed for supervised learning [6]. Improvements of fuzzy-ARTMAP were proposed in various learning applications [7, 8]. Furthermore, fuzzy-ART(MAP) has inspired various min-max neural networks [2, 18]. An interesting probabilistic analysis of fuzzy-ARTMAP was proposed lately using martingales [1]. An extension of the fuzzy-ART(MAP) algorithm to a mathematical lattice data domain is the FLR algorithm [19]. Note that the FLR algorithm can be implemented as a neural network towards fuzzy lattice neurocomputing, or FLN for short [18, 23, 32]. On the other hand, SOM has been popular in signal- and other information-processing applications [26]. Various SOM extensions have been proposed including nonEuclidean metrics [31], weighting factors [13], etc. A different SOM extension is the generative topographic mapping (GTM) based on a constrained mixture of Gaussians [4]. However, conventional SOM as well as its extensions cannot cope with ambiguity. In response, SOM-based fuzzy c-means algorithms have been proposed [24]. Note that an early employment of SOM in fuzzy inference systems (FISs) appeared [37]. Lately, SOM extensions were presented for FIS analysis and design based on positive FINs [21, 29], the latter (FINs) are presented below.

1.3 Mathematical Background

This section covers lattice theory mathematics necessary for introducing novel enhancements later.

1.3.1 Crisp lattices

We present two equivalent definitions for a mathematical *lattice* [3], namely *order-based-* and *algebra-based-* definition, respectively. The former is based on the notion *partially ordered set*, or *poset* for short, defined in the Appendix.

By "a covers b" in a poset $(P, \leq)$ it is meant that $b < a$ but $b < x < a$ for no $x \in P$. Let $(P, \leq)$ be a poset with least element O. Every $x \in P$ which covers O, if such x exists, is called *atom*.

Let $(P, \leq)$ be a poset and $X \subseteq P$. An *upper bound* of X is a $a \in P$ with $x \leq a$, $\forall x \in X$. The *least upper bound* (l.u.b.), if it exists, is the unique upper bound contained in every upper bound. The l.u.b. is also called *supremum* or *lattice join* of X and denoted by $\sup X$ or $\vee X$. The notions *lower bound* of X and *greatest lower bound* (g.l.b.) of X are defined dually. The g.l.b. is also called *infimum* or *lattice meet* of X and denoted by $\inf X$ or $\wedge X$.

The *order-based* definition for a lattice follows.

Definition 1.1 *A lattice is a poset $(\mathsf{L}, \leq)$ any two of whose elements have both a greatest lower bound (g.l.b.), denoted by $x \wedge y$, and a least upper bound (l.u.b.), denoted by $x \vee y$. A lattice $(\mathsf{L}, \leq)$, or equivalently* crisp lattice, *is called* complete *when each of its subsets X has a l.u.b. and a g.l.b. in L.*

Setting $X = \mathsf{L}$ in definition 1.1 it follows that a nonvoid complete lattice contains both a *least* element and a *greatest* element denoted, respectively, by O and I. By definition, an *atomic lattice* $(\mathsf{L}, \leq)$ is a complete lattice in which every element is a joint of atoms.

The *algebra-based* definition for a lattice follows based on the notion *algebra* defined in the Appendix to this chapter.

Definition 1.2 *An algebra with two binary operations which satisfy L1-L4 is a lattice, and conversely.*

$$
\begin{array}{llll}
(L1) & x \wedge x = x & x \vee x = x & \text{(Idempotent)} \\
(L2) & x \wedge y = y \wedge x & x \vee y = y \vee x & \text{(Commutative)} \\
(L3) & x \wedge (y \wedge z) = (x \wedge y) \wedge z & & \text{(Associative)} \\
& x \vee (y \vee z) = (x \vee y) \vee z & & \\
(L4) & x \wedge (x \vee y) = x & x \vee (x \wedge y) = x & \text{(Absorption)}
\end{array}
$$

We remark that definition 1.2 is popular in applications of mathematical morphology [12, 34, 36]. This work employs mainly definition 1.1.

Both definitions 1.1 and 1.2 regard a *crisp* lattice, where the binary relation $x \leq y$ is either true or false. In particular, if $x \leq y$ (or, $y \leq x$) then x and y are called *comparable*; otherwise, x and y are called *incomparable* or *parallel*, symbolically $x \| y$. A simple crisp lattice example is shown next.

Example 1.1 *Let $\mathcal{P}(A) = \{\{\}, \{a\}, \{b\}, \{c\}, \{a, b\}, \{b, c\}, \{a, c\}, \{a, b, c\}\}$ be the power set of set $A = \{a, b, c\}$. It turns out that $(\mathcal{P}(A), \subseteq)$ is a complete lattice, ordered by set-inclusion, with* least *and* greatest *elements $O = \{\}$ and $I = \{a, b, c\}$, respectively. Figure 1.1 shows a Hasse diagram of lattice $(\mathcal{P}(A), \subseteq)$ such that a line segment connects two sets X (below) and Y (above) if and only if Y covers X. A* Hasse- *or, equivalently,* line- *diagram can be drawn only for a finite lattice.*

The inverse $\geq$ of an order relation $\leq$ is itself an order relation. More specifically, the order $\geq$ is called *dual* (order) of $\leq$, symbolically also $\leq^{\partial}$, or $\leq^{-1}$. Furthermore, the Cartesian product $(\mathsf{L}, \leq) = (\mathsf{L}_1, \leq_1) \times \dots \times (\mathsf{L}_N, \leq_N)$ of N *constituent* lattices $(\mathsf{L}_1, \leq_1), \dots, (\mathsf{L}_N, \leq_N)$ is a lattice [3]. In particular, if both $(a_1, \dots, a_N)$ and $(b_1, \dots, b_N)$ are in L then $(a_1, \dots, a_N) \leq (b_1, \dots, b_N)$ if and only if $a_i \leq b_i$, $i = 1, \dots, N$.

Of particular interest here is lattice $(\tau(\mathsf{L}), \leq)$, where $\tau(\mathsf{L})$ denotes the set of intervals in L (including also the empty set) partially-ordered by

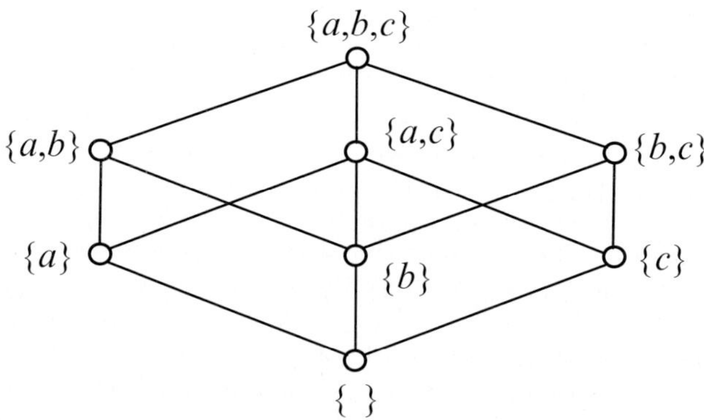

Fig. 1.1. Hasse diagram of the partially-ordered, complete lattice $(\mathcal{P}(A), \subseteq)$, where "$\mathcal{P}(A)$" is the powerset of set $A = \{a, b, c\}$ and "$\subseteq$" is the set-inclusion relation

set-inclusion — For a definition of an interval see in the Appendix. One way of dealing with lattice $(\tau(\mathsf{L}), \leq)$ is based on the product lattice $(\mathsf{L}^\partial \times \mathsf{L}, \leq^\partial \times \leq)$ [23].

1.3.2 Fuzzy lattices

The binary relation "$\leq$" in a crisp lattice can be fuzzified resulting in a *fuzzy lattice* as explained next — Note that a *fuzzy set* is denoted here by a pair (U, μ), where U is the *universe of discourse* and μ is a function $\mu : U \to [0, 1]$, namely *membership* function.

Definition 1.3 *A fuzzy lattice is a triple* $(\mathsf{L}, \leq, \mu)$, *where* $(\mathsf{L}, \leq)$ *is a crisp lattice and* $(\mathsf{L} \times \mathsf{L}, \mu)$ *is a fuzzy set such that* $\mu(x, y) = 1$ *if and only if* $x \leq y$.

Function μ in definition 1.3 is a *weak (fuzzy) partial order* relation in the sense that both $\mu(x, y) = 1$ and $\mu(y, z) = 1$ imply $\mu(x, z) = 1$, whereas if either $\mu(x, y) \neq 1$ or $\mu(y, z) \neq 1$ then $\mu(x, z)$ could be any number in $[0, 1]$. Fuzzification of a lattice can be pursued using either a *generalized zeta function* [25] or an *inclusion measure* function. The latter is defined next [19].

Definition 1.4 *Let* $(\mathsf{L}, \leq)$ *be a complete lattice with least element* O. *An inclusion measure is a map* $\sigma : \mathsf{L} \times \mathsf{L} \to [0, 1]$, *which satisfies the following conditions.*
$(IM0)$ $\sigma(x, O) = 0, x \neq O$
$(IM1)$ $\sigma(x, x) = 1, \forall x \in \mathsf{L}$
$(IM2)$ $x \wedge y < x \Rightarrow \sigma(x, y) < 1$
$(IM3)$ $u \leq w \Rightarrow \sigma(x, u) \leq \sigma(x, w)$ (Consistency Property)

For noncomplete lattices condition $(IM0)$ drops.

We remark that $\sigma(x, y)$ may be interpreted as a (fuzzy) degree of inclusion of x in y. Therefore, notations $\sigma(x, y)$ and $\sigma(x \leq y)$ are used interchangably. Alternative inclusion measure function definitions have been proposed by different authors [9]. If $\sigma : \mathsf{L} \times \mathsf{L} \to [0, 1]$ is an inclusion measure, in the sense of definition 1.4, then $(\mathsf{L}, \leq, \sigma)$ is a fuzzy lattice [18, 19].

1.3.3 Useful functions in a lattice

An inclusion measure can be defined in a crisp lattice $(\mathsf{L}, \leq)$ based on a *positive valuation* function (the latter is defined in the Appendix) as shown next.

Theorem 1.1 *If* $v : \mathsf{L} \to R$ *is a positive valuation function in a crisp lattice* $(\mathsf{L}, \leq)$ *then both functions (a)* $k(x, u) = v(u)/v(x \vee u)$, *and (b)* $s(x, u) = v(x \wedge u)/v(x)$ *are inclusion measures.*

We point out that a positive valuation in a crisp lattice $(\mathsf{L}, \leq)$ also defines a *metric* function $d : \mathsf{L} \times \mathsf{L} \to R_0^+$ given by $d(x, y) = v(x \vee y) - v(x \wedge y)$ — For a definition of a metric see in the Appendix to this chapter.

Given (i) a product lattice $(\mathsf{L}, \leq) = (\mathsf{L}_1, \leq_1) \times ... \times (\mathsf{L}_N, \leq_N)$, and (ii) both a positive valuation $v : \mathsf{L}_i \to \mathsf{R}$ and an *isomorphic* function $\theta_i : \mathsf{L}_i^{\partial} \to \mathsf{L}_i$ in a constituent lattice $(\mathsf{L}_i, \leq_i)$, $i = 1, ..., N$ — for a definition of an isomorphic function see in the Appendix to this chapter — then: (1) A positive valuation $v : \mathsf{L} \to \mathsf{R}$ is given by $v(x_1, ..., x_N) = v_1(x_1) + ... + v_N(x_N)$, (2) an isomorphic function $\theta : \mathsf{L}^{\partial} \to \mathsf{L}$ is given by $\theta(x_1, ..., x_N) = (\theta_1(x_1), ..., \theta_N(x_N))$, and (3) countably infinite *Minkowski metrics* d_p are given in L by

$$d_p(\mathbf{x}, \mathbf{y}) = [d_1^p(x_1, y_1) + ... + d_N^p(x_N, y_N)]^{1/p}, \tag{1.2}$$

where $p = 1, 2, ...$ and $d_i(x_i, y_i) = v_i(x_i \vee y_i) - v_i(x_i \wedge y_i)$, $x_i, y_i \in \mathsf{L}_i$, $i = 1, ..., N$. In the following, interest focuses on lattices stemming from the set R of real numbers.

1.3.4 Lattices stemming from R

Three different lattice examples are shown in Examples 1.2, 1.3, and 1.4 next including geometric interpretations on the plane.

Example 1.2 *Consider the set R of real numbers represented by a line (Fig. 1.2). It turns out that $(\mathsf{R}, \leq)$ is a noncomplete lattice including only comparable elements. Hence, lattice $(\mathsf{R}, \leq)$ is called* totally-ordered *or, equivalently,* chain. *Of particular interest is the complete sublattice $(\mathsf{I} = [0, 1], \leq)$.*

Example 1.3 *Lattices of interest are both $(\tau(\mathsf{R}), \leq)$ and $(\tau(\mathsf{I}), \leq)$, where $\tau(\mathsf{R})$ and $\tau(\mathsf{I})$ denote the set of (closed) intervals in R and I, respectively. Consider the set of* hyperrectangles *or, equivalently,* hyperboxes, *in the partially-ordered lattice R^N. It turns out that a hyperbox is a (lattice) interval in R^N. Moreover, $(\tau(\mathsf{R}^N), \leq)$ denotes the noncomplete lattice of hyperboxes in R^N. Note that $\tau(\mathsf{R}^N) = [\tau(\mathsf{R})]^N$. Of particular interest is complete lattice $(\mathsf{I}^N, \leq)$, namely* unit-hypercube. *The corresponding complete lattice of hyperboxes is denoted by $(\tau(\mathsf{I}^N), \leq) \equiv ([\tau(\mathsf{I})]^N, \leq)$. Figure 1.3 shows (hyper)boxes in lattice $(\tau(\mathsf{R}^N), \leq)$ for $N = 2$ (i.e. the plane). The unit-square is also shown.*

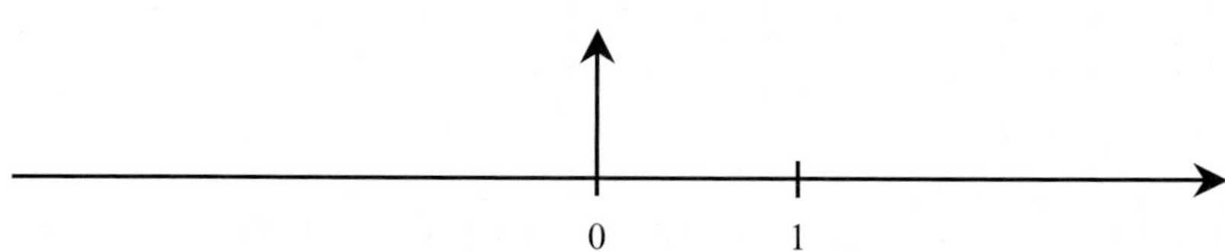

Fig. 1.2. The totally-ordered, noncomplete lattice $(\mathsf{R}, \leq)$ of real numbers. Note that lattice $(\mathsf{I}, \leq)$, where $\mathsf{I} = [0, 1]$, is a complete one

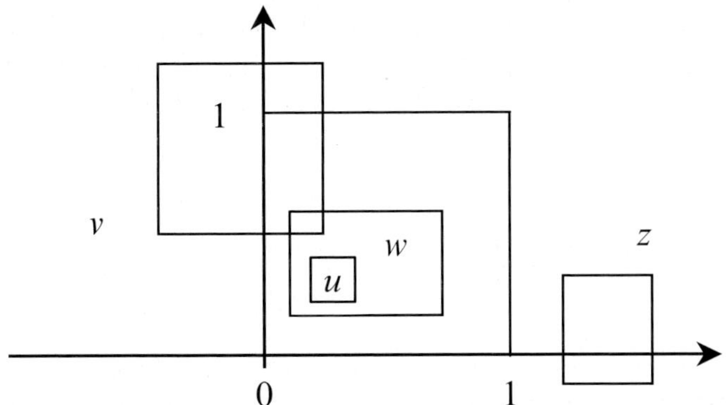

Fig. 1.3. Partially-ordered hyperboxes in the noncomplete lattice $(\mathsf{R}^N, \leq)$ are shown for $N = 2$ (i.e. the plane). The complete lattice unit-square is also shown. Box u is included in box w, i.e. $u \leq w$; all the other boxes are *incomparable*, e.g. $w\|z$, etc

The *diagonal* of a hyperbox in R^N is defined as follows.

Definition 1.5 *The* diagonal *of a hyperbox* $[a, b]$ *in* R^N, *where* $a, b \in R^N$ *with* $a \leq b$, *is defined as a nonnegative real function* $\mathrm{diag}_p : \tau(R^N) \to R_0^+$ *given by* $diag_p([a, b]) = d_p(a, b)$, $p = 1, 2, ...$

1.3.5 Lattices of generalized intervals

Definition 1.6 *(a) A* positive generalized interval *of height h is a map* $\mu_{a,b}^h$:
$$R \to \{0, h\} \ given \ by \ \mu_{a,b}^h(x) = \begin{cases} h, \ a \leq x \leq b \\ \\ 0, \ otherwise \end{cases}, \ where \ h \in (0, 1]. \ (b) \ A$$
negative generalized interval *of height h is a map* $\mu_{a,b}^h : R \to \{0, -h\}$ *given by*
$$\mu_{a,b}^h(x) = \begin{cases} -h, \ a \geq x \geq b \\ \\ 0, \ otherwise \end{cases}, \ where \ a > b \ and \ h \in (0, 1].$$

Note that a generalized interval is a "box" function, either positive or negative. In the interest of simplicity a generalized interval will be denoted as $[a, b]^h$, where $a \leq b$ $(a > b)$ for a positive (negative) generalized interval.

The set of positive (negative) generalized intervals of height h is denoted by $\mathsf{M}_+^h (\mathsf{M}_-^h)$. The set of generalized intervals of height h is denoted by M^h, i.e. $\mathsf{M}^h = \mathsf{M}_-^h \cup \mathsf{M}_+^h$. It turns out that the set M^h of generalized intervals is *partially ordered*; more specifically, M^h is a *mathematical lattice* [17, 18] with *lattice meet* and *lattice join* given, respectively, by $[a, b]^h \wedge [c, d]^h = [a \vee c, b \wedge d]^h$ and $[a, b]^h \vee [c, d]^h = [a \wedge c, b \vee d]^h$. Moreover, the corresponding lattice order relation $[a, b]^h \leq [c, d]^h$ in M^h is equivalent to "$c \leq a$".AND."$b \leq d$".

Example 1.4 *Figure 1.4 shows elements of lattice M^h. In particular, Fig. 1.4 shows all combinations for generalized intervals of height h as detailed in [17, 18, 20].*

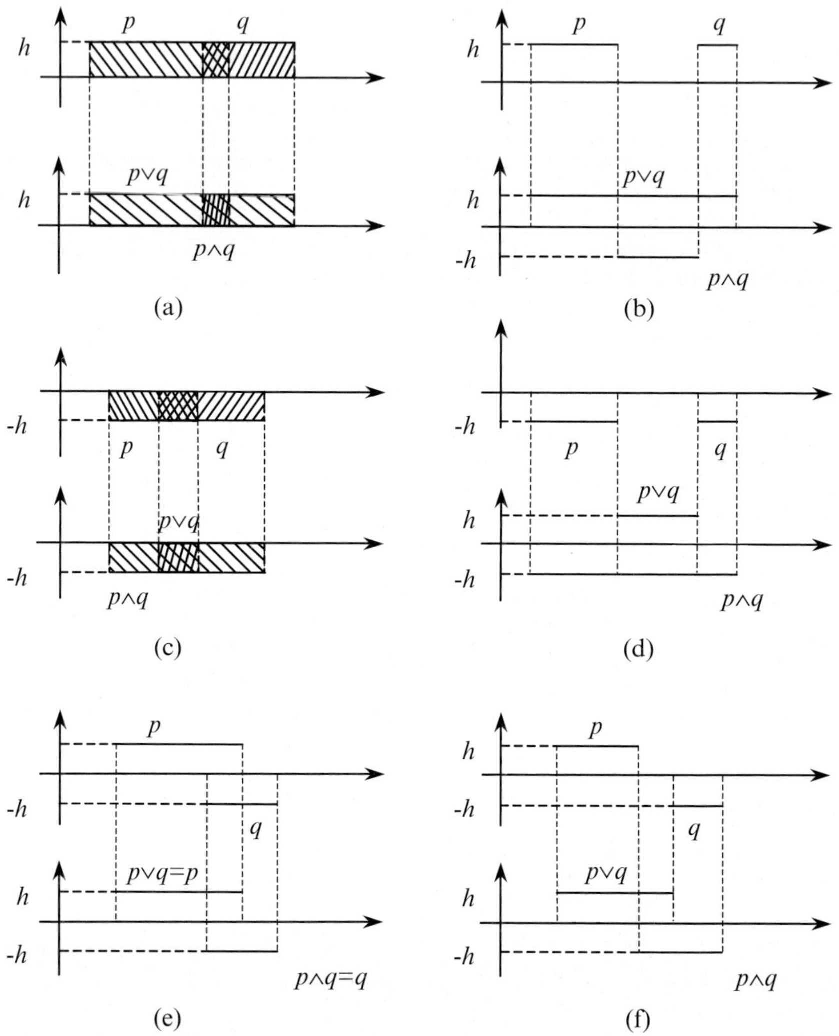

Fig. 1.4. Demonstrating the lattice- *join* $(p \vee q)$ and *meet* $(p \wedge q)$ for all different pairs (p, q) of generalized intervals of height h. Different fill-in patterns are used for partially overlapped generalized intervals. (a) "Intersecting" positive generalized intervals. (b) "Nonintersecting" positive generalized intervals. (c) "Intersecting" negative generalized intervals. (d) "Nonintersecting" negative generalized intervals. (e) "Intersecting" positive and negative generalized intervals. (f) "Nonintersecting" positive and negative generalized intervals

In the totally-ordered lattice R of real numbers any strictly increasing function $f_h : \mathsf{R} \to \mathsf{R}$ is a *positive valuation*, whereas any strictly decreasing function $\theta_h : \mathsf{R} \to \mathsf{R}$ is an *isomorphic* function. Given both f_h and θ_h, a positive valuation in lattice $(\mathsf{M}^h, \leq)$ is given by $v([a,b]^h) = f_h(\theta_h(a)) + f_h(b)$. Therefore, a metric between two generalized intervals is given by

$$d_h([a,b]^h, [c,d]^h) = [f_h(\theta_h(a \wedge c)) - f_h(\theta_h(a \vee c))] + [f_h(b \vee d) - f_h(b \wedge d)] \quad (1.3)$$

Choosing $\theta_h(x) = -x$ and f_h such that $f_h(x) = -f_h(-x)$ it follows $d_h([a,b]^h, [c,d]^h) = [f_h(a \vee c) - f_h(a \wedge c)] + [f_h(b \vee d) - f_h(b \wedge d)]$.

The set-union of all M^hs is the set M of *generalized intervals*, i.e. $\mathsf{M} = \underset{h \in (0,1]}{\cup} \mathsf{M}^h$. Our interest is in generalized intervals $[a,b]^h$ with $h \in (0,1]$ because the latter emerge from α-cuts of fuzzy numbers [18, 22]. It is interesting that different authors lately have considered the notion "α-fuzzy set" [33], the latter is identical to the notion "positive generalized interval" here.

The significance of a positive valuation function is demonstrated next.

Example 1.5 *Consider the positive generalized intervals $[-1,0]^1$ and $[3,4]^1$. Let $f_1(x) = x^3$ and $f_2(x) = (1 - e^{-x})/(1 + e^{-x})$ be the two strictly increasing functions shown in Fig. 1.5(a) and Fig. 1.5(b), respectively. Note that function $f_1(x)$ is steeply increasing, whereas function $f_2(x)$ is saturated. The computation of the (metric) distance $d_1([-1,0]^1, [3,4]^1)$ using $f_1(x)$ equals $d_1([-1,0]^1, [3,4]^1) = [f_1(3) - f_1(-1)] + [f_1(4) - f_1(-0)] = 65 + 27 = 92$. Whereas, the computation of the (metric) distance $d_1([-1,0]^1, [3,4]^1)$ using $f_2(x)$ equals $d_1([-1,0]^1, [3,4]^1) = [f_2(3) - f_2(-1)] + [f_2(4) - f_2(-0)] = 1.3672 + 0.9640 = 2.3312$. This example was meant to demonstrate that different positive valuation functions can drastically change the distance between two intervals. In practice, we often employ parametric positive valuations in order to introduce tunable nonlinearities by optimal parameter estimation.*

The space M^h of generalized intervals is a real *linear space* [18, 22] with

- *addition* defined as $[a,b]^h + [c,d]^h = [a+c, b+d]^h$.
- *multiplication* (by $k \in \mathsf{R}$) defined as $k[a,b]^h = [ka, kb]^h$.

A subset C of a linear space is called *cone* if for all $x \in C$ and a real number $\lambda > 0$ we have $\lambda x \in C$. It turns out that both M^h_+ and M^h_- are cones.

1.3.6 The lattice of Fuzzy Interval Numbers (FINs)

Consider the following definition.

Definition 1.7 *A Fuzzy Interval Number, or FIN for short, is a function $F : (0,1] \to \mathsf{M}$ such that (1) $F(h) \in \mathsf{M}^h$, (2) either $F(h) \in \mathsf{M}^h_+$ (positive FIN), or $F(h) \in \mathsf{M}^h_-$ (negative FIN) for all $h \in (0,1]$, and (3) $h_1 \leq h_2 \Rightarrow \{x : F(h_1) \neq 0\} \supseteq \{x : F(h_2) \neq 0\}$.*

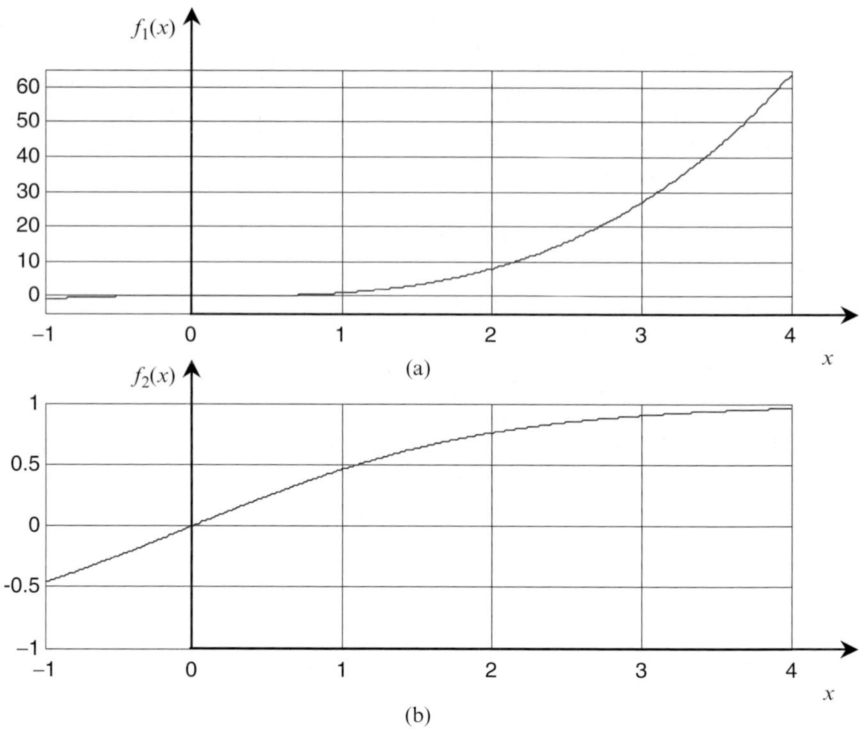

Fig. 1.5. Two positive valuation functions are shown on the domain $[-1, 4]$ including (a) The steeply increasing *cubic* function $f_1(x) = x^3$, and (b) The saturated *logistic* function $f_2(x) = (1 - e^{-x})/(1 + e^{-x})$

A FIN F can be written as the set union of generalized intervals; in particular, $F = \bigcup_{h \in (0,1]} \{[a(h), b(h)]^h\}$, where both interval-ends $a(h)$ and $b(h)$ are functions of $h \in (0, 1]$. The set of FINs is denoted by F. More specifically, the set of positive (negative) FINs is denoted by F_+ (F_-).

Example 1.6 *Figure 1.6 shows a positive FIN. The only restriction is that a FIN's membership function needs to be "convex".*

We define an *interval-FIN* as $F = \bigcup_{h \in (0,1]} \{[a(h), b(h)]^h\}$, where both $a(h)$ and $b(h)$ are constant, i.e. $a(h) = a$ and $b(h) = b$. In particular, for $a = b$ an interval-FIN is called *trivial-FIN*. In the aforementioned sense F_+ includes both (fuzzy) numbers and intervals.

We remark that a FIN is a mathematical object, which can be interpreted either as a possibility distribution (i.e. a fuzzy number) or as a probability distribution, etc. [17, 18, 22]. An ordering relation has been introduced in F as follows: $F_1 \leq F_2 \Leftrightarrow F_1(h) \leq F_2(h), \forall h \in (0, 1]$. It turns out that F is a mathematical lattice. The following proposition introduces a metric in F.

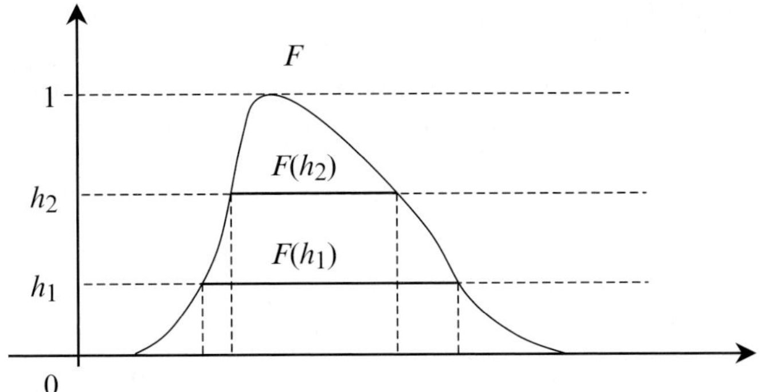

Fig. 1.6. A positive FIN $F = \bigcup_{h \in (0,1]} \{F(h)\}$ is the set-union of positive generalized intervals $F(h)$, $h \in (0, 1]$

Proposition 1.1 *Let F_1 and F_2 be FINs in the lattice F of FINs. A metric function $d_K : \mathsf{F} \times \mathsf{F} \to \mathsf{R}_0^+$ is given by*

$$d_K(F_1, F_2) = \int_0^1 d_h(F_1(h), F_2(h))dh \tag{1.4}$$

Based on d_K, a metric $D : \mathsf{F}^N \times \mathsf{F}^N \to \mathsf{R}_0^+$ can be defined between two N-dimensional FINs $\mathbf{F}_1 = [F_{1,1}, ..., F_{1,N}]^T$ and $\mathbf{F}_2 = [F_{2,1}, ..., F_{2,N}]^T$ as follows.

$$D(\mathbf{F}_1, \mathbf{F}_2) = \sqrt{\sum_{i=1}^{N} d_K^2(F_{1,i}, F_{2,i})} \tag{1.5}$$

We remark that formula (1.5) may involve a vector $\mathbf{x} = [x_1, ..., x_N]^T \in \mathsf{R}^N$ under the assumption that a vector entry x_i (number) is represented by the trivial-FIN $x_i = \bigcup_{h \in (0,1]} \{[x_i, x_i]^h\}$, $i = 1, ..., N$.

Addition and multiplication are extended from M^h to F as follows.

- The *product* kF_1, where $k \in \mathsf{R}$ and $F_1 \in \mathsf{F}$, is defined as $F_p : F_p(h) = kF_1(h)$, $h \in (0, 1]$.
- The *sum* $F_1 + F_2$, where $F_1, F_2 \in \mathsf{F}$ is defined as $F_s : F_s(h) = (F_1 + F_2)(h) = F_1(h) + F_2(h)$, $h \in (0, 1]$.

We remark that, on the one hand, the product kF_1 is always a FIN. On the other hand, when both F_1 and F_2 are in cone F_+ (F_-) then the sum $F_1 + F_2$ is in cone F_+ (F_-). However, if $F_1 \in \mathsf{F}_+$ and $F_2 \in \mathsf{F}_-$ then $F_1 + F_2$ might not be a FIN. The interest of this work is in positive FINs.

1.3.7 Practical FIN representation

From a practical viewpoint a FIN F is represented in the computer memory by a $L \times 2$ matrix $\begin{bmatrix} a_1 & b_1 \\ a_2 & b_2 \\ \vdots & \vdots \\ a_L & b_L \end{bmatrix}$ of real numbers, where L is a user-defined number of levels $h_1, h_2, ..., h_L$ such that $0 < h_1 \leq h_2 \leq ... \leq h_L = 1$; that is, FIN F equals $F = \bigcup_{i \in \{1,...,L\}} \{[a_i, b_i]^{h_i}\}$. In our experiments we usually use either $L = 16$ or $L = 32$ levels, spaced equally in the interval $[0, 1]$.

1.4 Enhancement of Both Fuzzy-ART and SOM

Based on the lattice-theoretic notions and tools presented previously, this section describes enhancements of both fuzzy-ART and SOM, namely *fuzzy lattice reasoning* (FLR) and *granular SOM* (grSOM), respectively, for supervised granular learning. We point out that the FLR is based on an *inclusion measure* function, whereas the grSOM is based on a *metric* function. Both aforementioned functions are used here in the lattice F_+ of positive FINs.

1.4.1 FLR: An enhancement of fuzzy-ART for classification

Algorithm FLR for training is presented next followed by algorithm FLR for testing. Both algorithms are applied on interval-FINs.

Algorithm FLR for training

FLR-0: A rule-base $RB = \{(u_1, C_1), ..., (u_L, C_L)\}$ is given, where $u_i \in \mathsf{F}_+^N$ is a hyperbox and $C_i \in \mathsf{C}$, $i = 1, ..., L$ is a class label — Note that C is a finite set of class labels.

FLR-1: Present the next input pair $(x_i, c_i) \in \mathsf{F}_+^N \times \mathsf{C}$, $i = 1, ..., n$ to the initially "set" RB.

FLR-2: If no more pairs are "set" in RB then store input pair (x_i, c_i) in RB; $L \leftarrow L + 1$; goto step FLR-1.

Else, compute the fuzzy degree of inclusion $k(x_i \leq u_l), l \in \{1, ..., L\}$ of input hyperbox x_i to all "set" hyperboxes $u_i, i = 1, ..., L$ in RB.

FLR-3: Competition among the "set" pairs in the RB: Winner is pair (u_J, C_J) such that $J \doteq arg \max_{l \in \{1,...,L\}} k(x_i \leq u_l)$. In case of multiple winners, choose the one with the smallest diagonal size.

FLR-4: The *Assimilation Condition*: Both (1) $diag(x_i \vee u_J)$ is less than a maximum user-defined threshold size D_{crit}, and (2) $c_i = C_J$.

FLR-5: If the *Assimilation Condition* is not satisfied then "reset" the winner pair (u_J, C_J); goto step FLR-2.
Else, replace the winner hyperbox u_J by the join-interval $x_i \vee u_J$; goto step FLR-1.

Algorithm FLR for testing

flr-0: Consider a rule-base $RB = \{(u_1, C_1), ..., (u_L, C_L)\}$.

flr-1: Present a hyperbox $x_0 \in \mathsf{F}_+^N$ to the rule base RB.

flr-2: Compute the fuzzy degree of inclusion $k(x_0 \leq u_l), l \in \{1, ..., L\}$ of hyperbox x_0 in all hyperboxes $u_i, i = 1, ..., L$ in the RB.

flr-3: Competition among the hyperboxes in RB: Winner is pair (u_J, C_J) such that $J \doteq arg \max_{l \in \{1,...,L\}} k(x_i \leq u_l)$.

flr-4: Hyperbox x_0 is classified to the class with label C_J.

By "hyperbox" above we mean "interval-FIN". We remark that the FLR has been described as a rule-based classifier [19], where a hyperbox h is assigned a class label thus corresponding to the following rule: If a point p is inside hyperbox h (let the latter by labeled by c) then p is in class c. For points outside all hyperboxes, as well as for points inside overlapping hyperboxes, inclusion measure k is used to assign a class. Note also that FLR has been implemented on a neural network architecture [18, 23].

There are inherent similarities as well as substantial differences between fuzzy-ART and FLR. In particular, both fuzzy-ART and FLR carry out learning rapidly in a single pass through the training data by computing hyperboxes in their data domain. Note that a computed hyperbox corresponds to the core of a fuzzy set and it can be interpreted as an information granule (cluster).

Advantages of FLR over fuzzy-ART include (1) *comprehensiveness*, (2) *flexibility*, and (3) *versatility* as summarized next [23]. (1) The FLR can handle intervals (granules), whereas fuzzy-ART deals solely with trivial intervals the latter are points in the unit-hypercube. (2) It is possible to optimize FLR's behavior by tuning an underlying positive valuation function v as well as an isomorphic function θ, whereas fuzzy-ART implicitly uses, quite restrictively, only $v(x) = x$ and $\theta(x) = 1 - x$. (3) The FLR can handle general lattice elements including points in the unit-hypercube, the latter is fuzzy-ART's sole application domain.

In addition, the FLR can deal with "missing" data as well as with "don't care" data in a constituent complete-lattice by replacing the aforementioned data by the *least* and the *greatest* element, respectively, in the corresponding constituent lattice [23].

Both of fuzzy-ART's *choice (Weber)* function and *match* function correspond to FLR's inclusion measure function "k". Moreover, fuzzy-ART's *complement coding* technique corresponds to a specific isomorphic function, namely $\theta(x) = 1 - x$. Apparently, choosing a different isomorphic function than $\theta(x) = 1 - x$ results in a different "coding" technique [18, 23].

Let f_i and θ_i be strictly-increasing and strictly-decreasing functions, respectively, in a constituent lattice R in R^N. A typical assumption for both fuzzy-ART and FLR is to select f_i and θ_i such that equation $v([a, b]) = 1 + diag_1([a, b])$ is satisfied [19]. On the one hand, two popular functions f_i and θ_i in the complete lattice unit-interval $[0, 1]$ are $f_i(x) = x$ and $\theta_i(x) = 1 - x$. On the other hand, two popular functions f_i and θ_i in the noncomplete lattice R are $f_i(x) = 1/(1 + e^{-\lambda(x - x_0)})$ and $\theta_i(x) = 2x_0 - x$.

Inclusion measure "k" above retains an *Occam razor* semantic interpretation as detailed in [19]. In particular, winner of the competition in steps FLR-3/flr-3 above is the hyperbox whose diagonal needs to be modified *the least* so as to "barely" include an input datum/hyperbox.

The FLR was interpreted lately as a reasoning scheme, which supports two different modes of reasoning, namely *Generalized Modus Ponens* and *Reasoning by Analogy* [19]. A novel interpretation is presented here as follows. Inclusion measure $k(p \leq q)$ is interpreted as the degree of truth of implication "$p \Rightarrow q$" involving the truth values p and q, respectively, of two propositions. Note that various mechanisms have been proposed in the literature for calculating the degree of truth of an implication "$p \Rightarrow q$" given the truth values p and q [14]. The basic difference here is that the truth values p and q of the two propositions involved in implication "$p \Rightarrow q$" take on values in a general complete lattice [11] rather than taking on values solely in the unit-interval $[0, 1]$. However, the truth of implication "$p \Rightarrow q$" takes on values in the unit-interval $[0, 1]$. More specifically, the truth of implication "$p \Rightarrow q$" is calculated as $k(p \Rightarrow q) = v(q)/v(p \vee q)$. In conclusion, the FLR carries out interactively tunable inferences. A couple of FLR drawbacks are described next.

Fuzzy-ART's *proliferation problem*, that is the proliferation of hyperboxes/clusters, is inherited to FLR. However, FLR is equipped with tools such as an inclusion measure as well as a metric function to reduce "in principle" the number of hyperboxes.

Another drawback of FLR, also inherited from fuzzy-ART, is that the learned clusters (in particular their total number, size, and location) depend on the order of presenting the training data. A potential solution is to employ an ensemble of FLR classifiers in order to boost performance stably [18].

1.4.2 grSOM: An Enhancement of SOM for Classification

Algorithm FLR is applicable in the space F_+ of (positive) FINs. Algorithm grSOM for learning (training) is presented next followed by algorithm grSOM for generalization (testing).

Algorithm grSOM for training

GR-0: Define the size L of a $L \times L$ grid of neurons. Each neuron can store both a N-dimensional FIN $W_{i,j} \in \mathsf{F}_+^N$, $i, j \in 1, ..., L$ and a class label $C_{i,j} \in \mathsf{C}$, where C is a finite set. Initially all neurons are *uncommitted*.

GR-1: Memorize the first training data pair $(x_1, C_1) \in \mathsf{F}_+^N \times \mathsf{C}$ by committing, randomly, a *neuron* in the $L \times L$ grid.
Repeat the following steps a user-defined number N_{epochs} of epochs, $p = 1, ..., N_{epochs}$.

GR-2: For each training datum $(x_k, C_k) \in \mathsf{F}_+^N \times \mathsf{C}$, $k = 1, ..., n$ "reset" all $L \times L$ grid neurons. Then carry out the following computations.

GR-3: Calculate the Minkowski metric $d_1(x_k, W_{i,j})$ between x_k and *committed* neurons $W_{i,j}$ $i, j \in \{1, ..., L\}$.

GR-4: Competition among the "set" (and, *committed*) neurons in the $L \times L$ grid: Winner is neuron (I, J) whose weight $W_{I,J}$ is the nearest to x_k, i.e. $(I, J) \doteq arg \min\limits_{i,j \in \{1,...,L\}} d_1(x_k, W_{i,j})$.

GR-5: *Assimilation Condition*: Both (1) Vector $W_{i,j}$ is in the neighborhood of vector $W_{I,J}$ on the $L \times L$ grid, and (2) $C_{I,J} = C_k$.

GR-6: If the *Assimilation Condition* is satisfied then compute a new value $W'_{i,j}$ as follows:
$$W'_{i,j} = \left[1 - \frac{h(k)}{1 + d_K(W_{I,J}, W_{i,j})}\right] W_{i,j} + \frac{h(k)}{1 + d_K(W_{I,J}, W_{i,j})} x_k.$$
Else, if the *Assimilation Condition* is not satisfied, "reset" the winner (I, J); goto GR-4.

GR-7: If all the $L \times L$ neurons are "reset" then commit an *uncommitted* neuron from the grid to memorize the current training datum (x_k, C_k). If there are no more *uncommitted* neurons then increase L by one.

Algorithm grSOM for testing

gr-0: Present $x_0 \in \mathsf{F}_+^N$ to a trained grSOM.

gr-1: Calculate the Minkowski metric $d_1(x_0, W_{i,j})$ for *committed* neurons $W_{i,j}$, $i, j \in \{1, ..., L\}$.

gr-2: Competition among the *committed* neurons in the $L \times L$ grid: Winner is neuron (I, J) such that $(I, J) \doteq arg \min\limits_{i,j \in \{1,...,L\}} d_1(x_0, W_{i,j})$.

gr-3: The class C_0 of x_0 equals $C_0 \doteq C_{I,J}$.

Function "$h(k)$", in the training phase above, reduces smoothly from 1 down to 0 with the epoch number k. The above algorithm is called *incremental-grSOM* [20]. It differs from another grSOM algorithm, namely *greedy-grSOM* [22], in that only the incremental-grSOM employs convex combinations of (positive) FINs. Both grSOM and SOM partition the data domain in Voronoi-regions, and each one of the aforementioned regions can also be interpreted as an information granule.

A fundamental improvement of grSOM over SOM is the sound capacity of grSOM to rigorously deal with nonnumeric data including both fuzzy numbers and intervals represented by FINs. However, the decision-making function of grSOM (as well as the corresponding function of SOM) does not admit a logical/linguistic interpretation. Rather, since the aforementioned function is an energy-type objective function, that is a *metric*, optimization is pursued during learning using *energy minimization* techniques.

1.5 Conclusion

This chapter was meant as a reference towards proliferating the employment of both fuzzy-ART and SOM in granular classification applications. Enhancements of fuzzy-ART as well as of SOM were presented, namely FLR and grSOM, respectively. FLR/grSOM is applicable in the lattice of fuzzy interval numbers, or FINs for short, including both (fuzzy) numbers and intervals. The FLR was interpreted as a reasoning scheme, whereas the grSOM was interpreted as an energy minimizer. The employment of mathematical lattice theory was instrumental for introducing useful tools.

Ample experimental evidence suggests that FLR and/or grSOM can comparatively improve classification performance [18, 19, 20, 21, 22, 23].

Future work will consider alternative granular inputs to modified FLR /grSOM classifiers including type-2 fuzzy set inputs, rough set inputs, etc. In addition, fuzzy logic reasoning applications [15, 28] will be pursued.

References

1. Andonie R, Sasu L (2006) Fuzzy ARTMAP with input relevances. IEEE Trans Neural Networks 17(4):929–941
2. Bargiela A, Pedrycz W, Hirota K (2004) Granular prototyping in fuzzy clustering. IEEE Trans Fuzzy Systems 12(5):697–709
3. Birkhoff G (1967) Lattice Theory. American Math Society, Colloquium Publications XXV
4. Bishop CM, Svensén M, Williams CKI (1998) GTM: The generative topographic mapping. Neural Computation 10(1):215–234
5. Carpenter GA, Grossberg S, Rosen DB (1991) Fuzzy ART: Fast stable learning and categorization of analog patterns by an adaptive resonance system. Neural Networks 4(6):759–771
6. Carpenter GA, Grossberg S, Markuzon N, Reynolds JH, Rosen DB (1992) Fuzzy ARTMAP: A neural network architecture for incremental supervised learning of analog multidimensional maps. IEEE Trans Neural Networks 3(5):698–713
7. Castro J, Georgiopoulos M, DeMara R, Gonzalez A (2005) Data-partitioning using the Hilbert space filling curves: Effect on the speed of convergence of Fuzzy ARTMAP for large database problems. Neural Networks 18(7):967–984
8. Castro J, Secretan J, Georgiopoulos M, DeMara RF, Anagnostopoulos G, Gonzalez AJ (2007) Pipelining of fuzzy-ARTMAP without match-tracking: Correctness, performance bound, and beowulf evaluation. Neural Networks 20(1):109–128
9. Cripps A, Nguyen N (2007) Fuzzy lattice reasoning (FLR) classification using similarity measures. This volume, Chapter 13

10. Georgiopoulos M, Fernlund H, Bebis G, Heileman GL (1996) Order of search in fuzzy ART and fuzzy ARTMAP: Effect of the choice parameter. Neural Networks 9(9):1541–1559
11. Goguen JA (1967) L-fuzzy sets. J Math Analysis and Applications 18:145–174
12. Graña M, Sussner P, Ritter G (2003) Innovative applications of associative morphological memories for image processing and pattern recognition. Mathware & Soft Computing 10(2-3):155–168
13. Hammer B, Villmann T (2002) Generalized relevance learning vector quantization. Neural Networks 15(8-9):1059–1068
14. Hatzimichailidis A, Kaburlasos VG, Papadopoulos B (2006) An implication in fuzzy sets. In: Proc World Congress Computational Intelligence (WCCI) FUZZ-IEEE Program pp 203–208
15. Igel C, Temme KH (2004) The chaining syllogism in fuzzy logic. IEEE Trans Fuzzy Systems 12(6):849–853
16. Kaburlasos VG (1992) Adaptive Resonance Theory with Supervised Learning and Large Database Applications. PhD Thesis, Univ Nevada, Reno
17. Kaburlasos VG (2004) FINs: Lattice theoretic tools for improving prediction of sugar production from populations of measurements. IEEE Trans Systems, Man and Cybernetics - B, Cybernetics 34(2):1017–1030
18. Kaburlasos VG (2006) Towards a Unified Modeling and Knowledge-Representation Based on Lattice Theory — Computational Intelligence and Soft Computing Applications, ser Studies in Computational Intelligence 27. Springer, Heidelberg, Germany
19. Kaburlasos VG, Athanasiadis IN, Mitkas PA (2007) Fuzzy lattice reasoning (FLR) classifier and its application for ambient ozone estimation. Intl J Approximate Reasoning 45(1):152–188
20. Kaburlasos VG, Chatzis V, Tsiantos V, Theodorides M (2005) granular Self-Organizing Map (grSOM) neural network for industrial quality control. In: Astola JT, Tabus I, Barrera J (eds) Proc SPIE Math Methods in Pattern and Image Analysis 5916 pp 59160J:1–10
21. Kaburlasos VG, and Kehagias A (2007) Novel fuzzy inference system (FIS) analysis and design based on lattice theory. IEEE Trans Fuzzy Systems 15(2):243–260
22. Kaburlasos VG, Papadakis SE (2006) Granular self-organizing map (grSOM) for structure identification. Neural Networks 19(5):623–643
23. Kaburlasos VG, Petridis V (2000) Fuzzy lattice neurocomputing (FLN) models. Neural Networks 13(10):1145–1170
24. Karayiannis NB, Bezdek JC (1997) An integrated approach to fuzzy learning vector quantization and fuzzy c-means clustering. IEEE Trans Fuzzy Systems 5(4):622–628
25. Knuth KH (2007) Valuations on lattices: Fuzzification and its implications. This volume, Chapter 15
26. Kohonen T (1995) Self-Organizing Maps, ser Information Sciences 30. Springer.
27. Lin TY (2005) Introduction to special issues on data mining and granular computing (editorial). Intl J Approximate Reasoning 40(1-2):1–2
28. Liu J, Ruan D, Xu Y, Song Z (2003) A resolution-like strategy based on a lattice-valued logic. IEEE Trans Fuzzy Systems 11(4):560–567
29. Papadakis SE, Kaburlasos VG (submitted) Efficient structure identification in TSK models based on a novel self-organizing map (SOM)

30. Pedrycz W (2005) Knowledge-Based Clustering — From Data to Information Granules. John Wiley & Sons, Hoboken, NJ
31. Peltonen J, Klami A, Kaski S (2004) Improved learning of Riemannian metrics for exploratory analysis. Neural Networks 17(8-9):1087–1100
32. Piedra-Fernández JA, Cantón-Garbín M, Guindos-Rojas F (2007) Application of fuzzy lattice neurocomputing (FLN) in ocean satellite images for pattern recognition. This volume, Chapter 11
33. Popescu M, Gader P, Keller JM (2006) Fuzzy spatial pattern processing using linguistic hidden Markov models. IEEE Trans Fuzzy Systems 14(1):81–92
34. Ritter GX, Gader PD (2006) Fixed points of lattice transforms and lattice associative memories. In: Hawkes P (ed) Advances in Imaging and Electron Physics 144:165–242. Elsevier, Amsterdam, The Netherlands
35. Ritter GX, Urcid G (2003) Lattice algebra approach to single-neuron computation. IEEE Trans Neural Networks 14(2):282–295
36. Sussner P, Valle ME (2006) Gray-scale morphological associative memories. IEEE Trans Neural Networks 17(3):559–570
37. Vuorimaa P (1994) Fuzzy self-organizing map. Fuzzy Sets Syst 66(2):223–231

Chapter 1 Appendix

A *poset* is a pair $(P, \leq)$, where P is a set and $\leq$ is a binary *partial order relation* defined next.

Definition 1.8 *A* partial order *relation satisfies the following laws.*
  $(PO1)$ $x \leq x$ (Reflexivity)
  $(PO2)$ $x \leq y$ *and* $y \leq x \Rightarrow x = y$ (Antisymmetry)
  $(PO3)$ $x \leq y$ *and* $y \leq z \Rightarrow x \leq z$ (Transitivity)

We remark that relation $<$ means both $\leq$ and $\neq$.

Definition 1.9 *An* algebra *is a pair* (S, F)*, where* S *is a non-empty set, and* F *is a set of operations* f_a*, each mapping a power* $S^{n(a)}$ *of* S *into* S *for some non-negative finite integer* $n(a)$*.*

We remark that each operation f_a assigns to every $n(a)$-ple $(x_1, \ldots, x_{n(a)})$ of elements of S, an element $f_a(x_1, \ldots, x_{n(a)})$ in S, the result of performing the operation f_a on the sequence $x_1, \ldots, x_{n(a)}$. In particular, if $n(a) = 1$, the operation f_a is called *unary*; if $n(a) = 2$, it is called *binary*, etc.

Definition 1.10 *An* interval $[a, b]$*, with* $a \leq b$ *in a poset* $(P, \leq)$*, is defined as the set* $[a, b] = \{x \in P : a \leq x \leq b\}$*.*

Definition 1.11 *A* positive valuation *in a crisp lattice* $(L, \leq)$ *is a real function* $v : L \to R$*, which satisfies both*
  $(PV1)$ $v(x) + v(y) = v(x \wedge y) + v(x \vee y)$*, and*
  $(PV2)$ $x < y \Rightarrow v(x) < v(y)$*.*

Definition 1.12 *A* metric *in a set* A *is a nonnegative real function* $d : A \times A \to R_0^+$*, which satisfies*
  $(D0)$ $d(x, y) = 0 \Rightarrow x = y$
  $(D1)$ $d(x, x) = 0$
  $(D2)$ $d(x, y) = d(y, x)$
  $(D3)$ $d(x, z) \leq d(x, y) + d(y, z)$ (Triangle Inequality)

If only conditions $D1$, $D2$, and $D3$ are satisfied then function d is called *pseudo-metric*. We remark that a *metric space* is a pair (A, d) including both a set A and a metric $d : A \times A \to R_0^+$.

Definition 1.13 *Let* P *and* Q *be posets. A map* $\psi : P \to Q$ *is called*
  (i) Order-perserving *(or, alternatively,* monotone*), if* $x \leq y$ *in* P *implies* $\psi(x) \leq \psi(y)$ *in* Q*.*
  (ii) Order-isomorphism *(or, simply,* isomorphism*), if both* $x \leq y$ *in* $P \Leftrightarrow \psi(x) \leq \psi(y)$ *in* Q *and* "ψ *is onto* Q"*.*

We remark that when there is an isomorphism from P to Q, then P and Q are called *isomorphic*, symbolically $P \cong Q$; moreover, the corresponding function ψ is called *isomorphic* (function).

Learning in Lattice Neural Networks that Employ Dendritic Computing

Gerhard X. Ritter[1] and Gonzalo Urcid[2]

[1] University of Florida, Dept. of Computer & Information Science & Engineering
Gainesville, Florida 32611-6120, USA
`ritter@cise.ufl.edu`
[2] National Institute of Astrophysics, Optics & Electronics, Department of Optics
Tonanzintla, Puebla 72000, Mexico
`gurcid@inaoep.mx`

Summary. Recent discoveries in neuroscience imply that the basic computational elements are the dendrites that make up more than 50% of a cortical neuron's membrane. Neuroscientists now believe that the basic computation units are dendrites, capable of computing simple logic functions. This paper discusses two types of neural networks that take advantage of these new discoveries. The focus of this paper is on some learning algorithms in the two neural networks. Learning is in terms of lattice computations that take place in the dendritic structure as well as in the cell body of the neurons used in this model.

2.1 Introduction

During the past two decades artificial neural networks based on lattice theory or employing various aspects of lattice theoretic computations have attracted considerable attention which is partially due to their useful applications in a variety of disciplines [2, 3, 5, 7, 9, 10, 11, 12, 19, 20, 22, 23, 24]. However, just like the various artificial neural networks currently in vogue such as multilayer perceptrons, radial basis function neural networks, and support vector machines, these lattice based neural networks bear little resemblance to biological neural networks. In biological neurons, the terminal axonal branches make contact with the soma and the many dendrites of other neurons. The sites of contact are the *synaptic sites* where the *synapse* takes place. The number of synapses on a *single* neuron in the cerebral cortex ranges between 500 and 200 000. Most of the synapses occur on the dendritic tree of the neuron, and it is here where the information is processed [4, 8, 16, 17]. Dendrites make up the largest component in both surface area and volume of the brain. Part of this is due to the fact that pyramidal cell dendrites span all cortical layers in all regions of the cerebral cortex [1, 4, 16]. Thus, when attempting

G.X. Ritter and G. Urcid: *Learning in Lattice Neural Networks that Employ Dendritic Computing*, Studies in Computational Intelligence (SCI) **67**, 25–44 (2007)
`www.springerlink.com`   © Springer-Verlag Berlin Heidelberg 2007

to model artificial brain networks that bear more than just a passing resemblance to biological brain networks, one cannot ignore dendrites (and their associated spines) which make up more than 50% of the neuron's membrane. For a more thorough background in dendritic computing we refer the reader to the references cited in [13].

In order to take advantage of recent advances in neurobiology and the biophysics of neural computation and to nudge the field of ANNs back to its original roots, we proposed a model of single neuron computation that takes into account the computation performed by dendrites [13]. Extrapolating on this model, we constructed a single-layer, feedforward neural network based on dendritic computing within the lattice domain [14]. This new neural network model was referred to as a *single layer morphological perceptron* (SLMP). However, a more appropriate name would be a *single layer lattice perceptron* (SLLP) since the basic neural computations are derived from the lattice algebra $(\mathbb{R}, +, \vee, \wedge)$, where $\mathbb{R}$ denotes the set of real numbers and $+$, $\vee$, and $\wedge$ denote the operations of addition, maximum, and minimum, respectively, of two numbers.

This chapter is organized as follows: Sect. 2.2 outlines the functionality of the SLLP; Sect. 2.3 discuss methods of training SLLPs. An example comparing SLLPs with some other current artificial neural networks is also given; in Sect. 2.4 we provide the basic algorithm for multi-class learning with SLLPs; Sect. 2.5 gives an extension of the SLLP to a lattice based associative memory with one hidden layer and provide a formulation for computing the weights of axonal terminal fibers. Concluding remarks to the material exposed here are given in Sect. 2.6. The chapter ends with a list of essential references to which the reader may turn for supplementary information.

2.2 The Dendritic Single Layer Lattice Perceptron

In the dendritic SLLP model, a set of n input neurons $N_1, \ldots, N_n$ accepts input $x = (x_1, \ldots, x_n) \in \mathbb{R}^n$. That is, the value x_i of the ith input neuron N_i need not be binary. An input neuron provides information through its axonal arborization to the dendritic trees of a set of m output neurons $M_1, \ldots, M_m$. Explicitely, the state value of a neuron $N_i(i = 1, \ldots, n)$ propagates through its axonal tree all the way to the terminal branches that make contact with the neuron $M_j(j = 1, \ldots, m)$. The weight of an axonal branch of neuron N_i terminating on the kth dendrite of M_j is denoted by w_{ijk}^ℓ, where the superscript $\ell \in \{0, 1\}$ distinguishes between *excitatory* ($\ell = 1$) and *inhibitory* ($\ell = 0$) input to the dendrite (see Fig. 2.1). The kth dendrite of M_j will respond to the total input received from the neurons $N_1, \ldots, N_n$ and will either accept or inhibit the received input. The computation of the kth dendrite of M_j is given by

$$\tau_k^j(x) = p_{jk} \bigwedge_{i \in I(k)} \bigwedge_{\ell \in L(i)} (-1)^{1-\ell}(x_i + w_{ijk}^\ell) \, , \tag{2.1}$$

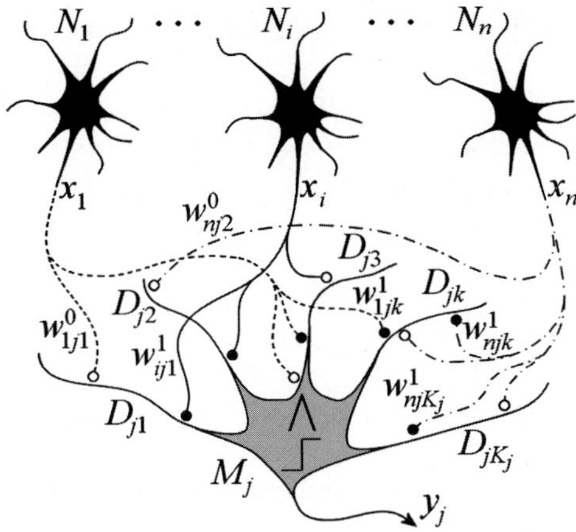

Fig. 2.1. The neural pathways in an SLLP. Terminations of excitatory and inhibitory fibers are marked with • and ∘, respectively. Symbol D_{jk} denotes dendrite k of M_j and K_j its number of dendrites. Neuron N_i can synapse D_{jk} with excitatory or inhibitory fibers, e.g. weights w_{1jk}^1 and w_{nj2}^0 respectively denote excitatory and inhibitory fibers from N_1 to D_{jk} and from N_n to D_{j2}

where $\boldsymbol{x} = (x_1, \ldots, x_n)$ denotes the input value of the neurons $N_1, \ldots, N_n$ with x_i representing the value of N_i, $I(k) \subseteq \{1, \ldots, n\}$ corresponds to the set of all input neurons with terminal fibers that synapse on the kth dendrite of M_j, $L(i) \subseteq \{0, 1\}$ corresponds to the set of terminal fibers of N_i that synapse on the kth dendrite of M_j, and $p_{jk} \in \{-1, 1\}$ denotes the excitatory ($p_{jk} = 1$) or inhibitory ($p_{jk} = -1$) response of the kth dendrite of M_j to the received input. Note that the number of terminal axonal fibers on N_i that synapse on a dendrite of M_j is at most two since $L(i)$ is a subset of $\{0, 1\}$.

The value $\tau_k^j(\boldsymbol{x})$ is passed to the cell body and the state of M_j is a function of the input received from all its dendrites. The total value received by M_j is given by

$$\tau^j(\boldsymbol{x}) = p_j \bigwedge_{k=1}^{K_j} \tau_k^j(\boldsymbol{x}) \;, \tag{2.2}$$

where K_j denotes the total number of dendrites of M_j and $p_j = \pm 1$ denotes the response of the cell body to the received dendritic input. Here again, $p_j = 1$ means that the input is accepted, while $p_j = -1$ means that the cell rejects the received input. Figure 2.1 illustrates the neural pathways from the sensor or input neurons to a particular output neuron M_j. The *next state* of M_j is then determined by an activation function f, namely $y_j = f[\tau^j(\boldsymbol{x})]$. In this exposition we restrict our discussion to the hard-limiter

$$f[\tau^j(\boldsymbol{x})] = \begin{cases} 1 & \Leftrightarrow & \tau^j(\boldsymbol{x}) \geq 0 \\ 0 & \Leftrightarrow & \tau^j(\boldsymbol{x}) < 0 \end{cases}. \tag{2.3}$$

2.3 Single Class Learning in SLLPs

Early learning algorithms were derived from the proofs of the following two theorems which provide insight into the computational capability of SLLPs.

Theorem 2.1 *If $X \subset \mathbb{R}^n$ is compact and $\varepsilon > 0$, then there exists a single layer lattice perceptron that assigns every point of X to class C_1 and every point $\boldsymbol{x} \in \mathbb{R}^n$ to class C_0 whenever $d(\boldsymbol{x}, X) > \varepsilon$.*

The expression $d(\boldsymbol{x}, X)$ in the statement of Theorem 2.1 refers to the distance of the point $\boldsymbol{x} \in \mathbb{R}^n$ to the set X. As a consequence, any compact configuration, as the one shown in Fig. 2.2(a), whether it is convex or non-convex, connected or not connected, or contains a finite or an infinite number of points, can be approximated within any desired degree of accuracy $\varepsilon > 0$ by an SLLP with one output neuron.

An SLLP can be extended to multiple output neurons in order to handle multiclass problems, just like its classical counterpart. However, unlike the well known *single layer perceptron* (SLP), which is a linear discriminator, the SLLP with multiple outputs can solve multiclass nonlinear classification problems. This computational capability of an SLLP with multiple output neurons is attested by Theorem 2.2 below, which is a generalization of Theorem 2.1 to multiple sets. Suppose $X_1, \ldots, X_m$ denotes a collection of disjoint compact subsets of $\mathbb{R}^n$. The goal is to classify, $\forall j = 1, \ldots, m$, every point of X_j as a point belonging to class C_j and not belonging to class C_i whenever $i \neq j$. For each $p \in \{1, \ldots, m\}$, define $Y_p = \bigcup_{j \neq p}^{m} X_j$. Since each Y_p is compact and $Y_p \cap X_p = \emptyset$, the Euclidean distance between the two sets X_p and Y_p is greater than zero; i. e., $d(X_p, Y_p) > 0$. Then, set $\varepsilon_p = d(X_p, Y_p)$ for all $p = 1, \ldots, m$ and let $\varepsilon_0 = \frac{1}{2}\min\{\varepsilon_1, \ldots, \varepsilon_p\}$.

Theorem 2.2 *If $\{X_1, \ldots, X_m\}$ is a collection of disjoint compact subsets of $\mathbb{R}^n$ and ε is a positive number with $\varepsilon < \varepsilon_0$, then there exists a single layer*

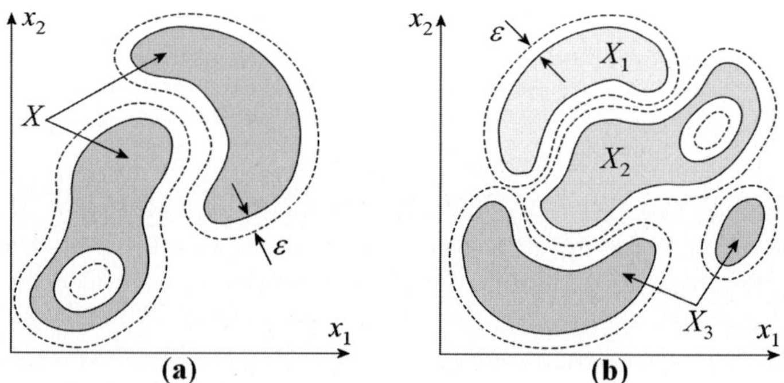

Fig. 2.2. (a) Compact set X and (b) collection of disjoint compact sets X_1, X_2, X_3, each with a banded region of thickness ε (dashed). Theorems 2.1 and 2.2 guarantee the existence of SLLPs able to classify sets X and, respectively, X_1, X_2, X_3, within a desired ε accuracy

lattice perceptron that assigns each point $\boldsymbol{x} \in \mathbb{R}^n$ to class C_j whenever $\boldsymbol{x} \in X_j$ and $j \in \{1, \ldots, m\}$, and to class $C_0 = \neg \bigcup_{j=1}^m C_j$ whenever $d(\boldsymbol{x}, X_i) > \varepsilon, \forall i = 1, \ldots, m$. Furthermore, no point $\boldsymbol{x} \in \mathbb{R}^n$ is assigned to more than one class.

Figure 2.2(b) illustrates the conclusion of Theorem 2.2 for the case $m = 3$. Based on the proofs of these two theorems, which can be found in [13], we constructed training algorithms for SLLPs [13, 14, 15]. For faster computation, instead of the Euclidean distance, the SLLP algorithms work with the Chebyshev distance as defined next.

Definition 2.1 *The* Chebyshev *distance (also known as checkerboard distance between two patterns $\boldsymbol{x}, \boldsymbol{y} \in \mathbb{R}^n$ is given by*

$$d(\boldsymbol{x}, \boldsymbol{y}) = \max_{1 \leq i \leq n} |x_i - y_i| \,. \tag{2.4}$$

During the learning phase, the output neurons grow new dendrites while the input neurons expand their axonal branches to terminate on the new dendrites. The algorithms always converge and have rapid convergence rate when compared to backpropagation learning in traditional perceptrons.

Training can be realized in one of two main strategies, which differ in the way the separation surfaces in pattern space are determined. One strategy is based on elimination, whereas the other is based on merging. In the former approach, a hyperbox is initially constructed large enough to enclose all patterns belonging to the same class, possibly including foreign patterns from other classes. This large region is then carved to eliminate the foreign patterns. Training completes when all foreign patterns in the training set have been eliminated. The elimination is performed by computing the intersection of the regions recognized by the dendrites, as expressed in (2.2) for some neuron $M_j : \tau^j(\boldsymbol{x}) = p_j \bigwedge_{k=1}^{K_j} \tau_k^j(\boldsymbol{x})$.

The latter approach starts by creating small hyperboxes around individual patterns or small groups of patterns all belonging to the same class. Isolated boxes that are identified as being close according to a distance measure are then merged into larger regions that avoid including patterns from other classes. Training is completed after merging the hyperboxes for all patterns of the same class. The merging is performed by computing the union of the regions recognized by the dendrites. Thus, the total net value received by output neuron M_j is computed as:

$$\tau^j(\boldsymbol{x}) = p_j \bigvee_{k=1}^{K_j} \tau_k^j(\boldsymbol{x}) \,. \tag{2.5}$$

The two strategies are equivalent in the sense that they are based on the same mathematical framework and they both result in closed separation surfaces around patterns. The equivalence can be attested by examining the equations employed to compute the total net values in the two approaches (2.2) and (2.5), and remarking that the maximum of any K values $a_1, a_2, \ldots, a_K$, can be equivalently written as a minimum: $\bigvee_{k=1}^{K} a_k = -\bigwedge_{k=1}^{K}(-a_k)$. Thus, if the output value y_j at neuron M_j is computed in terms of minimum as $y_j = f[p_j \bigwedge_{k=1}^{K_j} \tau_k^j(\boldsymbol{x})]$, then y_j can be equivalently computed in terms of maximum as $y_j = f[-p_j \bigvee_{k=1}^{K_j}(-\tau_k^j(\boldsymbol{x}))]$.

The major difference between the two approaches is in the shape of the separation surface that encloses the patterns of a class, and in the number of dendrites that are grown during training to recognize the region delimited by that separation surface. Since the elimination strategy involves removal of pieces from an originally large hyperbox, the resulting region is bigger than the one obtained with the merging strategy. The former approach is thus more general, while the latter is more specialized. This observation can guide the choice of the method for solving a particular problem. Additional developments related with the material presented in this section appear in [6, 18]; also in this book, see Chap. 3 by Barmpoutis and Ritter on *Orthonormal Basis Lattice Neural Networks*.

Figure 2.3 illustrates two possible partitionings of the pattern space $\mathbb{R}^2$ in terms of intersection (a) and, respectively, union (b), in order to recognize the solid circles ($\bullet$) as one class C_1. In part (a) the C_1 region is determined as the intersection of three regions, each identified by a corresponding dendrite. The rectangular region marked D_1 is intersected with the complement of the region marked D_2 and the complement of the region marked D_3. An excitatory dendrite recognizes the interior of an enclosed region, whereas an inhibitory dendrite recognizes the exterior of a delimited region. Thus, the corresponding dendrite D_1 is excitatory, while dendrites D_2 and D_3 are inhibitory. If we assign $j = 1$ in (2.1) and (2.2) as representing the index

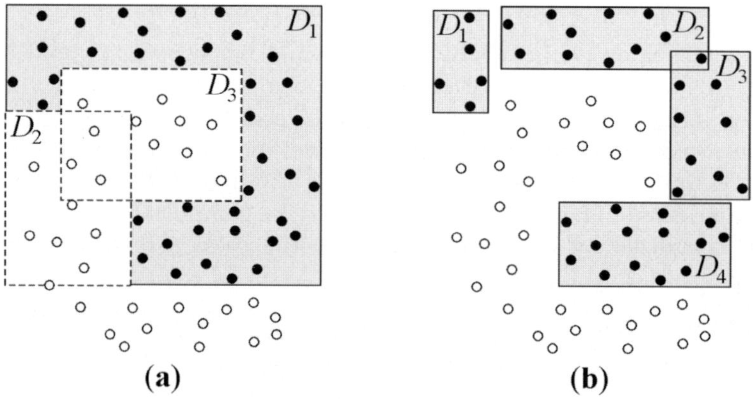

(a)   (b)

Fig. 2.3. Two partitionings of the pattern space $\mathbb{R}^2$ in terms of intersection (a) and union (b), respectively. The solid circles ($\bullet$) belong to class C_1, which is recognized as the shaded area. Solid and dashed lines enclose regions learned by excitatory and, respectively, inhibitory dendrites

of the output neuron for class C_1, then the output value is computed as $y_1 = f[\tau^1(\boldsymbol{x})] = f[p_1 \bigwedge_{k=1}^{3} \tau_k^1(\boldsymbol{x})]$, where $\tau_k^1(\boldsymbol{x})$ is computed as in (2.1). There, the responses p_{1k} are $p_{11} = 1$ and $p_{12} = p_{13} = -1$. The response of the cell body is $p_1 = 1$.

In part (b) the C_1 region is determined as the union of four regions, each identified by a corresponding dendrite. This time, all dendrites $D_1, \ldots, D_4$ are excitatory, so their responses will be $p_{1k} = 1, k = 1, \ldots, 4$. Using (2.1) and (2.5) we obtain the output value $y_1 = f[\tau^1(\boldsymbol{x})] = f[p_1 \bigvee_{k=1}^{4} \tau_k^1(\boldsymbol{x})]$, where $p_1 = 1$. As noted above, the output value can be equivalently computed with minimum instead of maximum as $y_1 = f[\tau^1(\boldsymbol{x})] = f[p_1^* \bigwedge_{k=1}^{4} p_{1k}^* \tau_k^1(\boldsymbol{x})]$, i.e. by conjugating the responses $p_1^* = -p_1$ and $p_{1k}^* = -p_{1k}$.

2.3.1 Training based on Elimination

A training algorithm that employs elimination is given below. In the training or learning set $\{(\mathbf{x}^\xi, c_\xi) : \xi = 1, \ldots, m\}$, the vector $\boldsymbol{x}^\xi = (x_1^\xi, \ldots, x_n^\xi) \in \mathbb{R}^n$ denotes an exemplar pattern and $c_\xi \in \{0, 1\}$ stands for the corresponding class number in a *one-class* problem, i.e., $c_\xi = 1$ if $\boldsymbol{x}^\xi \in C_1$, and $c_\xi = 0$ if $\boldsymbol{x}^\xi \in C_0 = \neg C_1$. Numbered steps are prefixed by S and comments are provided within brackets.

Algorithm 2.1. SLLP TRAINING BY ELIMINATION

S1. Let $k = 1$, $P = \{1, \ldots, m\}$, $I = \{1, \ldots, n\}$, $L = \{0, 1\}$; **for** $i \in I$ **do**

$$w_{ik}^1 = - \bigwedge_{c_\xi = 1} x_i^\xi \; ; \; w_{ik}^0 = - \bigvee_{c_\xi = 1} x_i^\xi$$

[Initialize parameters and auxiliary index sets; set weights for first dendrite (hyperbox enclosing class C_1); k is the dendrite counter.]

S2. Let $p_k = (-1)^{\mathrm{sgn}(k-1)}$; **for** $i \in I$, $\ell \in L$ **do** $r_{ik}^\ell = (-1)^{(1-\ell)}$; **for** $\xi \in P$ **do**

$$\tau_k(\mathbf{x}^\xi) = p_k \bigwedge_{i \in I} \bigwedge_{\ell \in L} r_{ik}^\ell (x_i^\xi + w_{ik}^\ell)$$

[Compute response of current dendrite; $\mathrm{sgn}(x)$ is the signum function.]

S3. For $\xi \in P$ **do** $\tau(\mathbf{x}^\xi) = \bigwedge_{j=1}^{k} \tau_j(\mathbf{x}^\xi)$
[Compute total response of output neuron M.]

S4. If $f(\tau(\mathbf{x}^\xi)) = c_\xi \; \forall \xi \in P$ **then** let $K = k$; **for** $k = 1, \ldots, K$ and
     for $i \in I, \ell \in L$ **do** print network parameters: $k, w_{ik}^\ell, r_{ik}^\ell, p_k$ STOP
[If with k generated dendrites learning is successful, output final weights and input-output synaptic responses of the completed network; K denotes the final number of dendrites grown in neuron M upon convergence (the algorithm ends here). If not, training continues by growing additional dendrites (next step).]

S5. Let $k = k + 1$, $I = I' = X = E = H = \emptyset$, and $D = C_1$
[Add a new dendrite to M and initialize several auxiliary index sets; initially, set D gets all class 1 patterns.]

S6. Let choice$(\mathbf{x}^\gamma) \in C_0$ such that $f(\tau(\mathbf{x}^\gamma)) = 1$
[Select a misclassified pattern from class C_0; γ is the index of the misclassified pattern and function choice(x) is a random or sort selection mechanism.]

S7. Let

$$\mu = \bigwedge_{\xi \neq \gamma} \{ \bigvee_{i=1}^{n} |x_i^\gamma - x_i^\xi| : \mathbf{x}^\xi \in D \}$$

[Compute the minimum Chebyshev distance from the selected misclassified pattern $\mathbf{x}^\gamma$ to all patterns in set D; since D is updated in step S12, μ changes during the generation of new terminals on the same dendrite.]

S8. Let $I' = \{ i : |x_i^\gamma - x_i^\xi| = \mu,\ \mathbf{x}^\xi \in D,\ \xi \neq \gamma \}$ and
$X = \{ (i, x_i^\xi) : |x_i^\gamma - x_i^\xi| = \mu,\ \mathbf{x}^\xi \in D,\ \xi \neq \gamma \}$
[Keep indices and coordinates of patterns in set D that are on the border of the hyperbox centered at $\mathbf{x}^\gamma$.]

S9. For $(i, x_i^\xi) \in X$, **if** $x_i^\xi < x_i^\gamma$ **then** $w_{ik}^1 = -x_i^\xi$, $E = \{1\}$ and
if $x_i^\xi > x_i^\gamma$ **then** $w_{ik}^0 = -x_i^\xi$, $H = \{0\}$
[Assign weights and input values for new axonal fibers in the current dendrite that provide correct classification for the misclassified pattern.]

S10. Let $I = I \cup I'$; $L = E \cup H$
[Update index sets I and L with only those input neurons and fibers needed to classify $\boldsymbol{x}^\gamma$ correctly.]

S11. Let $D' = \{ \mathbf{x}^\xi \in D : \forall i \in I,\ -w_{ik}^1 < x_i^\xi$ and $x_i^\xi < -w_{ik}^0 \}$
[Keep C_1 exemplars $\boldsymbol{x}^\xi$ $(\xi \neq \gamma)$ that do not belong to the recently created region in step S9; auxiliary set D' is used for updating set D that is reduced during the creation of new possible axonal fibers on the current dendrite.]

S12. If $D' = \emptyset$ **then** return to S2 **else** let $D = D'$ and loop to step S7
[Check if there is no need to wire more axonal fibers to the current dendrite. Going back to S2 means that the current dendrite is done, no more terminal fibers need to be wired in it but the neuron response is computed again to see if learning has been achieved; returning to S7 means that the current dendrite needs more fibers.]

   To illustrate the results of an implementation of the training algorithm based on elimination, we employed a data set from [21], where it was used to test a simulation of a *radial basis function network* (RBFN). The data set consists of two nonlinearly separable classes of 10 patterns each, where the class of interest, C_1, comprises the patterns depicted with solid circles ($\bullet$). All patterns were used for both training and test. Figure 2.4 compares the class

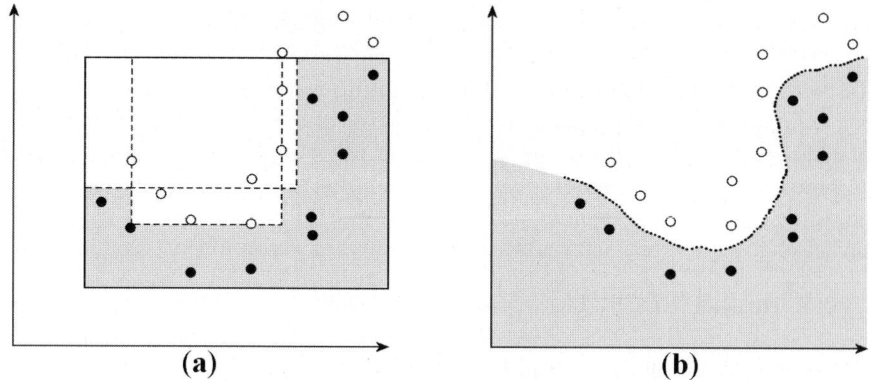

Fig. 2.4. The closed class C_1 region (shaded) learned by an **SLLP** with dendritic structures using the elimination algorithm (a), in comparison to the open region learned by an **MLP** (b), both applied to the data set from [21]. During training, the **SLLP** grows only 3 dendrites, one excitatory and two inhibitory (dashed). Compare (a) to the output in Fig. 2.5 of the merging version of the **SLLP** training algorithm

C_1 regions learned by an **SLLP** with dendritic structures using the elimination-based algorithm (a) and, respectively, by a backpropagation **MLP** (b).

The first step of the algorithm creates the first dendrite, which sends an excitatory message to the cell body of the output neuron if and only if a point of $\mathbb{R}^2$ is in the rectangle (solid lines) shown in Fig. 2.4(a). This rectangle encloses the entire training set of points belonging to class C_1. Subsequent steps of the algorithm create two more dendrites having inhibitory responses. These dendrites will inhibit responses to points in the carved out region of the rectangle as indicated by the dashed lines in Fig. 2.4(a). The only "visible" region for the output neuron will now be the dark shaded area of Fig. 2.4(a).

The three dendrites grown in a single epoch during training of the **SLLP** are sufficient to partition the pattern space. In contrast, the **MLP** created the open surface in Fig. 2.4(b) using 13 hidden units and 2000 epochs. The **RBFN** also required 13 basis functions in its hidden layer [21]. The separation surfaces drawn by the **SLLP** are closed, which is not the case for the **MLP**, and classification is 100% correct, as guaranteed by the theorems listed at the beginning of this section. Additional comments and examples in relation to Algorithm 2.1 are discussed in [13].

2.3.2 Training based on Merging

A training algorithm based on merging for a **SLLP** is given below. The algorithm constructs and trains an **SLLP** with dendritic structures to recognize the patterns belonging to the class of interest C_1. The remaining patterns in the training set are labeled as belonging to class $C_0 = \neg C_1$. The input to the algorithm consists of a set of training patterns $X = \{x^1, \ldots, x^k\} \subset \mathbb{R}^n$

with associated desired outputs y^ξ with $y^\xi = 1$ if and only if $\boldsymbol{x}^\xi \in C_1$, and $y^\xi = 0$ if $\boldsymbol{x}^\xi$ is not in class C_1. In the form provided below, the algorithm will construct small hyperboxes about training patterns belonging to C_1 and then *merge* these hyperboxes that are close to each other in the pattern space. For this reason, the number of dendrites of the output neuron M created by the algorithm will exceed the cardinality of the number of training patterns belonging to class C_1. Besides having one dendrite per pattern in class C_1, M may also have dendrites for some unique pattern pairs of patterns in C_1.

Algorithm 2.2. SLLP Training by Merging

S1. **Let** $k = 0$ and set all n-dimensional pattern pairs $\{\mathbf{x}^\xi,\ \mathbf{x}^\nu\}$ from C_1 as unmarked; **let**

$$d_{\min} = \bigwedge_{\xi \neq \gamma} d(\mathbf{x}^\xi, \mathbf{x}^\gamma) = \bigwedge_{\xi \neq \gamma} \{ \bigvee_{i=1}^{n} |x_i^\xi - x_i^\gamma| : \mathbf{x}^\xi \in C_1, \mathbf{x}^\gamma \in C_0 \};$$

let $\varepsilon = 0.5 d_{\min}$; **for** $i = 1, \ldots, n$ set $\alpha_i^1, \alpha_i^0 < \varepsilon$

[k is a dendrite counter, the α's are $2n$ accuracy factors where $d_{\min}$ is the minimal Chebyshev interset distance between classes C_1 and C_0; ε is a tolerance parameter for merging two hyperboxes.]

S2. **For** $\boldsymbol{x}^\xi \in C_1$ **do** step S3

S3. **Let** $k = k + 1$; **for** $i = 1, \ldots, n$ **do** $w_{ik}^1 = -(x_i^\xi - \alpha_i^1)$, $w_{ik}^0 = -(x_i^\xi + \alpha_i^0)$
[Create a dendrite on the neural body of the output neuron M and for each input neuron N_i grow two axonal fibers on the kth dendrite by defining their weights.]

S4. **For** $\boldsymbol{x}^\xi \in C_1$ **do** steps S5 to S10 [Outer loop.]

S5. **Mark** each unmarked pattern pair $\{\boldsymbol{x}^\xi, \boldsymbol{x}^\nu\}$, where $\mathbf{x}^\xi$ is fixed by step S4 and **do** steps S6 to S10
[The pattern pair can be chosen in the order $\boldsymbol{x}^\nu$ appears in the training set or at random.]

S6. **If** $d(\boldsymbol{x}^\xi, \boldsymbol{x}^\nu) < d_{\min} + \varepsilon$, **do** steps S7 to S10 **else** loop to step S5

S7. **Identify** a region $\mathcal{R}$ in pattern space that connects patterns $\boldsymbol{x}^\xi$ and $\boldsymbol{x}^\nu$; **for** $i = 1, \ldots, n$ define a set of merging parameters ε_i
such that $0 < \varepsilon_i \leq d_{\min} + \varepsilon$
[This ensures a user defined merging size for each dimension. If $|x_i^\xi - x_i^\nu| \geq \varepsilon_i$, region $\mathcal{R}$ will be bounded by the hyperplanes x_i^ξ and x_i^ν or by the hyperplanes $0.5(x_i^\xi + x_i^\nu - \varepsilon_i)$ and $0.5(x_i^\xi + x_i^\nu + \varepsilon_i)$ otherwise.]

S8. **If** $\mathbf{x}^\gamma \in C_0$ and $\mathbf{x}^\gamma \notin \mathcal{R}$ **then** do step S9 **else** do step S10
[$\mathcal{R}$ is the merged region determined in step S7.]

S9. Let $k = k + 1$; **for** $i = 1, \ldots, n$ **do**

    if $|x_i^\xi - x_i^\gamma| \geq \varepsilon_i$

        then $w_{ik}^1 = -\min\{x_i^\xi, x_i^\gamma\}$, $w_{ik}^0 = -\max\{x_i^\xi, x_i^\gamma\}$,

        else $w_{ik}^1 = -0.5(x_i^\xi + x_i^\gamma - \varepsilon_i)$, $w_{ik}^0 = -0.5(x_i^\xi + x_i^\gamma + \varepsilon_i)$

[Create a new dendrite and axonal branches that recognize the new region.]

S10. If there are unmarked pattern pairs remaining, loop to step **S5**; **if** all class C_1 pattern pairs $\{\boldsymbol{x}^\xi, \boldsymbol{x}^\nu\}$ with the fixed $\boldsymbol{x}^\xi$ have been marked, **then** loop to step **S4**; when outer loop exits, **let** $K = k$ and STOP

According to this algorithm, the output neuron M will be endowed with K dendrites. If $|C_1|$ denotes the number of training patterns belonging to class C_1, then $K - |C_1|$ corresponds to the number of dendrites that recognize regions that connect a pattern pair, while $|C_1|$ of the dendrites recognize regions around individual class one patterns. We need to point out that this training algorithm as well as our previously mentioned algorithm starts with the creation of hyperboxes enclosing training points. In this sense there is some similarity between our algorithms and those established in the fuzzy min-max neural networks approach, which also uses hyperboxes [19, 20]. However, this is also where the similarity ends, as all subsequent steps are completely different. Furthermore, our approach does not employ fuzzy set theory.

Figure 2.5 illustrates the results of an implementation of the SLLP training algorithm based on merging, applied to the same data set as in Fig. 2.4. Again, all patterns were used for both training and test. During training 19 excitatory dendrites are grown, 10 for regions around each pattern from class C_1 and 9 more to merge the individual regions. The separation surface is closed and recognition is 100% correct, as expected.

There is one more region in Fig. 2.5, drawn in dashed line, corresponding to an inhibitory dendrite, and its presence is explained as follows. The merging algorithm outlined above creates regions that are sized to avoid

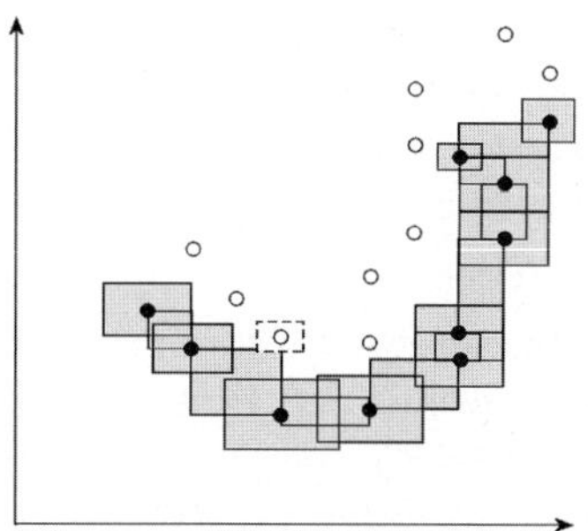

Fig. 2.5. The class C_1 region (shaded) learned by an SLLP with dendritic structures using the merging-based algorithm, applied to the data set from [21]. During training, the SLLP grows 20 dendrites, 19 excitatory and 1 inhibitory (dashed). Compare to the results in Fig. 2.4(a) obtained with the elimination version of the algorithm

touching patterns belonging to class C_0 or approaching them closer than a certain distance. In a more general version, larger hyper-boxes are allowed to be constructed. In this case, an additional step would identify foreign patterns that are approached or touched, and create inhibitory dendrites to eliminate those patterns. This more general approach is being used in the experiment of Fig. 2.5 and explains the presence of the inhibitory dendrite whose region is depicted with dashed line.

2.4 Multiple Class Learning in SLLPs

For better clarity of the description, the training algorithms described so far were limited to a single non-zero class, which corresponds to a single output neuron of the **SLLP** with dendritic structures. Below we present a straightforward generalization to multiple classes, which will invoke either one of the two procedures (elimination or merging) discussed in Sect. 2.3 as a subroutine.

The generalized algorithm consists of a main loop that is iterated m times, where m represents the number of non-zero classes and also the number of output neurons of the resulting **SLLP**. Within the loop, the single-class procedure is invoked. Thus, one output neuron at a time is created and trained to classify the patterns belonging to its corresponding class. The algorithm proceeds as follows.

Algorithm 2.3. Multi-class SLLP Training

S1. For $j = 1, \ldots, m$ **do** steps **S2** through **S4**
[j is a non-zero class index.]

S2. Generate the output neuron M_j
[Create a new output neuron for each non-zero class.]

S3. For each pattern x^ξ of the training set **do**
  If x^ξ is labeled as belonging to class C_j
   then temporarily reassign x^ξ as belonging to C_1
   else temporarily reassign x^ξ to class C_0
[The assignment is for this iteration only; original pattern labels are needed in subsequent iterations.]

S4. Call a single-class procedure to train output neuron M_j on the
  training set modified to contain patterns of only one non-zero class
[Algorithm 2.1 (elimination) or 2.2 (merging) can be used here.]

The straightforward generalization presented above suffers from a potential problem. The resulting **SLLP** partitions the pattern space in regions that might partially overlap. It is desirable that the learned regions be disjoint. Otherwise, a test pattern located in an area of overlap will erroneously be classified as belonging to more than one class. Theorem 2.2 guarantees the existence of an

SLLP with multiple output neurons that is able to classify m classes disjointly. Therefore, the generalization of the algorithm can be modified to prevent overlap between classes.

One way of modifying the algorithm would consist of taking into account current information during training about the shape of the regions learned so far, and using this information when growing new dendrites and assigning synaptic weights. An alternative would be to draw regions of controlled size based on minimum interset distance, in such a way that two regions that belong to different classes cannot touch. A similar idea was mentioned in the training algorithm based on merging.

Yet another approach to prevent overlap of different class regions would involve the augmentation of the SLLP with an additional layer of neurons. The former output layer of the SLLP will thus become the hidden layer of a two-layer lattice perceptron with dendritic structures. The role of the supplemental layer is basically to change the neural values of the hidden nodes, where several can be active simultaneously, into values where at most one output neuron may be active (may fire) at a time.

It is worthwhile mentioning that multiple layers are not required to solve a nonlinear problem with a lattice perceptron, as is the case for classical perceptrons. The theorems in Sect. 2.3 prove that a single layer is sufficient. The two-layer SLLP described in the previous paragraph simply provides a conceivable manner to prevent ambiguous classification in the straightforward generalization of the training algorithm. Existence of single layer lattice perceptrons that are able to solve a multi-class problem with no class overlap is guaranteed.

2.5 Dendritic Lattice Associative Memories

One goal in the theory of associative memories is for the memory to recall a stored pattern $y \in \mathbb{R}^m$ when presented a pattern $x \in \mathbb{R}^n$, where the pattern association expresses some desired pattern correlation. More precisely, suppose $X = \{x^1, \ldots, x^k\} \subset \mathbb{R}^n$ and $Y = \{y^1, \ldots, y^k\} \subset \mathbb{R}^m$ are two sets of pattern vectors with desired association given by the diagonal $\{(x^\xi, y^\xi) : \xi = 1, \ldots, k\}$ of $X \times Y$. The goal is to store these pattern pairs in some memory $\mathcal{M}$ such that for $\xi = 1, \ldots, k$, $\mathcal{M}$ recalls y^ξ when presented with the pattern x^ξ. If such a memory $\mathcal{M}$ exists, then we shall express this association symbolically by $x^\xi \rightarrow \mathcal{M} \rightarrow y^\xi$. Additionally, it is generally desirable for $\mathcal{M}$ to be able to recall y^ξ even when presented with a somewhat corrupted version of x^ξ.

A modification of the lattice based perceptron with dendritic structure leads to a novel associative memory that is distinct from matrix correlation memories and their various derivatives. This new *dendritic lattice associative memory* or DLAM can store any desirable number of pattern associations, has perfect recall when presented with an exemplar pattern and is extremely robust in the presence of noise. For this new hetero-associative memory one defines a set of sensory (input) neurons $N_1, \ldots, N_n$ that receive input $\boldsymbol{x}$ from the space $\mathbb{R}^n$ with N_i receiving input x_i, the ith-coordinate of $\boldsymbol{x}$. If, as before, $X = \{\boldsymbol{x}^1, \ldots, \boldsymbol{x}^k\} \subset \mathbb{R}^n$ represents the set of exemplar patterns, then the input neurons will propagate their input values x_i to a set of k hidden neurons $H_1, \ldots, H_k$, where each H_j has exactly one dendrite. Every input neuron N_i has exactly two axonal fibers terminating on the dendrite of H_j. The weights of the terminal fibers of N_i terminating on the dendrite of H_j are given by

$$w_{ij}^\ell = \begin{cases} -(x_i^j - \alpha_j) & \Leftrightarrow \ell = 1 \\ -(x_i^j + \alpha_j) & \Leftrightarrow \ell = 0 \end{cases}, \tag{2.6}$$

where $i = 1, \ldots, n$ and $j = 1, \ldots, k$, and for each $j \in \{1, \ldots, k\}$, the number $\alpha_j > 0$ is a user defined *noise parameter* associated with the pattern x^j. This noise parameter needs to satisfy

$$\alpha_j < \frac{1}{2}\min\{d(\boldsymbol{x}^j, \boldsymbol{x}^\gamma) : \gamma \in K(j)\} = \frac{1}{2}d_{\min}, \tag{2.7}$$

where $K(j) = \{1, \ldots, k\} \setminus \{j\}$ and $d(\boldsymbol{x}^j, \boldsymbol{x}^\gamma)$ denotes the Chebyshev distance between patterns $\boldsymbol{x}^j$ and $\boldsymbol{x}^\gamma$ (see Def. 2.1).

For a given input $\boldsymbol{x} \in \mathbb{R}^n$, the dendrite of the hidden unit H_j computes

$$\tau^j(\boldsymbol{x}) = \bigwedge_{i=1}^{n} \bigwedge_{\ell=0}^{1} (-1)^{1-\ell}(x_i + w_{ij}^\ell). \tag{2.8}$$

The state of the neuron H_j is determined by the hard-limiter activation function

$$f(z) = \begin{cases} 0 & \Leftrightarrow z \geq 0 \\ -\infty & \Leftrightarrow z < 0 \end{cases}. \tag{2.9}$$

The output of H_j is given by $f[\tau^j(\boldsymbol{x})]$ and is passed along its axonal fibers to m output neurons $M_1, \ldots, M_m$. The activation function defined by (2.9) is a hard-limiter in the algebra $\mathcal{A} = (\mathbb{R}_{-\infty}, \vee, +)$ since the *zero* of $\mathcal{A}$ is $-\infty$ (for the operation $\vee$) and the *unit* of $\mathcal{A}$ corresponds to 0. This mirrors the hard-limiter in the algebra $(\mathbb{R}, +, \times)$ defined by $f(z) = 0$ if $z < 0$ and $f(z) = 1$ if $z \geq 0$, since in this algebra the zero is 0 and the unit is 1.

Similar to the hidden layer neurons, each output neuron M_h, where $h = 1, \ldots, m$, has one dendrite. However, each hidden neuron H_j has exactly one excitatory axonal fiber and no inhibitory fibers terminating on the dendrite of M_h. Figure 2.6 illustrates this dendritic network model. The excitatory fiber of H_j terminating on M_h has synaptic weight $v_{jh} = y_h^j$. Note that, for an auto-associative **DLAM**, $m = n$ and $v_{jh} = x_h^j$.

The computation performed by M_h is given by

$$\tau^h(s) = \bigvee_{j=1}^{k} (s_j + v_{jh}), \tag{2.10}$$

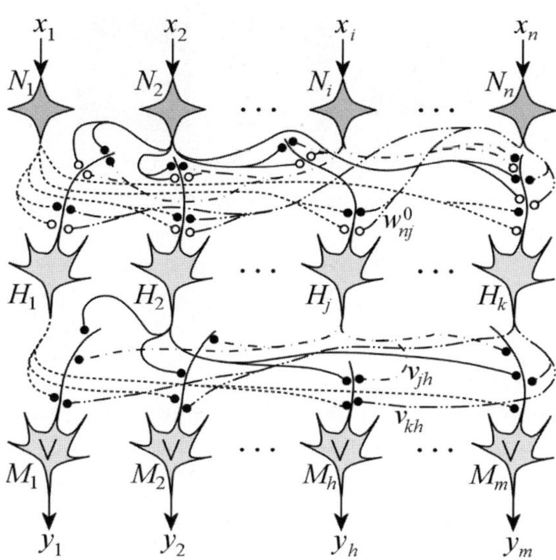

Fig. 2.6. The topology of a DLAM. This associative memory is a feedforward fully connected neural network; all axonal branches from input neurons N_i synapse via two fibers (excitatory and inhibitory) on all hidden neurons H_j, which in turn connect to all output nodes M_h via excitatory fibers

where s_j denotes the output of H_j, namely $s_j = f[\tau^j(\boldsymbol{x})]$, with f defined in (2.9). The activation function g for each output neuron M_h is the linear identity function $g(z) = z$.

Each neuron H_j will have the output value $s_j = 0$ if and only if $\mathbf{x}$ is an element of the hypercube $B^j = \{\boldsymbol{x} \in \mathbb{R}^n : x_i^j - \alpha_j \le x_i \le x_i^j + \alpha_j\}$ and $s_j = -\infty$ whenever $\boldsymbol{x} \in \mathbb{R}^n \setminus B^j$. The output of the network having the weights set according to (2.6) will be $\boldsymbol{y} = (y_1, \ldots, y_m)' = (y_1^j, \ldots, y_m^j)' = \boldsymbol{y}^j$ if and only if $\boldsymbol{x} \in B^j$. That is, whenever $\boldsymbol{x}$ is a corrupted version of $\boldsymbol{x}^j$ with each coordinate of $\boldsymbol{x}$ not exceeding the allowable noise level α_j, then $\boldsymbol{x}$ will be associated with $\boldsymbol{x}^j$. If the amount of noise exceeds the level α_j, then the network rejects the input by yielding the output vector $(-\infty, \ldots, -\infty)'$. Obviously, each uncorrupted pattern $\boldsymbol{x}^\xi$ will be associated with $\boldsymbol{x}^\xi$.

It is important to remark that the α_j values computed with (2.7) are *input independent* and establish maximum allowable noise bounds on input patterns in order to guarantee their perfect recall. In addition, the α_j parameter for the jth exemplar pattern depends only on distances against the other exemplar patterns. An alternative scheme considers a single *input dependent* α value that takes into account the distances between an arbitrary noisy input and each exemplar pattern. Let $\tilde{\boldsymbol{x}} \in \mathbb{R}^n$ denote a noisy version of an exemplar pattern $\boldsymbol{x} \in X$, then the value of α used for the weights in (2.6) is given by

$$\alpha = \min_{1 \le \gamma \le k} \{d(\tilde{\boldsymbol{x}}, \boldsymbol{x}^\gamma)\} = \tilde{d}_{\min} \; ; \; \tilde{\boldsymbol{x}} \notin X \, , \; \boldsymbol{x}^\gamma \in X. \tag{2.11}$$

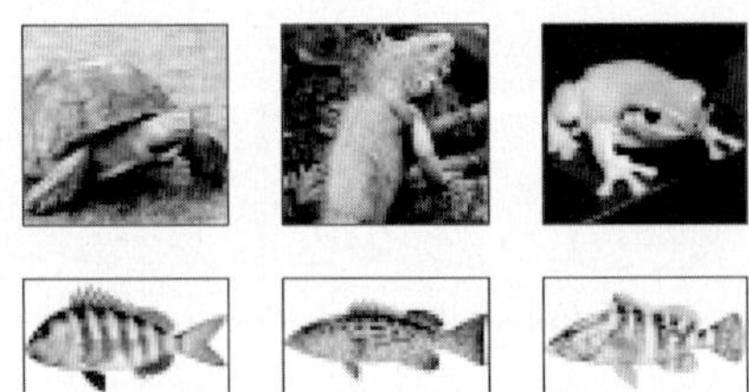

Fig. 2.7. Top row depicts the images $\boldsymbol{p}^1, \boldsymbol{p}^2, \boldsymbol{p}^3$, which were converted into the prototype patterns of the set $X = \{\boldsymbol{x}^1, \boldsymbol{x}^2, \boldsymbol{x}^3\}$; bottom row shows the images used to generate the corresponding association patterns from the set $Y = \{\boldsymbol{y}^1, \boldsymbol{y}^2, \boldsymbol{y}^3\}$

The modified **DLAM** has the advantage that it can work with inputs corrupted by *unspecified* amounts of random noise and still preserves robustness.

2.5.1 Image Retrieval from Noisy Inputs with a **DLAM**

To illustrate the performance of this associative memory, we use a visual example consisting of the associated image pairs $P = \{\boldsymbol{p}^1, \boldsymbol{p}^2, \boldsymbol{p}^3\}$ and $Q = \{\boldsymbol{q}^1, \boldsymbol{q}^2, \boldsymbol{q}^3\}$ shown in Fig. 2.7. Each $\boldsymbol{p}^\xi$ is a 50×50 pixel 256-gray scale image, whereas each $\mathbf{q}^\xi$ is a 30×50 pixel 256-gray scale image, where $\xi = 1, 2, 3$. Using the standard row-scan method, each pattern image $\boldsymbol{p}^\xi$ and $\boldsymbol{q}^\xi$ was converted into an associative pair of pattern vectors $\boldsymbol{x}^\xi = (x_1^\xi, \ldots, x_{2500}^\xi)'$ and $\boldsymbol{y}^\xi = (y_1^\xi, \ldots, y_{1500}^\xi)'$. Thus, for this particular example, we have $X = \{\boldsymbol{x}^1, \boldsymbol{x}^2, \boldsymbol{x}^3\} \subset \mathbb{R}^{2500}$ and $Y = \{\boldsymbol{y}^1, \boldsymbol{y}^2, \boldsymbol{y}^3\} \subset \mathbb{R}^{1500}$.

The patterns illustrated in the top row of Fig. 2.8 were obtained by distorting 100% of the vector components of each $\boldsymbol{x}^j$ within a noise level α_j, chosen to satisfy the inequality in (2.7). Specifically, we set

$$\alpha_1 = 0.49\min[d(\boldsymbol{x}^1, \boldsymbol{x}^2), d(\boldsymbol{x}^1, \boldsymbol{x}^3)] \approx 97\,,$$
$$\alpha_2 = 0.49\min[d(\boldsymbol{x}^2, \boldsymbol{x}^1), d(\boldsymbol{x}^2, \boldsymbol{x}^3)] \approx 92\,, \tag{2.12}$$
$$\alpha_3 = 0.49\min[d(\boldsymbol{x}^3, \boldsymbol{x}^1), d(\boldsymbol{x}^3, \boldsymbol{x}^2)] \approx 92\,.$$

The values were obtained by truncation to the nearest integer. Figure 2.8 shows that our model achieves perfect recall association when presented the very noisy versions of the exemplar patterns.

In the case of the modified **DLAM** that employs the α bound defined in (2.11), further experiments were performed on the amphibian images considered as a set of auto-associations. Different kinds of distortions, such as *random dilative*, *random erosive*, and *random mixed* noise with probabilities of $0.2, 0.3, 0.4$, were tested. The noise is introduced in the exemplar patterns without being conditioned or constrained to a pre-established value. For the set $X = \{\boldsymbol{x}^1, \boldsymbol{x}^2, \boldsymbol{x}^3\}$ of exemplar patterns (amphibians), a matrix D can be defined that contains in the upper right portion the different distances obtained from computation; for convenience, all entries are initialized to ∞.

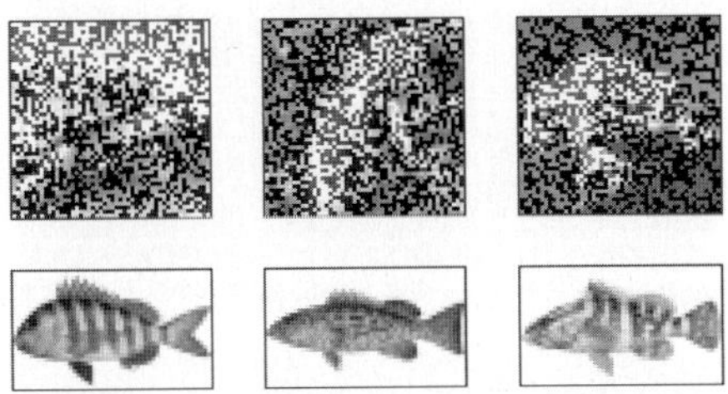

Fig. 2.8. Top row shows the corrupted prototype patterns $\tilde{x}^1, \tilde{x}^2, \tilde{x}^3$; bottom row illustrates that the memory makes the correct association when presented with the corrupted input

Table 2.1. Test cases for the modified DLAM

Test	Pattern	Noise type	D^*	α-range	γ
1	x^1	40%-dilative	$(128, 253, 247)$	$[128, 247)$	1
2	x^2	30%-dilative	$(245, 128, 252)$	$[128, 245)$	2
3	x^3	20%-dilative	$(240, 233, 128)$	$[128, 233)$	3
4	x^1	40%-erosive	$(128, 211, 242)$	$[128, 211)$	1
5	x^2	30%-erosive	$(224, 128, 227)$	$[128, 224)$	2
6	x^3	20%-erosive	$(241, 220, 128)$	$[128, 220)$	3
7	x^1	40%-mixed	$(128, 253, 247)$	$[128, 247)$	1
8	x^2	30%-mixed	$(249, 128, 253)$	$[128, 249)$	2
9	x^3	20%-mixed	$(247, 229, 128)$	$[128, 229)$	3

Given a noisy input, the checkerboard distance matrix is recomputed using the distorted input pattern $\tilde{x}$. In this example, the D matrix has the following entries

$$D = \begin{pmatrix} d_{12} & d_{13} & d_{14} \\ \infty & d_{23} & d_{24} \\ \infty & \infty & d_{34} \end{pmatrix} \quad ; \quad D^* = (d_{14}, d_{24}, d_{34}), \tag{2.13}$$

where the third column registers the distances of each exemplar pattern to the noisy one (the fourth pattern). For brevity, D^* will denote the transpose of the third column of matrix D. Table 2.1 lists some noisy patterns with their respective distances to the three exemplar patterns and Fig. 2.9 displays the exemplars as well as the noisy versions used as test cases.

The 1st column numbers each test, the 2nd column specifies the corrupted exemplar pattern, the 3rd column describes the type of random noise, the 4th column gives the Chebyshev distance vector of the noisy input with respect to each exemplar pattern, the 5th column gives the valid recollection α-*range*, and the last column gives the pattern index showing perfect recall.

Any value of $\alpha \in [\tilde{d}_{\min}, \eta)$, where η is the next minimum distance, i.e., the second closer exemplar pattern to $\tilde{x}$, can be used for weight assignment. If $\alpha < \tilde{d}_{\min}$ the recalled pattern is always zero (black image) and if $\alpha \geq \eta$, the recalled pattern turns to be a mixture with other exemplar patterns. Observe

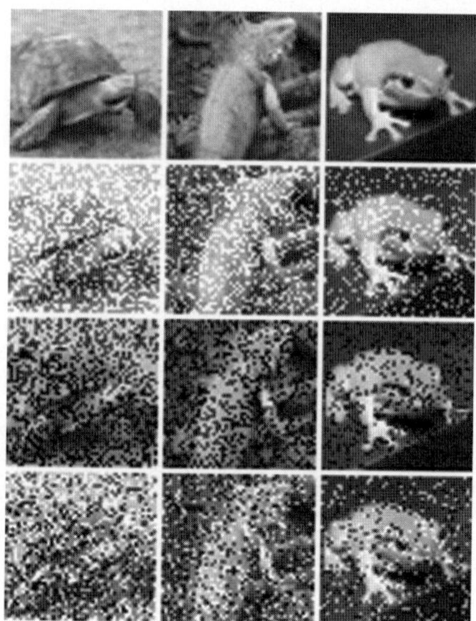

Fig. 2.9. 1st row, exemplar patterns; row 2, dilative noisy versions (test cases 1, 2, 3); row 3, erosive noise versions (test cases 4, 5, 6); row 4, patterns contaminated with mixed random noise (test cases 7, 8, 9)

that, if the input independent α_j values in (2.12) are used the DLAM fails since $\alpha_j < \widetilde{d}_{\min}$ for $j = 1, 2, 3$.

An earlier approach to lattice associative memories is based on matrix minimax algebra whose description the reader will find in Chap. 5 of this book, *Noise Masking for Pattern Recall using a Single Lattice Matrix Associative Memory* by Urcid and Ritter. See also Chap. 8, *Morphological and Certain Fuzzy Morphological Associative Memories for Classification and Prediction*, where the use of fuzzy set theory in lattice associative memories has been developed and applied successfully by Sussner and Valle.

2.6 Conclusions

We presented a single layer feed-forward neural network (SLLP) for pattern recognition and a two layer associative memory with both of these networks taking into account the dendritic processes of neurons. Emphasis was on learning. The theory and examples presented make it obvious that the SLLP has several advantages over both traditional single layer perceptrons and perceptrons with hidden layers. For instance, the SLLP needs no hidden layers when solving nonconvex problems. There are no convergence problems and speed of learning far exceeds traditional back propagation methods. Questions may arise as to whether dendrites merely represent hidden layers. Such questions are valid in light of the fact that, theoretically, a two hidden layer perceptron can also classify any compact region. In comparison to hidden neurons in regular perceptrons, dendrites use no activation functions while hidden neurons usually employ sigmoidal activation functions. Also, with hidden layers, the number of hidden neurons are predetermined before training. In the model presented, dendrites and axonal terminal fibers are grown automatically as the neuron learns a specific task. This is analogous to cortical neurons which grow dendrites and synaptic connections during learning. Furthermore, no error remains after training. All pattern vectors of the training set will always be correctly identified after training stops.

To be precise, learning does not really take place in the lattice associative memory. Here dendrites and synaptic connections are hard-wired and weights are directly computed from the associated pattern vectors. However, the dendritic lattice approach has major advantages over traditional memories such as Hopfield or Kohonen matrix correlation memories and their various modifications. Again, there are no convergence problems. Storage capacity of dendritic lattice associative memories described here is unlimited; for any set of k associations, only k hidden neurons are required and the network can learn new pattern associations by adding one hidden neuron with appropriate connections to the input/output layer neurons for each new association. Associations for uncorrupted input is always perfect and, as illustrated, the memory is extremely robust in the presence of noisy inputs.

References

1. Eccles JC (1977) The Understanding of the Brain, McGraw-Hill, New York, NY
2. Kaburlasos VG (2003) Improved fuzzy lattice neurocomputing for semantic neural computing. In: Proc Inter Joint Conf on Neural Networks pp 1850–1855
3. Keller JM, Hunt DJ (1985) Incorporating fuzzy membership functions into the perceptron algorithm. IEEE Trans Pattern Analysis and Machine Intelligence 7:693–699
4. Koch C, Segev I (1989) Methods in Neuronal Modeling: From Synapses to Networks. MIT Press, Cambridge, MA
5. Marks RJ, Oh S, Arabshahi P et al (1992) Steepest descent adaption of min-max fuzzy if-then rules. In: Proc Inter Joint Conf on Neural Networks 3:471–477
6. Myers DS (2006) The synaptic morphological perceptron. In: Ritter GX, Schmalz MS, Barrera J, Astola JT (eds) Proc SPIE Math of Data/Image Pattern Recognition, Compr and Encryp 6315 pp 63150B:1–11
7. Petridis V, Kaburlasos VG (1998) Fuzzy lattice neural networks: a hybrid model for learning. IEEE Trans on Neural Networks 9:877–890
8. Rall W, Segev I (1987) Functional possibilities for synapses on dendrites and dendritic spines. In: Edelman GM, Gall EE, Cowan WM (eds) Synaptic Function. Wiley, New York, NY
9. Ritter GX, Sussner P (1997) Associative memories based on lattice algebra. In: Proc IEEE Inter Conf on Sys, Man & Cyber pp 3570–3575
10. Ritter GX, Sussner P, Diaz de Leon JL (1998) Morphological associative memories. IEEE Trans on Neural Networks 9:281–293

11. Ritter GX, Diaz de Leon JL, Sussner P (1999) Morphological bidirectional associative memories. Neural Networks 12:851–867
12. Ritter GX, Urcid G, Iancu L (2003) Reconstruction of noisy patterns using morphological associative memories. J Math Im & Vision 19:95–111
13. Ritter GX, Urcid G (2003) Lattice algebra approach to single neuron computation. IEEE Trans on Neural Networks 14(2):282–295
14. Ritter GX, Iancu L, Urcid G (2003) Morphological perceptrons with dendritic structure. In: Proc FUZZ-IEEE pp 1296–1301
15. Ritter GX, Iancu L (2003) Single layer feedforward neural network based on lattice algebra. In: Proc IJCNN pp 2887–2892
16. Segev I (1998) Dendritic processing. In: Arbib M (ed) Handbook of Brain Theory and Neural Networks. MIT Press, Cambridge, MA
17. Sejnowski TJ, Qian N (1992) Synaptic integration by electro-diffusion in dendritic spines. In: McKenna T, Davis J, Zornetzer SF (eds) Single Neuron Computation. Academic Press, New York, NY
18. Urcid G (2005) Transformations of neural inputs in lattice dendrite computation. In: Proc SPIE Math Methods in Pattern and Image Analysis 5916:201–212
19. Simpson PK (1992) Fuzzy min-max neural networks - part 1: classification. IEEE Trans on Neural Networks 3: 776–786
20. Simpson PK (1993) Fuzzy min-max neural networks - part 2: clustering. IEEE Trans on Fuzzy Systems 1:32–45
21. Wasnikar VA, Kulkarni AD (2000) Data mining with radial basis functions. In: Dagli CH et al (eds) Intelligent Engineering Systems Through Artificial Neural Networks. ASME Press, New York, NY
22. Won Y, Gader PD, Coffield P (1997) Morphological shared-weight neural networks with applications to automatic target recognition. IEEE Trans on Neural Networks 8:1195–1203
23. Zhang XH, Wong F, Wang PZ (1993) Decision-support neural networks and its application in financial forecast. Bull of Studies and Exchanges on Fuzziness Applications 54:40–49
24. Zhang XH, Hang C, Tan S, Wang PZ (1996) The min-max function differentiation and training of fuzzy neural networks. IEEE Trans on Neural Networks 7:1139–1150

Orthonormal Basis Lattice Neural Networks

Angelos Barmpoutis[1] and Gerhard X. Ritter[2]

[1] University of Florida, Gainesville, FL 32611, USA
  abarmpou@cise.ufl.edu
[2] University of Florida, Gainesville, FL 32611, USA
  ritter@cise.ufl.edu

Summary. Lattice based neural networks are capable of resolving some difficult non-linear problems and have been successfully employed to solve real-world problems. In this chapter a novel model of a lattice neural network (LNN) is presented. This new model generalizes the standard basis lattice neural network (SB-LNN) based on dendritic computing. In particular, we show how each neural dendrite can work on a different orthonormal basis than the other dendrites. We present experimental results that demonstrate superior learning performance of the new Orthonormal Basis Lattice Neural Network (OB-LNN) over SB-LNNs.

3.1 Introduction

The artificial neural model which employs lattice based dendritic computation has been motivated by the fact that several researchers have proposed that dendrites, and not neurons, are the elementary computing devices of the brain, capable of implementing logical functions [1, 16]. Inspired by the neurons of the biological brain, a lattice based neuron that possesses dendritic structures was developed and is discussed in detail in [13, 14, 17].

Several applications of LNNs have been proposed, due to their high capability of resolving some difficult non-linear problems. LNNs were employed in applications for face and object localization [3, 9], Auto-Associative memories [10, 18, 19], color images retrieval and restoration [20] etc. Furthermore, various models of fuzzy lattice neural networks (FLNN) were studied in [5, 6] and some of their applications in the area of text classification and classification of structured data domains were presented in [7, 8].

Despite the high capabilities of LNNs, training dendritic networks such as Lattice Based Morphological Perceptrons with Dendritic Structure [13], results in creating huge neural networks with a large number of neurons; the size of the trained network sometimes is comparable to the size of the training data.

In this work a new model of LNNs is proposed. In this model each neural dendrite can work on a different orthonormal basis than the other dendrites.

A. Barmpoutis and G.X. Ritter: *Orthonormal Basis Lattice Neural Networks*, Studies in Computational Intelligence (SCI) **67**, 45–58 (2007)
www.springerlink.com   © Springer-Verlag Berlin Heidelberg 2007

The orthonormal basis of each dendrite is chosen appropriately in order to optimize the performance of the Orthonormal Basis Lattice Neural Network (OB-LNN). OB-LNNs have some useful properties such as automatic compression of the size of the neural network and they show significantly better learning capabilities than the standard basis LNNs. Validation experimental results in synthetic and real datasets are presented and demonstrate superior learning performance of OB-LNNs over SB-LNNs.

The rest of the chapter is organized into the following sections: In Sect. 3.2, we make a brief review of the LNN model. In Sect. 3.3, the Orthonormal Basis Lattice Neural Network is presented. This section is divided in two parts. In the first part the model of OB-LNNs is presented. This is followed by an algorithm for training such a neural network. Finally, in Sect. 3.4, validation experimental results are presented which demonstrate superior learning performance of OB-LNNs over standard basis LNNs.

3.2 Lattice Neural Networks

The primary distinction between traditional neural networks and LNNs is the computation performed by the individual neuron. Traditional neural networks use a multiply accumulate neuron with thresholding over the ring $(\Re, \times, +)$ given by the formula

$$\tau_j(x) = \sum_{i=1}^{n} x_i w_{ij} - \theta_j \tag{3.1}$$

where $\tau_j(x)$ is the total input to the jth neuron, x_i are the values of the input neurons connected with the jth neuron, and w_{ij} are their weights. Finally θ_j are bias weights.

In the case of a lattice based neuron, lattice operators $\vee$ (maximum) and $\wedge$ (minimum) are used. These operators and the addition $(+)$ form the rings $(\Re \cup -\infty, \vee, +)$ and $(\Re \cup \infty, \wedge, +)$. The computation performed by a neuron is given by the formula

$$\tau_j(x) = p_j \vee_{i=1}^{n} r_{ij}(x_i + w_{ij}) \tag{3.2}$$

or

$$\tau_j(x) = p_j \wedge_{i=1}^{n} r_{ij}(x_i + w_{ij}) \tag{3.3}$$

where $\tau_j(x)$ is the total input to the jth neuron, x_i are the values of the input neurons connected with the jth neuron, and w_{ij} are their weights. Parameters r_{ij} take $+1$ or -1 value if the ith input neuron causes excitation or inhibition to the jth neuron. P_j takes also $+1$ or -1 value if the type of the output response is excitatory or inhibitory. A more detailed presentation of the theory of lattice based morphological neurons and how their networks work can be found in [11, 14].

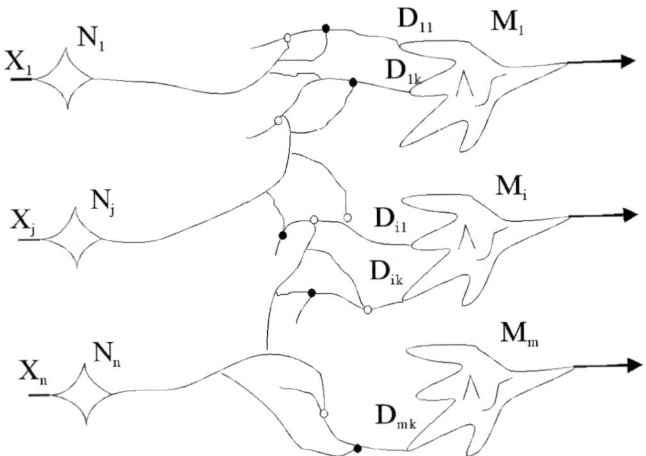

Fig. 3.1. An artificial neural network model which employs lattice based dendritic computations

Using the above computational framework, a lattice neural network can be constructed using layers of lattice based neurons which are connected to neurons of other layers. Each lattice based neuron consists of dendrites which connect the neuron with the previous layers neurons. Figure 3.1 shows a lattice neural network with an input layer and a lattice based neuron layer. Each lattice neuron M_j consists of dendrites D_{ij}. The neurons of the input layer are connected to the next layer via the dendrites. The black and white circles denote excitatory and inhibitory connection respectively. Each dendrite can be connected with an input neuron at most twice (with one inhibitory and one excitatory connection).

The computation performed by a single (the kth) dendrite can be expressed using lattice operators by the formula:

$$\tau_k(x) = p_k \wedge_{i \in I} \wedge_{l \in L} (-1)^{1-l}(x_i + w_{ik}^l) \tag{3.4}$$

where I is a subset of 1,...,n which corresponds to the set of all input neurons N_i with terminal fibers that synapse on the kth dendrite of the current lattice based neuron. L is a subset of 1,0 and the other parameters are the same with those used in (3.2) and (3.3).

The geometrical interpretation of the computation performed by a dendrite is that every single dendrite defines a hyperbox. The borders of this hyperbox form the decision boundaries in a particular location of the input space. The left part of Fig. 3.2 shows a hyperbox that separates data points of two different classes. Here the input space is the plane of real numbers ($\Re^2$) and the hyperbox is a rectangle. This hyperbox can be defined by a single dendrite via its weight values w_{ij}.

Decision boundaries with more complex shapes, can be formed by using more dendrites. Furthermore boundaries which separate more than two classes

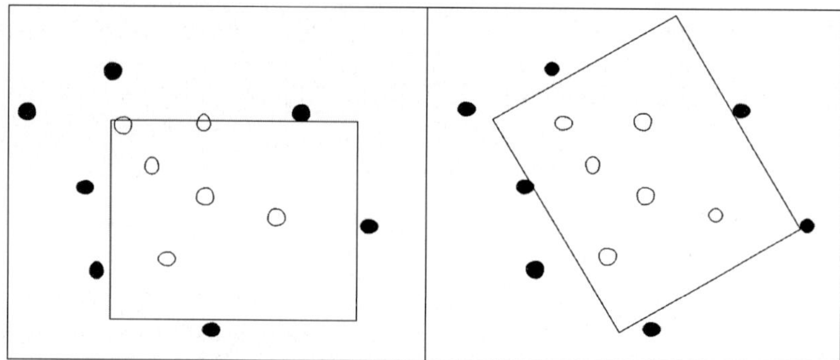

Fig. 3.2. Decision boundaries of a SB-LNN dendrite (left) and a OB-LNN dendrite (right)

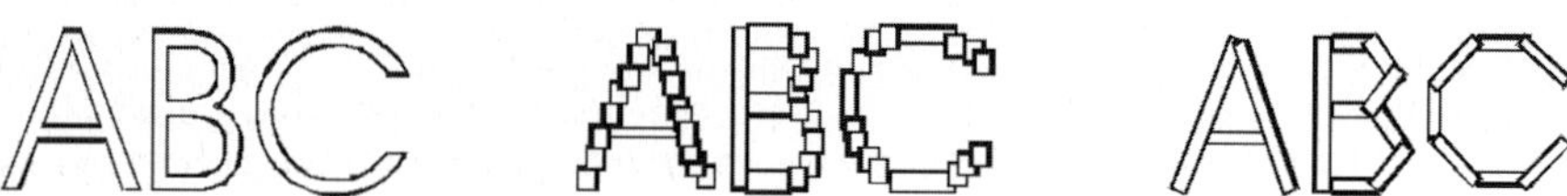

Fig. 3.3. This is an illustration of the decision boundaries that can be formed by a SB-LNN (middle) and an OB-LNN (right). The desired decision boundaries form the ABC letters (left)

can be formed, if more lattice based neurons are employed. The left part of Fig. 3.3 shows an example of decision boundaries with more complex shape, forming the letters ABC. In the middle of the same figure we can see how a group of hyperboxes (2-dimensional boxes in this case) can approximate the decision boundaries. Each box can be defined by a dendrite. Complicated figures require a large number of dendrites, in order to achieve satisfactory approximation of the decision boundaries.

Note that here the word hyperbox has a more general meaning, since some of the bounding planes of the hyperboxes may have infinite length. The word hyperbox as used in this article includes *open* hyperboxes. For example, in $\Re^1$ a half-open interval of form $[\alpha, \infty)$ or $(-\infty, \alpha]$ are half-open *boxes*, while in $\Re^2$ a rectangle as well as the convex area bounded by one, two, or three mutually orthogonal lines are also considered to be hyperboxes. Following similar reasoning, two parallel lines can be also considered as a hyperbox, etc.

An algorithm for training a lattice neural network with dendritic structure can be found in [11] and method for learning LNNs can be found in [17]. A comparison of various training methods for LNNs that employ dendritic computing is presented in [12]. The next section is divided into two parts. In the first part the model of Orthonormal Basis lattice neural network is presented. This is followed by an algorithm for training such a neural network.

3.3 Orthonormal Basis Lattice Neural Networks

In this section, the Orthonormal Basis Lattice Neural Network is presented. This section is divided into two parts. In the first part the model of OB-LNNs is presented. This is followed by an algorithm for training such a neural network.

3.3.1 The OB-LNN model

As it was discussed earlier, the geometrical interpretation of the computation performed by a dendrite is that every single dendrite defines a hyperbox in the space of input values. These hyperboxes are oriented parallel to the Cartesian axis of the space of input values. The left part of Fig. 3.2 presents a decision hyperbox defined by a dendrite. Its edges are parallel to the x and y axis of the input space ($\Re^2$). Due to this constraint about the orientation of the hyperboxes, the decision boundaries formed by lattice neural networks are not smooth and box patterns are annoyingly visible along the boundaries (Fig. 3.3 middle).

To overcome these disadvantages, a new type of dendrite can be defined, which is able to form a hyperbox parallel to an arbitrary orthonormal system. The right part of Fig. 3.2 presents such a hyperbox, whose orientation is no longer parallel to the Cartesian axis of the input space. A neural network consisting of lattice based neurons with such dendrites would be able to produce smoother decision boundaries (Fig. 3.3 right). In Fig. 3.3 the decision boundaries formed by such an Orthonormal Basis Lattice Neural Network are compared with these formed by a Standard Basis LNN. Notice that the number of hyperboxes (thus the number of dendrites as well) required by the OB-LNN is much smaller than the number of those required by a SB-LNN.

Another advantage of the Orthonormal Basis LNNs is that they store information about the local orientation of the classes. This is demonstrated with an example in Fig. 3.4. Suppose that the samples of a class form the shape of the letter A. A neural network with Orthonormal Basis Dendtrites

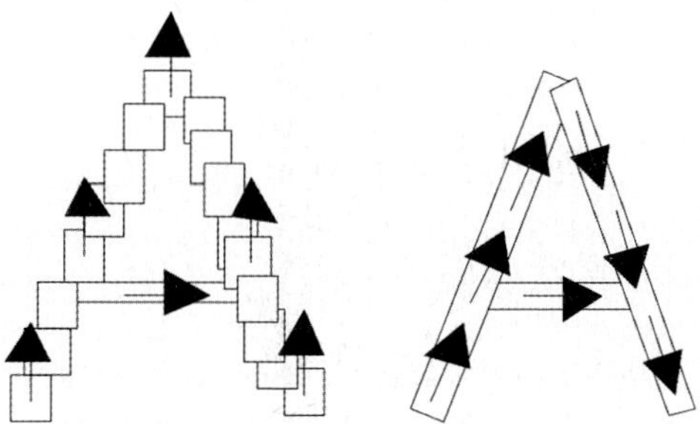

Fig. 3.4. Another advantage of OB-LBNN (right) is that they also store information about the local orientation of the classes

can approximate this shape forming mainly 3 hyperboxes. The orientation of each hyperbox contains information about the local orientation of this class (Fig. 3.4 right). This useful information cannot be obtained by the standard basis LNN that was trained for the same purpose (Fig. 3.4 left). This property is better illustrated in Sect. 3.4 (Fig. 3.6), where training results of standard and orthonormal basis LNNs are presented.

The computation performed by the kth Orthonormal Basis dendrite can be expressed using lattice operators, changing slightly (3.4) as follows

$$\tau_k(\mathbf{X}) = p_k \wedge_{i \in I} \wedge_{l \in L} (-1)^{1-l} \left[(\mathbf{R}_k \mathbf{X})_i + w_{ik}^l \right] \tag{3.5}$$

where $\mathbf{X}$ is the input value vector $(x_1, x_2, ..., x_I)^T$, $\mathbf{R}_k$ is a square matrix whose columns are unit vectors forming an orhonormal basis, and $(\mathbf{R}_k \mathbf{X})_i$ is the ith element of the vector $\mathbf{R}_k \mathbf{X}$. Each dendrite now works in its own orthonormal basis defined by the matrix $\mathbf{R}_k$. The weights of the dendrite act on the elements of vector $\mathbf{R}_k \mathbf{X}$, hence the weights act on the rotated by the orthonormal basis $\mathbf{R}_k$ space.

Note that Standard Basis Lattice Neural Networks are a sub group of the Orthonormal Basis Lattice Neural Networks where the matrix $\mathbf{R}_k$ is the identity matrix. In this case it is obvious $(\mathbf{R}_k \mathbf{X})_i = x_i$, thus (3.5) becomes equal to (3.4).

3.3.2 Training OB-LNNs

The training of an Orthonormal Basis Lattice Neural Network is based on finding the best possible values for the weights and $\mathbf{R}_k$ matrices. In other words, in order to train an OB-LNN, one must train its dendrites.

The training methods which can be used for training SB-LNNs can also be used for training OB-LNNs with an appropriate modification in order to adopt the fact that each dendrite works on its own orthonormal basis. The training procedure of an orthonormal basis dendrite can be treated as a maximization problem. The quantity that we are trying to maximize is the volume of the hyperbox which is defined by the dendrite. Due to this volume maximization process the OB-LNN training algorithm generally produces more compressed neural networks compared to those in the standard basis case. This property is expressed by lemma 1, which is discussed at the end of this section.

By fixing the matrix $\mathbf{R}_k$, the weights of the dendrite can be estimated by the training procedure described in [11]. The weights define the hyperbox; therefore its volume can be directly calculated from the weights. We can repeat the process by varying appropriately the matrix $\mathbf{R}_k$, until the volume of the hyperbox reaches a maximum.

The matrix $\mathbf{R}_k$ is a rotation matrix, i.e. it rotates the vector $\mathbf{X}$. A variation of this matrix $d\mathbf{R}$ is also a rotation matrix. A new rotation matrix $\mathbf{R}'_k$ can be obtained by multiplying $\mathbf{R}_k$ with matrix $d\mathbf{R}$, $(\mathbf{R}'_k = \mathbf{R}_k d\mathbf{R})$. A variation matrix $d\mathbf{R}$ can be easily constructed by using the following equation:

$$dR = exp(tS) \tag{3.6}$$

where $\mathbf{S}$ is a randomly generated skew symmetric matrix, and t is a scalar value. Note that the exponential is the matrix exponential. Note also that the matrix exponential of a skew symmetric matrix is always a rotation matrix. The smaller the absolute value of t is, the smaller the variation, which is caused by the matrix $d\mathbf{R}$, is. A brief review of the properties of matrix exponential and skew-symmetric matrices can be found in the appendix at the end of this article.

By using the above, any cost minimization method can be used in order to minimize the negative of the hyperboxes volume (or to maximize its volume) in steps 2 and 7 of Algorithm 1. Simulated annealing [15] and greedy searching for experiments in 2D are the methods used in the experiments presented in Sect. 3.4, in order to find the matrix $\mathbf{R}_k$ that gives the maximum hyperboxes volume.

The training algorithm of an OB-LNN morphological perceptron is summarized below. This algorithm is an extension, in the space of OB-LNN, of the training algorithm for morphological perceptron proposed in [11]. The algorithm is presented for the case of 2 classes only, but it can be easily extended to problems with a larger number of classes. A more detailed description of the training algorithm in the case of SB-LNN is presented in [11].

---

input : N training samples $\mathbf{X}_i$, and N outputs $d_i = 0$ or 1 for class C1 or C2, respectively, $i = 1, ..., N$

output: The number of generated dendrites L, their weights $\mathbf{W}_j$ and their orthonormal basis $\mathbf{R}_j$, $j = 1, ..., L$

Step 1: $L \leftarrow 1$;

Step 2: Find the size $\mathbf{W}$ and orientation $\mathbf{R}$ of the smallest possible hyperbox containing all the samples of C1 ;

Step 3: Assign the result of step 2 to $\mathbf{W}_1$ and $\mathbf{R}_1$;

Step 4: If there are misclassified points of C2, go to step 5 ;
else go to step 10 ;

Step 5: Pick arbitrarily a misclassified point ξ of C2 ;

Step 6: $L \leftarrow L + 1$;

Step 7: Find the size $\mathbf{W}$ and orientation $\mathbf{R}$ of the biggest hyperbox that contains ξ, but it does not contain any point of C1 ;

Step 8: Assign the result of step 7 to $\mathbf{W}_L$ and $\mathbf{R}_L$;

Step 9: Go to step 4 ;

Step 10: Terminate the algorithm and report the results ;

Algorithm 1: Training an orthonormal basis lattice neural network

---

It can be easily shown that the time complexity of this algorithm is equal to $O_{OBLNN} = O_{SBLNN} \times O_{Min}$ where O_{SBLNN} is the time complexity of the training algorithm in the case of SB-LNN and O_{Min} is the time complexity of the minimization process used in steps 2 and 7. Therefore, in order to obtain improved learning capability we loose in speed performance. The improvement in speed performance will be one of the research topics in our future work. Practically we can use efficiently this algorithm for problems defined in 2 dimensions, e.g. image processing related problems and problems defined in 3 dimensions, e.g. point set processing and 3D volume image processing problems.

More specifically, in 2-dimensional problems the orthonormal basis is defined by a 2×2 rotation matrix $\mathbf{R}$. For the storage of this matrix we need to store only 1 real number, which is either the only lower triangular element of a 2×2 skew-symmetric matrix (see in the Appendix to this chapter), or the rotation angle of the orthonormal system around the origin. In the 3-dimensional case, the orthonormal basis is defined by a 3×3 rotation matrix, which can be stored using only 3 real number storage units. Similarly to the 2D case these 3 numbers are the three lower triangular elements of a 3×3 skew-symmetric matrix $\mathbf{A}$ such that $\mathbf{R} = exp(\mathbf{A})$. Generally in the n-dimensional case, we need $n(n-1)/2$ real number storage units in order to store the orthonormal basis.

The difference between the training algorithm of an orthonormal basis lattice neural network and the training algorithm of a standard basis lattice neural network, is in the processes performed in steps 2 and 7. For the standard basis case these two steps find only the size $\mathbf{W}$ and not the orientation $\mathbf{R}$ of the new dendrite. The standard orthonormal basis (expressed by the identity matrix) is employed for every dendrite. As a result of this difference between the orthonormal basis and the standard basis algorithms, the orthonormal basis algorithm generally produces more compressed neural networks, i.e. with smaller number of dendrites compared to those in the standard basis case. This property is discussed in more details by the following lemma and its proof.

Lemma 1. *The number of dendrites generated by training an orthonormal basis lattice neural network L_O is smaller or equal to the number of dendrites obtained by training on the same dataset a standard basis lattice neural network L_S ($L_O \leq L_S$).*

The proof of the lemma involves understanding of the process performed in step 7 of Algorithm 1. Without loss of generality we will assume that in step 5 the same misclassified point ξ is arbitrarily selected for both orthonormal basis and standard basis lattice neural network training. Let assume that for one particular ξ the SB-LNN algorithm creates a hyperbox of volume A. The OB-LNN algorithm will create a hyperbox with the maximum possible volume, say B. Therefore $A \leq B$, and the equality holds in the case that the standard basis hyperbox happens to have the maximum possible volume. Hence the number of the training points included in volume A must be smaller or equal to those included in volume B. As a consequence the number of iterations performed by the orthonormal basis training algorithm is smaller or equal to those performed by the standard basis training algorithm. This proves the lemma since $L_O \leq L_S$.

3.4 Experimental Results

In this section, validation experimental results are presented which demonstrate superior learning performance of OB-LNNs over SB-LNNs. The experiments were performed using synthetic 2-dimentional datasets and the well known Iris flower dataset.

The first dataset was synthetically generated and it forms two 2D spirals. The points of the two spirals obey the equations

$$[x_1(\theta), y_1(\theta)] = [2\theta cos(\theta)/\pi, 2\theta sin(\theta)/\pi] \tag{3.7}$$

and

$$[x_2(\theta), y_2(\theta)] = [-x_1(\theta), -y_1(\theta)] \tag{3.8}$$

respectively. Several versions of this datasets were generated with a) 130, b) 258, c) 514, d) 770, e) 1026, f) 1538 samples. The largest dataset (2538 samples) was used for the testing dataset. The samples of the smallest dataset (130 samples) are presented in Fig. 3.5. Small circles denote the samples of the one spiral, and small crosses denote the samples of the other one.

Two neural networks were trained in the previously described datasets: a) a Standard Basis Lattice Neural Network and b) the proposed Orthonormal Basis Lattice Neural Network. Both networks were lattice based morphological perceptrons with dendritic structure [11] so that their performance could be compared directly. In the case of the OB-LNN perceptron, the dendrites were

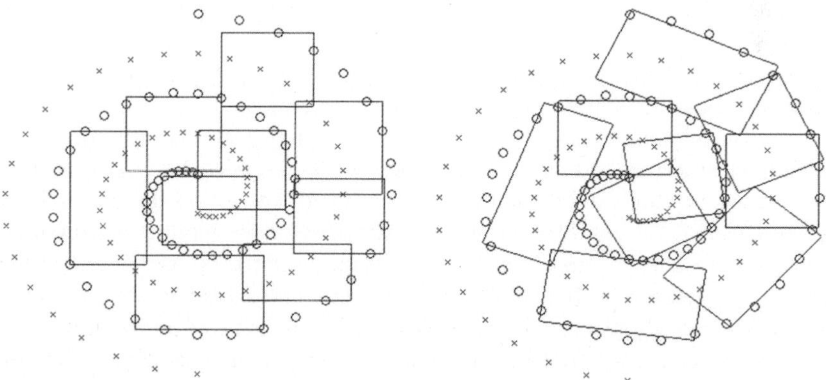

Fig. 3.5. This figure shows some of the hyperboxes formed by a SB-LNN (left) and a OB-LNN(right). The 130 training samples are denoted by circles and crosses and they are forming two spirals. The hyperboxes presented here for comparison, are the 9 smallest of each case

Table 3.1. This Table presents classification errors and number of dendrites needed for the training of an OB-LNN and a SB-LNN, using different sizes of training samples (see column 1). The 2nd and 4th columns present the number of dendrites needed for the correct classification of the training samples. The 3rd and 5th columns show the percentage of misclassified samples using always 1538 testing samples. The last column shows the quantity equals to one minus the ratio of the 3rd column over the 5th column

Training samples	OB-LNN Dendrites	OB-LNN Error	SB-LNN Dendrites	SB-LNN Error	1-Ratio of errors
130	12	14.0%	14	18.34%	23.45%
258	14	6.5%	17	9.69%	32.20%
514	15	3.2%	19	4.55%	28.57%
770	16	2.3%	19	3.20%	26.88%
1026	15	1.5%	20	2.54%	40.94%

Orthonormal Basis dendrites, which were trained in order to maximize the volume of the hyperboxes that they formed.

Table 3.1 presents the classification errors of the trained neural networks for different sizes of the training datasets (column 1). The size of the testing dataset for all the experiments was 1538 samples. In all cases, the classification errors made by the OB-LNN are significantly smaller than these made by the SB-LNN. This conclusively demonstrates superior training performance of the proposed neural network over the standard basis lattice neural networks.

Furthermore, the number of the dendrites, which are required to form the decision boundaries between some populations, measures the learning ability of a neural network. Table 3.1 also presents the final number of dendrites required by the lattice neural networks in order to classify correctly all the training samples (columns 2 and 4). In all cases, the number of dendrites trained by the OB-LNN is smaller than the number of dendrites trained by a standard basis LNN. This means that the OB-LNN compresses automatically its size.

Figure 3.5 shows most of the hyperboxes formed by the dendrites of the Standard basis (left) and the Orthonormal basis LNNs (right). These hyperboxes were formed by the training process using 130 training samples. By observing this figure we can see the differences between the decision boundaries formed by the OB-LNN and those formed by the SB-LNN. The hyperboxes generated by the Orthonormal basis LNN have bigger volume (area in the 2D domain) than those generated by the standard basis LNN. In the case of OB-LNN each hyperbox is rotated appropriately because of the fact that it is working on a different orthonormal basis than the other dendrites.

As it was discussed earlier in Sect. 3.3, each hyperbox contains information about the local orientation of the classes. This property is illustrated in Fig. 3.6. In this figure the decision boundaries between the two spirals generated by a SB-LNN (left) and an OB-LNN (right) are presented. On each hyperbox several ellipses are plotted. The sizes of the principal axes of each

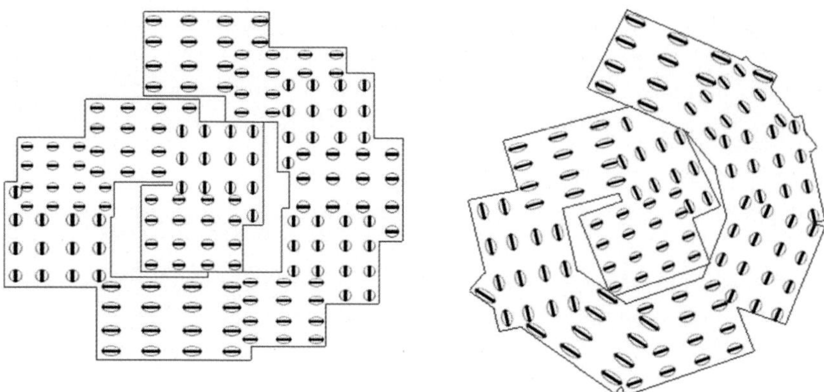

Fig. 3.6. Decision boundaries formed by a SB-LNN (left) and a OB-LNN (right).
On each hyperbox several ellipses are plotted. The sizes of the principal axes of each
ellipse are proportional to the sizes (length and width) of the relative hyperbox

Table 3.2. This Table presents classification errors of the three neural networks for
different amounts of training samples

Training samples	OB-LNN	SB-LNN	Perceptron
50%	8.33%	10.67%	13.33%
60%	6.41%	10.00%	12.21%
70%	4.20%	6.33%	7.34%
80%	3.12%	3.67%	6.54%
90%	2.01%	3.33%	6.38%
100%	0%	0%	3.67%

ellipse are proportional to the size (length and width) of the relative hyperbox.
The dominant axis of each ellipse is also plotted, forming a vector field.

Observing the vector field generated by the SB-LNN (top) and this generated by the OB-LNN (bottom), one can conclude that the hyperboxes of
the OB-LNN contain information about the local orientation of the classes.
In the case of SB-LNN this property cannot be generally observed.

Another set of experiments was held by using the Iris flower dataset. The
Iris Flower dataset is a popular multivariate dataset that was introduced by
R.A. Fisher as an example for discriminant analysis. The data reports on
four characteristics of the three species of the Iris Flower, sepal length, sepal
width, petal length, and petal width. The goal of a discriminant analysis is to
produce a simple function that, given the four measurements, will classify a
flower correctly.

Several experiments were performed by using randomly different amounts
of the Iris flower samples as testing samples (1st column of Table 3.2). The
following three neural networks were trained: a) an OB-LNN perceptron,
b) a SB-LNN perceptron and c) a multilayer perceptron (MLP) with one
hidden layer. Several different architectures were used for the MLP with 1

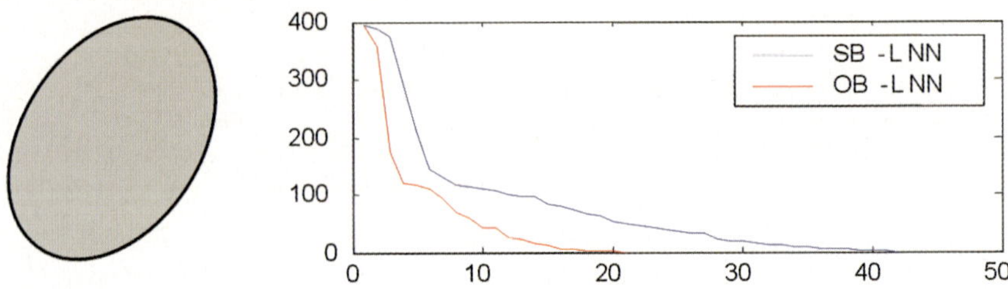

Fig. 3.7. Left: This dataset forms an ellipse. Right: Plot of the misclassified points over the number of dendrites required by a OB-LNN and a SB-LNN during the training process

hidden layer, and the results of Table 3.2 are the best obtained. The whole Iris data set were used as the testing data set. The classification errors of the three neural networks are presented in Table 3.2. In all cases, the classification errors made by the OB-LNN are significantly smaller.

Finally, another experiment was also held in order to compare the sizes of the trained neural networks. A synthetic dataset was generated forming two classes; one within an ellipse (Fig. 3.7 left) and the other one outside of it. The sample points of the two classes were picked up randomly using uniform distribution.

The same two lattice neural networks with dendritic structure were used: a) an OB-LNN perceptron and b) a SB-LNN perceptron. The right plate of Fig. 3.7 shows the plot of the number of misclassified sample points over the number of dendrites generated by the two neural networks during the training process. The final number of dendrites required by the OB-LNN in order to classify correctly all the training samples is significantly smaller than the number of dendrites trained by a standard basis LNN. This also conclusively demonstrates superior learning performance of OB-LNNs over SB-LNNs.

3.5 Conclusion

A novel model, namely Orthonormal Basis Lattice Neural Network, was presented. Comparisons of the proposed model with the standard basis model of Lattice neural networks were shown. Validation experimental results were also presented demonstrating the advantages of the proposed model. Our future work will be focused on applying this model in applications for face and object localization, Auto-Associative memories and color images retrieval and restoration, in which areas the standard basis lattice neural networks have been applied [3, 9, 10, 18, 19, 20]. Extension of the proposed model in the area of Fuzzy Lattice Neurocomputing [5, 6] and improvement in speed performance will also be some of our future research topics.

References

1. Eccles J (1977) The Understanding of the Brain. McGraw-Hill, New York
2. Gallier J, Xu D (2003) Computing exponentials of skew-symmetric matrices and logarithms of orthogonal matrices. Intl J of Robotics and Autom 18:10–20
3. Grana M, and Raducanu B (2001) Some applications of morphological neural networks. In: Proc Intl Joint Conf Neural Networks 4:2518–2523
4. Horn RA, Johnson CR (1991) Topics in Matrix Analysis. Cambridge University Press
5. Kaburlasos V, Petridis V (2000) Fuzzy lattice neurocomputing (FLN) models. Neural Networks 13(10):1145–1170
6. Petridis V, Kaburlasos V (1998) Fuzzy lattice neural network (FLNN): a hybrid model for learning. IEEE Transactions on Neural Networks 9(5):877–890
7. Petridis V, Kaburlasos V (2001) Clustering and classification in structured data domains using fuzzy lattice neurocomputing (FLN). IEEE Transactions on Knowledge and Data Engineering 13(2):245–260
8. Petridis V, Kaburlasos V, Fragkou P, Kehagias A (2001) Text classification using the σ-FLNMAP neural network. In: Proc Intl Joint Conference on Neural Networks 2:1362–1367
9. Raducanu B, Grana M (2001) Morphological neural networks for vision based self-localization. In: Proc IEEE Intl Conference on Robotics and Automation 2:2059–2064
10. Ritter GX, Iancu L (2004) A morphological auto-associative memory based on dendritic computing. In: Proc IEEE Intl Joint Conference on Neural Network 2:915–920
11. Ritter GX, Iancu L, Urcid G (2003) Morphological perceptrons with dendritic structure. In: Proc IEEE Intl Conference on Fuzzy Systems 2:1296–1301
12. Ritter GX, Schmalz M (2006) Learning in lattice neural networks that employ dendritic computing. In: Proc IEEE World Congress on Computational Intelligence pp 7–13
13. Ritter GX, Sussner P (1996) An introduction to morphological neural networks. In: Proc Intl Conference on Pattern Recognition 4:709–717
14. Ritter GX, Urcid G (2007) Learning in lattice neural networks that employ dendritic computing. This volume, Chapter 2
15. Saul A, Teukolsky S, Flannery B, Press W, Vetterling B (2003) Numerical Recipes Example Book (C++). Cambridge University Press
16. Segev I (1988) Dendritic processing. In: Arbib MA (ed) The Handbook of Brain Theory and Neural Networks pp 282–289. MIT press
17. Sussner P (1998) Morphological perceptron learning. In: Proc IEEE ISIC/CIRA/ISAS Joint Conference pp 477–482
18. Sussner P (2004) Binary autoassociative morphological memories derived from the kernel method and the dual kernel method. In: Proc Intl Joint Conference on Neural Networks 1:236–241
19. Sussner P (2004) A fuzzy autoassociative morphological memory. In: Proc Intl Joint Conference on Neural Networks 1:326–331
20. Yun Z, Ling Z, Yimin Y (2004) Using multi-layer morphological neural network for color images retrieval. In: Proc World Congress on Intelligent Control and Automation 5:4117–4119

Chapter 3 Appendix

In this section we review briefly some of the mathematical preliminaries related to the methods presented in this article.

The matrix exponential of a square $n \times n$ matrix $\mathbf{A}$ is defined as

$$exp(\mathbf{A}) = \sum_{i=0}^{\infty} \frac{\mathbf{A}^i}{i!} = \mathbf{I} + \mathbf{A} + \frac{\mathbf{A}^2}{2} + \frac{\mathbf{A}^3}{6} + \text{...} \tag{3.9}$$

where $\mathbf{I}$ is the $n \times n$ identity matrix. The summation of the infinite terms of (3.9) does converge and the obtained result is a $n \times n$ positive definite matrix. Similarly to the matrix exponential we can define the matrix logarithm as the inverse operation. The matrix logarithm of a $n \times n$ positive definite real valued matrix $\mathbf{A}$ gives a $n \times n$ real valued matrix. Both exp and log matrix operations are invariant to rotations, i.e. for every $n \times n$ orthogonal matrix $\mathbf{v}$ we have $exp(\mathbf{v}\mathbf{A}\mathbf{v}^T) = \mathbf{v}exp(\mathbf{A})\mathbf{v}^T$ and $log(\mathbf{v}\mathbf{A}\mathbf{v}^T) = \mathbf{v}log(\mathbf{A})\mathbf{v}^T$.

A skew-symmetric matrix $\mathbf{A}$ is a $n \times n$ matrix whose negative is also the transpose of itself (i.e. $\mathbf{A}^T = -\mathbf{A}$). A skew-symmetric matrix has all its diagonal elements zero, and the rest of the elements satisfy the property $A_{i,j} = -A_{j,i}$.

The matrix exponential of a skew-symmetric matrix is an orthogonal matrix $\mathbf{R} = exp(\mathbf{A})$, where $\mathbf{R}$ is an orthogonal matrix and it is also a positive definite matrix because it is expressed as the matrix exponential of a matrix. The zero $n \times n$ matrix is an example of a skew-symmetric matrix. By evaluating (3.9) it turns out that the matrix exponential of the zero matrix is the $n \times n$ identity matrix (i.e. $\mathbf{I} = exp(\mathbf{0})$), which is also an orthogonal matrix. The $n \times n$ identity matrix $\mathbf{I}$ can be seen as the zero rotation matrix of the n-dimensional space. Generally, the matrix operation $exp(\lambda\mathbf{A})$, where λ is a scalar which is close to zero, produces an orthogonal matrix which is close to the identity and rotates slightly the n-dimensional space.

$$\lim_{\lambda \to 0} exp(\lambda\mathbf{A}) = \mathbf{I} \tag{3.10}$$

A random small rotation of the n-dimensional space can be generated from Eqn. (3.10) by using a randomly generated skew-symmetric matrix $\mathbf{A}$ and a scalar λ which is close to zero. A random skew-symmetric matrix can be generated by using the multivariate Gaussian distribution for the lower triangular elements of the matrix. If we set the rest of the elements (diagonal and upper triangular) to be zero, we can obtain a randomly generated skew-symmetric matrix by subtracting the transpose of this lower triangular matrix from itself.

Computing exponential of a $n \times n$ skew-symmetric matrix has time complexity $O(n^3)$. A detailed discussion on computing exponentials of skew-symmetric matrices and logarithms of orthogonal matrices was presented in [2]. Finally, a more detailed study on the matrix exponential, skew-symmetric matrices and the related theory can be found in [4].

4

Generalized Lattices Express Parallel Distributed Concept Learning

Michael J. Healy and Thomas P. Caudell

University of New Mexico   `mjhealy@ece.unm.edu`
Albuquerque, New Mexico 87131   `tpc@ece.unm.edu`

Summary. Concepts have been expressed mathematically as propositions in a distributive lattice. A more comprehensive formulation is that of a generalized lattice, or category, in which the concepts are related in hierarchical fashion by lattice-like links called concept morphisms. A concept morphism describes how an abstract concept can be used within a more specialized concept in more than one way as with "color", which can appear in "apples" as either "red", "yellow" or "green". Further, "color" appears in "apples" because it appears in "red", "yellow" or "green", which in turn appear in "apples", expressed via the composition of concept morphisms. The representation of such concept relationships in multi-regional neural networks can be expressed in category theory through the use of categories, commutative diagrams, functors, and natural trasformations. Additionally, categorical model theory expresses the possible worlds described by concepts. The analysis of morphisms between the possible worlds highlights the importance of reciprocal connections in neural networks.

4.1 Introduction

The contributions in this volume discuss lattice theory from different perspectives. Some relate it to neural network algorithms in which the partial order relation ($\leq$) and the meet/join operations ($\wedge$, $\vee$), regarded as min/max operations, are part of the computational model. There is a history of lattice theory in neural network theoretical models [12, 18, 28], and in some the meet/join operations have other interpretations. One of these involves symbolic languages for modeling the semantics of neural computation. Symbolic languages including formal logics have appeared over the years in attempts to obtain a mathematically precise expression of the semantics of neural networks [2, 3, 6, 12, 13, 17, 19, 24, 26, 27]. In addition to classical propositional and first-order logic [3], the logics used have included fuzzy logic [24], geometric logic [12], and non-monotonic logics [26]. In the semantic model of [12], the input data for a neural network represent paired entities from two domains. The network's long-term adaptation is explained as the learning of

M.J. Healy and T.P. Caudell: *Generalized Lattices Express Parallel Distributed Concept Learning*, Studies in Computational Intelligence (SCI) **67**, 59–77 (2007)
`www.springerlink.com`   © Springer-Verlag Berlin Heidelberg 2007

a geometric logic theory about each domain while simultaneously learning an inference relation between the theories that expresses a mapping between their domains. Because of the form of geometric logic used, this yields lattices coupled by a very special type of lattice homomorphism. In the process of learning, the network modifies its connection weights based upon input data, and this effectively combines domain knowledge extracted from the data with pre-existing knowledge to derive new theories and new inferences between their formulas. Each initial theory consists of a set of predicates representing observable features of the entities, and proceeds through the adaptation process to include derived formulas and their relationships, and these and the modified inferencing express the information gained.

Vickers applies geometric logic, a non-Boolean logic similar to intuitionistic logic ([29, 30]), to domain theory. The logic is applied as a "logic of finite observations". This derives from the interpretation of a statement Q in the logic as affirmable: It cannot be falsified, only affirmed through demonstration (observation or proof). The axioms of a geometric theory specify basic knowledge about some domain. As with most logics, the propositions, predicates, the formula-constructing operations—in this case conjunction and disjunction along with the existential quantifier $\exists$—- and the proof theory of the logic, resulting in entailments such as $\exists x P(x) \vdash \exists x Q(x)$, are used to reason about any domain to which the theory applies. In the logic of finite observations, however, negation $\neg$, implication formulas $P(x) \to Q(x)$ and the universal quantifier $\forall$ are used only in formulating the axioms. Further, as with intuitionistic logic, the Law of the Excluded Middle (LEM, $\forall x (P(x) \vee \neg P(x))$) is not an axiom of geometric logic: It must be either established as valid when it applies or assumed as an axiom of a theory where its use is desired. The analyst has the task of formulating axioms for a theory sufficiently detailed to express all relevant, pre-existing knowledge about the domain of investigation and then applying the geometric operations to make inferences about new data.

Propositional geometric logic is a special case of this. For a propositional theory formulated in geometric logic, a lattice can be constructed from the axioms, other propositions, and the entailments, where $\vdash$ becomes the lattice order relation $\leq$. Inference is then carried out based solely upon the lattice operations of meet ($\wedge$) and join ($\vee$), interpreted as conjunction and disjunction. The full logic with predicates, quantifiers and open formulas, on the other hand, requires a structure more general than a lattice— it requires category theory.

Yet, single-argument predicates of the form $P(x)$ are used within a lattice in [12]. The explanation for this is that the argument x represents the observations in a domain, not individual entities discussed within the domain theory, and the predicate is affirmed when an observation applies to it. This notation is a convenience for conducting an analysis over multiple domains while maintaining the utmost simplicity. The different domains are obtained by coupling networks with systems of interconnects to form a composite or multi-regional network. The example examined in [12] is that of a LAPART network [13], a coupling of two ART-1 networks in such a way that the composite network can learn inferences across theories in addition to learning the individual theories. From this, a new theory can be formed that combines the newly-formed sub-network theories together with the newly-formed implications between their formulas, all based upon observation. A composite domain accompanies the composite theory, having the form of a product—if the domains are thought of as sets, the composite in this case is a cartesian product. A finding of this analysis is that most of the structure supporting the learning of coupled domain theories is missing from the example architecture.

The domain theories in propositional geometric logic yield upper-complete distributive lattices, having infinite as well as finite joins and having finite meets (the lattices have infinite meets, but they are non-geometric, so are not used). Mathematically, an advantage of this use of geometric logic is that a theory in the logic directly corresponds to a topology for its domain. This makes it possible to perform model-theoretic analyses in terms of topological spaces and continuous functions (although this is not quite point-set topology, since a slight modification to the set union operation must be made). Thus, the application of geometric propositional logic combines the advantages of logical inference, lattice theory, topology, and domain theory in describing the semantics of neural networks. This facilitates the design of improved neural networks by noting which entailments, lattice meets and joins, and continuous structures are missing from an existing architectural model. The analysis is not restricted to binary-input neural networks because graded values can be represented in a quantized form (see, for example, [13]). With all its advantages, however, this kind of analysis reveals only a part of the gap between present-day architectures and the full potential of neural networks.

More recently, we have been experimenting with a much more comprehensive semantic theory for neural networks based upon category theory [16]. Here, the main application of category theory is not to represent the predicate version of geometric logic and assign formulas with predicates and quantifiers to network nodes. Instead, neural network nodes are still regarded as representing closed logical fragments, but these are whole domain theories and the theories have quantities of more than one type. They are associated with network nodes based upon the nodes' input connection pathways traced back to the input nodes, which are assigned theories that describe the properties of the input features. Also, the logic in which the theories are expressed is left to the analyst's choice; it can be geometric, intuitionistic, classical first-order, or fuzzy, and so forth. When adopted in full, this strategy enables the expression of neural network semantics in the framework of categorical logic and categorical model theory, resulting in a deeper and more useful analysis. One example of this is an experiment with an application of a neural network to multi-spectral imaging, in which an existing architecture was re-designed to produce images with a significantly higher quality [15]. Here, we provide a brief recounting of the categorically-based semantic theory.

4.2 Lattices and Categories

A category can be thought of as a system of mathematical structures of some kind, concrete or abstract, together with the relationships between them that express that type of structure [1, 7, 20, 21, 25]. Each relationship, called a morphism or arrow, has the form $f: a \longrightarrow b$ with a *domain* object a and a *codomain* object b. A lattice is a special case of a category in which the morphisms are the relations $a \leq b$. In a *category* C, each pair of arrows $f: a \longrightarrow b$ and $g: b \longrightarrow c$ (with a head-to-tail match, where the codomain b of f is also the domain of g as indicated) has a *composition* arrow $g \circ f : a \longrightarrow c$ whose domain a is the domain of f and whose codomain c is the codomain of g. In a lattice, of course, this expresses the transitivity of $\leq$. Composition satisfies the familiar associative law, so that in triples which have a head-to-tail match by pairs, $f: a \longrightarrow b$, $g: b \longrightarrow c$ and $h: c \longrightarrow d$, the result of composition is order-independent, $h \circ (g \circ f) = (h \circ g) \circ f$. Also, for each object a, there is an *identity morphism* $\mathtt{id}_a: a \longrightarrow a$ (in a lattice, $a \leq a$) such that the identities $\mathtt{id}_a \circ g = g$ and $f \circ \mathtt{id}_a = f$ hold for any arrows $f: a \longrightarrow b$ and $g: c \longrightarrow a$.

Unlike in a lattice, there can be many morphisms in either or both directions between a pair of objects a and b. The category **Set** of sets and functions (with composition of functions) is a familiar example, for given arbitrary sets α and β, there can be many functions with either set as domain or codomain. With a multiplicity of morphisms existing between two objects in a typical category, and given the notion of composition, the notion of a *commutative diagram* takes on great significance. A diagram in a category C is simply a collection of objects and morphisms of C (the domain and codomain objects of a morphism are always included with it). In a commutative diagram, any two morphisms with the same domain and codomain, where at least one of the morphisms is the composition of two or more diagram morphisms, are equal. *Initial and terminal objects* (the bottom and top elements in a lattice) are also important when they exist in a category C. An initial object i is the domain of a unique morphism $f: i \longrightarrow a$ with every object a of C as codomain. A terminal object t has every object a of C as the domain of a unique morphism $f: a \longrightarrow t$ with t as codomain.

The principle of duality is a fundamental notion in category theory. The dual or opposite C^{op} of a category C has the same objects, and the arrows and compositions $g \circ f$ reversed, $f^{\mathrm{op}} \circ g^{\mathrm{op}}$. The *dual of a statement* in category theory is the statement with the words "domain" and "codomain", "initial" and "final", and the compositions reversed. If a statement is true of a category C, then its dual is true of C^{op}; if a statement is true of all categories, the dual statement is also true of all categories because every category is dual to its dual. Roughly speaking, "half the theorems of category theory are obtained for free", since proving a theorem immediately yields its dual as an additional theorem (see any of [1, 21, 25]).

In addition to the widely-used notion of duality, category theory provides a mathematically rigorous notion of "isomorphism", a term which is often used in a loose, intuitive sense. One sometimes hears a statement such as "the two [concepts, data types, program constructs, etc.] are in some sense isomorphic". If the entities under discussion can be formalized as objects in a category, one can make such statements with mathematical rigor. If a, b are objects of a category C such that there exist arrows $f : a \longrightarrow b$ and $g : b \longrightarrow a$ with $f \circ g = \mathtt{id}_b$ and $g \circ f = \mathtt{id}_a$, then the morphism f is called an *isomorphism* (as is g also) and g is called its *inverse* (and f is called the inverse of g), and the two objects are said to be isomorphic. The property of an identity morphism ensures that isomorphic objects in a category are interchangeable in the sense that they have the same relationships with all objects of the category. It is easily shown that all initial objects in a category are isomorphic, and the same holds for terminal objects (see the above definitions for initial and terminal objects).

Let Δ be a diagram in a category C, shown in Fig. 4.1 with objects a_1, a_2, a_3, a_4, a_5 and morphisms $f_1 : a_1 \longrightarrow a_3$, $f_2 : a_1 \longrightarrow a_4$, $f_3 : a_2 \longrightarrow a_4$, $f_4 : a_2 \longrightarrow a_5$. Also shown are two cone-like structures, K' and K''. Each of these *cocones* extends Δ, forming a commutative diagram: For example, K' adds an object b' and morphisms $g'_i : a_i \longrightarrow b'$ $(i = 1, \ldots, 5)$ such that $g'_3 \circ f_1 = g'_1 = g'_4 \circ f_2$ and $g'_4 \circ f_3 = g'_2 = g'_5 \circ f_4$. The cocones for Δ are objects in a category $\mathbf{coc}_\Delta$ whose morphisms are morphisms of C from one cocone apical object to the other, $h : b' \longrightarrow b''$, having the property that the legs of the codomain cocone K'' factor through the legs of the domain K' and the morphism h,

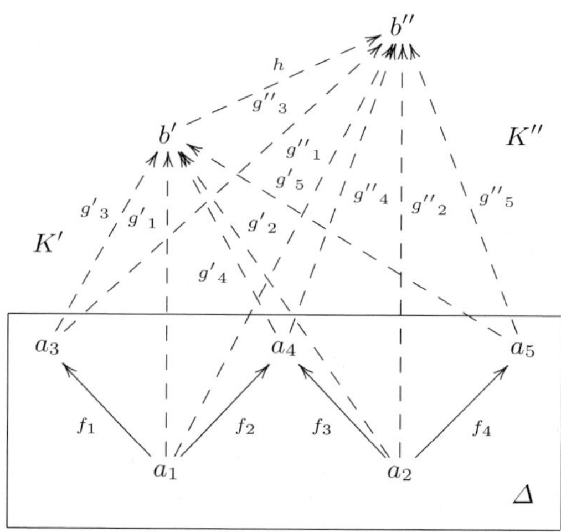

Fig. 4.1. A cocone morphism $h : K' \longrightarrow K''$ in $\mathbf{coc}_\Delta$ is a morphism $h : b' \longrightarrow b''$ in C between the apical objects b' and b'' of cocones K' and K'', respectively, that is a factor of each leg morphism $g''_i : a_i \longrightarrow b''$ of K'', with $g''_i = h \circ g'_i$

$$g''_i = h \circ g'_i \quad (i = 1, \ldots, 5).$$
(4.1)

With morphisms so defined, the composition of cocone morphisms follows directly. A *colimit for the diagram* Δ is an initial object K in the category $\mathbf{coc}_\Delta$. That is, for every other cocone K' for Δ, there exists a unique cocone morphism $h\colon K \longrightarrow K'$. The original diagram Δ is called the *base diagram* for the colimit and the diagram $\overline{\Delta}$ formed by adjoining K to Δ is called its *defining diagram*. By *initiality*, all colimits for a given base diagram are isomorphic. A *coproduct* is the colimit for a discrete diagram, one having objects but no morphisms among them except the identity morphism for each diagram object, which is always assumed to be present in a diagram. In **Set**, for example, coproducts are disjoint unions of the component sets.

Limits are the dual notion to colimits, obtained by "reversing the arrows" and interchanging "initial" and "terminal". By duality, a limit for a diagram Δ, if one exists, is a terminal cone, whose morphisms are directed *into* the diagram. A limit for a discrete diagram is called a *product*, and the leg morphisms are called projections. The familiar cartesian product of sets is an example in the category **Set**. Limits and colimits are useful constructs where they are available, and colimits have a history of use in categorical logic and computer science ([11, 32]). A theorem in category theory can be used to derive an algorithm for calculating limits in any category that contains limits for all of its diagrams, and similarly for colimits by dualization (see The Limit Theorem in [25]).

4.3 A Category of Concepts

We express the semantics of a system in terms of a distributed system of concepts about the system's environment as well as the system itself. This manner of describing a system can be thought of as a knowledge representation that is implicit in the system's disposition to sample its environment and act upon it. The concept system, of which the knowledge representation is a part, is an ontology for the system's environment as experienced through the system inputs, both external to the system and possibly internally-generated as well. The ontology is expressed mathematically as a category **Concept** whose objects are symbolic descriptions (concepts) of domains of items (sensed entities, events, situations, and their parts). The concepts are represented incrementally in an adaptive system based upon sensor input, and, where the capabilities exist, are based upon the interplay of sensor input with efference copy (where motor command outputs are used also as system inputs) and with feedback from higher-level processing to levels of processing more directly associated with the sensors. For example, the concept representations in a neural network are formed through modification of connection weights, which in some networks exist in both feedforward and feedback connection pathways. Determining the knowledge representation capability of a given neural architecture, and designing architectures with a desired capacity, are major goals of analysis using the semantic theory.

Through feedback, sensor inputs can be filtered and adaptively-formed concept representations learned from them can be made consistent with the already-existing knowledge representation store in, say, a cognitive system. We refer to concept representations formed at or near the sensor level through this two-way processing as *percepts*, even when they are not associated with conscious awareness or a biological organism. Concept representations formed through further processing at a "higher" level are, as a consequence, based in perception. This view of concept representation based in perception has important implications for theories of cognition and intelligence. Indeed, it is consistent with converging evidence and theory in cognitive neuroscience [4, 8, 9, 10, 22, 31]. Toward the end of this chapter, we discuss categorical model theory and suggest that it indicates the desirability of feedback in a neural network. It also provides a mathematical foundation for perceptual knowledge representations.

As explained in Sect. 4.5, mathematical rigor is maintained in this formulation by describing structure-preserving mappings from **Concept** to a category representing the neural system's computational structure, thus showing how concepts, their morphisms, and the consequences of composition including commutative concept diagrams are represented in the system computations. A **Concept** morphism $s\colon T \longrightarrow T'$ is an association of the description constituting concept T with a subconcept, or logical part, of the description constituting concept T', with the property that the axioms of the domain T of the morphism are mapped to axioms or theorems of the codomain T'. We use an already-available mathematical convenience, a category of formal logic theories and theory morphisms ([7, 11, 23]) for the category **Concept**. The category has an overall hierarchical structure, since the more abstract theories are represented within the more specialized ones and the morphisms convey this. We use the terms "theory" and "concept" interchangeably.

As shown in the next section, colimits provide a means of combining concepts to derive concepts of greater complexity. Because they are more complex, these concepts are also more specific, or specialized, than any of the concepts in their diagrams. In the lattice formulation for geometric propositional logic, the lattice meet $\wedge$ operation plays an analogous role, since it forms a proposition of greater specificity than those being combined. However, because colimits are based upon diagrams in a category of theories, more can be expressed in this manner than with lattices. The same can be said for limits relative to the lattice join operation $\vee$. Limits form simpler, hence, more abstract, concepts than those in their base diagrams. But asserting that a *category* of theories is more expressive than a lattice begs the question of whether the theory category is not itself a lattice. After all, a lattice is a category, and in that sense a category is a sort of generalized lattice. How is the concept category not a lattice?

The key point is that a theory category, like most categories, has many "edges" between a given pair of "nodes"; that is, there can be many ways in which an object a is related to an object b (and, in many categories, b can be related to a as well). This is a consequence of the complexity allowed in the morphisms, which is two-fold: First, a theory morphism must map all symbols (sorts, operations, constants) of the domain to symbols of the same type in the codomain. Second, each axiom of the domain theory, when transformed by symbol substitution, must become an axiom or theorem of the codomain theory.

Consider a concept of color stated as a theory, as follows:

```
Concept T
   sorts Colors, Items
   op has_color: Items*Colors -> Boolean
   const c: Colors
   Axiom color-is-expressed is
      exists (it: Items) (has_color (it, c))
end
```

The statement line `sorts Colors, Items` introduces the basic sorts, or "logical containers", of T; logic in this form is called *sorted*. The explicit typing of variables and constants in sorted theories such as T is a convenient alternative to using predicates to qualify the quantities in every formula, and it allows operations to be interpreted as total (as opposed to partial) functions whose domains are expressed in the theory as sorts. The line `op has_color: Items*Colors -> Boolean` specifies an operation `has_color`. This operation acts as a predicate with items and colors as arguments, where `Items*Colors` is a product sort—a "container" for ordered pairs of items and colors. The `Boolean` sort contains the truth values `true` and `false`. It is part of a theory of logical operations that is implicitly included in every concept (it is an initial object of the concept category). Notice that T also contains a constant `c` of sort `Colors`; this can represent, for example, a feature always observed via one sensor element of a system. The axiom `color-is-expressed` states that the color `c` has some item that expresses it (in a sorted predicate calculus, simply asserting a formula as shown is the same as stating that it is true). This color is arbitrary, having no definition, but as a constant it is regarded as a definite color nevertheless. In actuality, T does not have enough content to constitute a theory about color, but it will suffice for this simple example.

To show how such an abstract theory might be used, let there be morphisms $s^{(i)}: T \longrightarrow T^{(i)}$ $(i \in \{1, 2, 3\})$, where T', T'' and $T^{(3)}$ are concepts representing three specific colors, `red`, `yellow` and `green`. For example:

```
Concept T'
   sorts Colors, Items
   op has_color: Items*Colors -> Boolean
   const red: Colors
   Axiom red-is-expressed is
      exists (it: Items) (has_color (it,red))
end
```

Theories T'' and $T^{(3)}$ are identical except that red is replaced by yellow and green, respectively. The morphism $s' : T \longrightarrow T'$ has the form

```
Morphism s':   Colors       ↦ Colors
               Items        ↦ Items
               has_color    ↦ has_color
               c            ↦ red
```

Morphisms s'' and $s^{(3)}$ are identical except for a color-symbol change. The axiom re-naming (red-is-expressed replaces color-is-expressed) need not be included in stating a morphism, since axiom names are simply a cosmetic touch. Also, it is customary to omit symbols that remain unchanged when stating a morphism, in this case Colors, Items and has_color. Continuing, T', T'' and $T^{(3)}$ are the domains for morphisms $t^{(i)} : T^{(i)} \longrightarrow T^{(4)}$ ($i \in \{1, 2, 3\}$) which have a common codomain, a theory that uses the three colors:

```
Concept T⁽⁴⁾
  sorts Colors, Apples
  op has_color: Apples*Colors -> Boolean
  const red: Colors
  const yellow: Colors
  const green: Colors
  Axiom some-apple-colors is
    exists (x, y, z: Apples)
      (has_color (x,red))
      and (has_color (y, yellow))
      and (has_color (z, green))
end
```

The morphism $t' : T' \longrightarrow T^{(4)}$, for example, has the form:

```
Morphism t':   Items         ↦ Apples
```

(The mappings of Colors to Colors, has_color to has_color, and red to red are not shown, as is customary.) Notice that the axiom of each of the theories T', T'' and $T^{(3)}$, transformed by the symbol substitution from t', t'' and $t^{(3)}$, respectively, maps to a theorem of $T^{(4)}$: For example, exists (x: Apples) (has_color (x,red)) is an immediate consequence of the axiom some-apple-colors of $T^{(4)}$.

Finally, the compositions $t' \circ s'$, $t'' \circ s''$ and $t^{(3)} \circ s^{(3)}$ are three distinct morphisms with domain T and codomain $T^{(4)}$. For example, $t' \circ s' : T \longrightarrow T^{(4)}$ is as follows, by tracing the symbol mappings:

```
Morphism t' ∘ s':  Items         ↦ Apples
                   c             ↦ red
```

Therefore, the category **Concept** is not a lattice. Of course, there would have been only one morphism from T to $T^{(4)}$ had T not included the constant c: Color. However, most theories are much more complex than the ones in this example, increasing the possibilities for morphisms.

Another distinguishing feature of a category of theories is that colimits and limits are more expressive than joins and meets, for they express the contents of their base diagrams, not just the concepts. For example, colimits "paste together" or "blend" the concepts in a diagram along shared concepts as indicated in the diagram morphisms.

4.4 Colimits — An Example

The concept of a triangle can be derived as a colimit for a diagram involving concepts about points and lines (the alternative of defining triangles in terms of angles would complicate the example). We can start with a rather abstract concept, a theory that defines lines in terms of undefined quantities, or primitives, called points. The definition is expressed using a predicate on with two arguments, a point and a line, and is true just in case the point "lies on" the line (see [5] for a discussion of geometries based upon this definition).

```
Concept T1
  sorts Points, Lines
  const p1: Points
  const p2: Points
  const p3: Points
  op on: Points*Lines -> Boolean
  Axiom Two-points-define-a-line is
    forall(x, y:Points)
      ((x not= y) implies
        (exists L:Lines)
          (on (x, L) and on (y, L) and
            ((forall m:lines) (on (x, m) and
              on (y, m)) implies (m = L) ))
end
```

Notice that T_1 also contains constants representing three arbitrary points p1, p2 and p3. Three other concepts T_2, T_3 and T_4 share T_1 except with different names for the point constants in each. The latter concepts also include a line constant and associate two of the point constants with it via the on predicate. This is specified with an additional axiom (name omitted) which also states that the two point constants denote distinct points. For example:

```
Concept T2
  sorts Points, Lines
  const pa1: Points
  const pa2: Points
  const paext: Points
  const La: Lines
```

```
 op on: Points*Lines --> Boolean
 Axiom Two-points-define-a-line is
   forall(x, y:Points)
     ((x not= y) implies
       (exists L:Lines)
         (on (x, L) and on (y, L) and
           ((forall m:lines) (on (x, m) and
             on (y, m)) implies (m = L) ))
 on (pa1, La) and on (pa2, La)
   and (pa1 not= pa2)
end
```

Concepts T_3 and T_4 are identical, except with the names pa1, pa2, paext, and La replaced with pb1, pb2, pbext, Lb in T_3 and pc1, pc2, pcext, Lc in T_4. A morphism $s_1 \colon T_1 \longrightarrow T_2$ maps the sort symbols Points and Lines and the on predicate symbol to the corresponding symbols in T_2, which happen to be identical. We reformulate all statements of T_1 by term replacement in accordance with the symbol mapping to form their image statements in T_2. As a consequence, the axiom of T_1 relating points to lines maps to itself as an axiom of T_2. The point constants p1, p2 and p3 map to the point constants pa1, pa2 and paext. In T_2, pa1 and pa2 are associated with the line La via the on predicate, and paext is intended as a point "external to" La.

Morphism s_1:
       p1 $\mapsto$ pa1
       p2 $\mapsto$ pa2
       p3 $\mapsto$ paext

Morphisms $s_2 \colon T_1 \longrightarrow T_3$ and $s_3 \colon T_1 \longrightarrow T_4$ are similar to s_1 but with different point constant targets pb1, pb2, pbext and pc1, pc2, pcext:

Morphism s_2:
       p1 $\mapsto$ pbext
       p2 $\mapsto$ pb1
       p3 $\mapsto$ pb2

Morphism s_3:
       p1 $\mapsto$ pc2
       p2 $\mapsto$ pcext
       p3 $\mapsto$ pc1

In T_3, it is the images pb1 of p2 and pb2 of p3 that are associated with the line constant, lb, while the image pbext of p1 is the "external" point. The associations are similarly reordered in T_4; the point-to-line associations in each concept can be seen by noticing which points are the targets of those of T_1 under the appropriate morphism and applying term substitution. A colimit for the diagram $\varDelta$ with objects T_1, T_2, T_3, T_4 and morphisms s_1, s_2, s_3 (which always exists in the category **Concept**) has a cocone as shown in Fig. 4.2, with apical object T_5 and leg morphisms $\ell_1 \colon T_1 \longrightarrow T_5$, $\ell_2 \colon T_2 \longrightarrow T_5$, $\ell_3 \colon T_3 \longrightarrow T_5$, and $\ell_4 \colon T_4 \longrightarrow T_5$. With $\varDelta$ as the base diagram, the defining diagram of the colimit, $\overline{\varDelta}$, is commutative, with

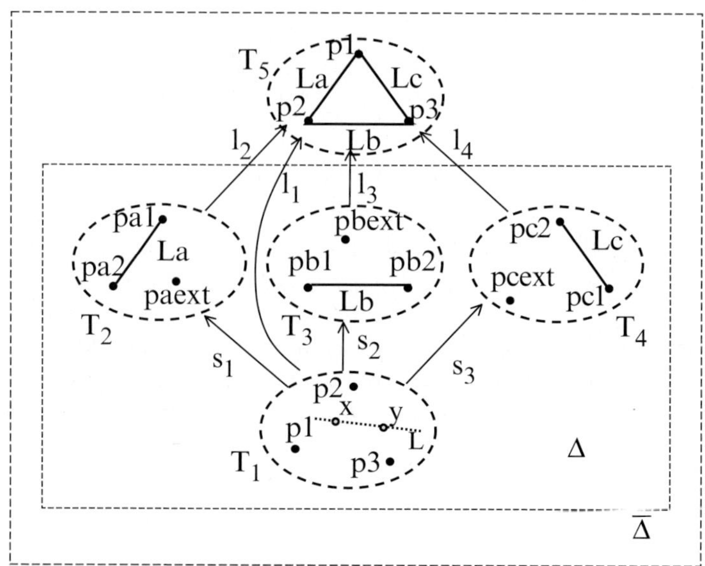

Fig. 4.2. A pictorial illustration of the colimit base and defining diagrams Δ and $\bar{\Delta}$. The contents of the concepts involved are pictured along with the diagrammatic structure. Solid dots and lines signify point and line constants. Concept T_1 has no line constant, so it is shown containing a dashed line with two open dots, representing the axiom relating points and lines which is present in all the concepts

$$\ell_1 = \ell_2 \circ s_1 = \ell_3 \circ s_2 = \ell_4 \circ s_3. \tag{4.2}$$

The resulting colimit object, T_5, is as follows:

```
Concept T5
  sorts Points, Lines
  const p1: Points
  const p2: Points
  const p3: Points
  const La: Lines
  const Lb: Lines
  const Lc: Lines
  op on: Points*Lines -> Boolean
  Axiom Two-points-define-a-line is
    forall(x, y:Points)
      ((x not= y) implies
        (exists L:Lines)
          (on (x, L) and on (y, L) and
            ((forall m:lines) (on (x, m) and
              on (y, m)) implies (m = L) ))
```

```
  on (p1, La) and on (p2, La) and
    (p1 not= p2)
  on (p2, Lb) and on (p3, Lb) and
    (p2 not= p3)
  on (p3, Lc) and on (p1, Lc) and
    (p3 not= p1)
end
```

As a consequence of the commutativity of the defining diagram $\overline{\Delta}$ of the colimit, its apical concept T_5 is a "blending" or "pasting together" of T_2, T_3 and T_4 along their common sub-concept T_1. That is, for the equality (4.2) to hold, separate symbols of T_2, T_3 and T_4 that are images of the same symbol of T_1 under the three diagram Δ morphisms s_1, s_2 and s_3 must merge into a single symbol in the colimit apical concept T_5. To make this clear, each symbol in T_5 that is a merging of symbols has been assigned the name of the T_1 symbol underlying the merging. Thus, symbols such as `Points`, `Lines` and `on` appear in T_5, and appear only once, since they are mapped to themselves by each of the morphisms s_1, s_2 and s_3. The point constants `p1, p2, p3` also appear. However, in T_5, each one represents a merging of two point constants from T_2, T_3 and T_4 and as a consequence appears in the definition of two different lines. In two of these concepts, the image of each single point constant appears in the definition of a line, but as a different point on a different line in each of the two concepts. In the third concept, it appears as an "external" point, not on the line named in that concept. For example, `p1` in T_1 is mapped to `pa1` in T_2 via s_1, to `pbext` in T_3 via s_2, and to `pc2` in T_4 via s_3. In T_5, therefore, it forms the point `p1` at the intersections of lines `La` and `Lc`, and lies external to line `Lb`. Because of the *initiality* of the colimit cocone, any other cocone for Δ is the codomain of a unique cocone morphism whose domain is the cocone containing T_5. Therefore, T_5 adds no extraneous information to that in Δ, and any other apical concept of a cocone for Δ is either isomorphic with T_5 or extends it without duplicating it.

Because it is more complex, the colimit apical object is more specific, or more specialized, than any of the concepts in its base diagram. Because it expresses the "pasting together" of the base diagram concepts around their shared sub-concepts, it expresses the concept relationships (morphisms) as well as the concepts in its base diagram. Because of the Colimit Theorem, the calculation of concept colimits can be automated. For example, the apical object T_5 and leg morphisms ℓ_1, ℓ_2, ℓ_3, and ℓ_4 above can be derived automatically from the objects and morphisms of the base diagram, Δ. Other examples are given in [32] for an application of engineering software synthesis via category theory. Where they are available, limits can be calculated also, yielding less complex or abstract theories and their attendant morphisms. This facilitates the generalization of learned concepts to new contexts by extracting useful invariants from them.

The ability to calculate concept colimits and limits suggests the ability to "flesh out" an ontology in an incremental fashion. This process can begin with a collection of concepts and morphisms describing the most basic properties of observable quantities and also any desired assumptions about the environment and the operation of a system within it. More specialized theories can be calculated as colimits, and more abstract ones as limits, through re-use of pre-existing concepts and morphisms. Taken together, the discussions in this section and the previous one suggest the use of category theory in the study of knowledge systems, learning, and the semantics of distributed, adaptive systems such as neural networks. Conducting these studies while utilizing the mathematical rigor of category theory requires the availability of categories that express the system computations together with structure-preserving mappings from **Concept** to these categories. Any such mappings must be capable of conveying only part of the ontology expressed in **Concept**, for two reasons: (1) **Concept** is infinite and realizable systems are finite, and (2) in an adaptive system, only a part of the ontology will have been learned at any one time. Another issue arises in systems with multiple sources of inputs working in parallel, such as neural networks with more than one sensor. Simultaneous inputs from the different sensors obtain different types of information about the same events. Fusing the information across sensors requires a multi-component system that can make several knowledge representations act as one.

4.5 Structural Mappings and Systems

A *functor* $F : C \longrightarrow D$, with domain category C and codomain category D, associates to each object a of C a unique image object $F(a)$ of D and to each morphism $f : a \longrightarrow b$ of C a unique morphism $F(f) : F(a) \longrightarrow F(b)$ of D. Moreover, F preserves the compositional structure of C, as follows. Let $\circ_C$ and $\circ_D$ denote the separate composition operations in categories C and D, respectively. For each composition $g \circ_C f$ defined for morphisms of C, $F(g \circ_C f) = F(g) \circ_D F(f)$, and for each identity morphism of C, $F(\mathrm{id}_a) = \mathrm{id}_{F(a)}$. It follows that the images of the objects and morphisms in a commutative diagram of C form a commutative diagram in D. This means that any structural constraints expressed in C are translated into D and, hence, F is a structure-preserving mapping. Functors can be many-to-one, and by this means much of the structure of the domain category of a functor can be "compressed", so that not all the structure is represented explicitly in the codomain. Functors can also be *into* mappings, so that the codomain need not be entirely utilized in representing the domain; this leaves room for a functor to be used in defining an extension of a given category, or in representing a structure in the context of one with a greater complexity.

We have proposed elsewhere (for example, [15, 16]) that a neural network at a given stage of learning can be associated with a category $\mathbf{N}_{A,w}$ that expresses both its connectivity and potential state changes in response to its next input. Here, A represents the architectural design including dynamic properties (the dynamics are represented only in summary fashion as rules for state changes), while w is the current connection-weight array. An analysis of the network determines whether it is possible to define a functor $M:\textbf{Concept}\longrightarrow \mathbf{N}_{A,w}$. This allows the knowledge-representation capabilities of an existing architecture to be evaluated with mathematical rigor, and can lead to insights for possible design improvements or for the design of entirely new architectures. Since functors are many-to-one mappings, concepts which have not been acquired at some stage of learning or simply cannot be represented by a given neural network can be included in the analysis.

A further advantage of the categorical approach is the notion of *natural transformations*, of the form $\alpha:F\longrightarrow G$ with domain functor $F:C\longrightarrow D$ and codomain functor $G:C\longrightarrow D$. A natural transformation α consists of a system of D-morphisms α_a, one for each object a of C, such that the diagram in D shown in Fig. 4.3 commutes for each morphism $f:a\longrightarrow b$ of C. That is, the morphisms $G(f)\circ\alpha_a:F(a)\longrightarrow G(b)$ and $\alpha_b\circ F(f):F(a)\longrightarrow G(b)$ are actually one and the same, $G(f)\circ\alpha_a=\alpha_b\circ F(f)$. In a sense, the two functors have their morphism images $F(f):F(a)\longrightarrow F(b)$, $G(f):G(a)\longrightarrow G(b)$ "stitched together" by other morphisms α_a,α_b existing in D, indexed by the objects of C. In a multi-sensor system, with each sensor having its own dedicated processing network and the networks connected into a larger system, natural transformations express *knowledge coherence*. They accomplish this by unifying the knowledge representations of the subnetworks of a neural network, which can include subnetworks for sensor association, planning and other functions (see [14] for an example network). Figure 4.4 illustrates this notion for two functors $M_1:\textbf{Concept}\longrightarrow \mathbf{N}_{A,w}$ and $M_2:\textbf{Concept}\longrightarrow \mathbf{N}_{A,w}$ representing sensor-specific processing knowledge and $M_3:\textbf{Concept}\longrightarrow \mathbf{N}_{A,w}$ representing knowledge fusion in an association region of a multi-regional network. Natural transformations $\gamma_1:M_1\longrightarrow M_3$ and $\gamma_2:M_2\longrightarrow M_3$ provide knowledge coherence between the separate knowledge representations.

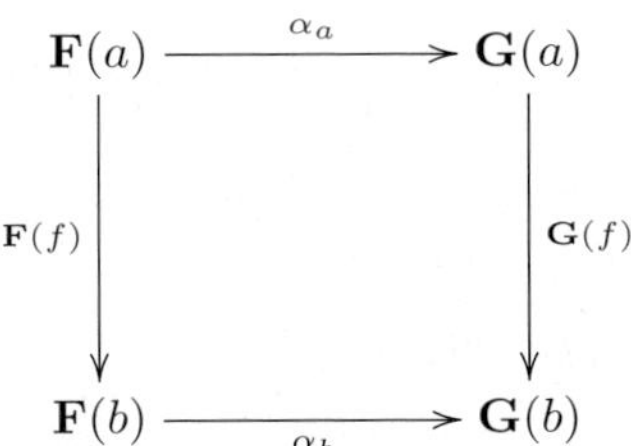

Fig. 4.3. A commutative diagram associated with a natural transformation. The morphisms $G(f)\circ\alpha_a:F(a)\longrightarrow G(b)$ and $\alpha_b\circ F(f):F(a)\longrightarrow G(b)$ are one and the same, $G(f)\circ\alpha_a=\alpha_b\circ F(f)$

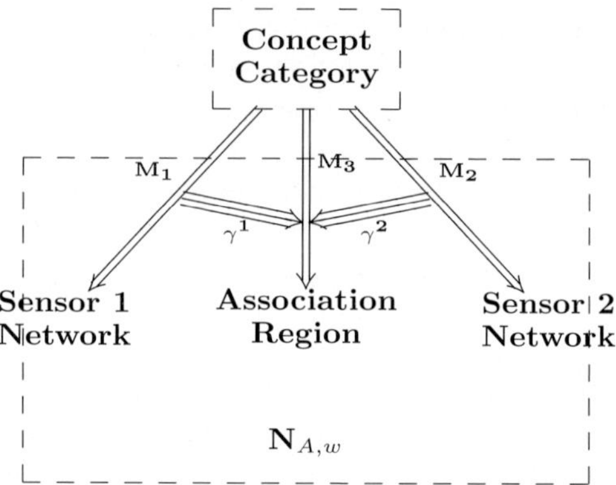

Fig. 4.4. Functors map the hierarchy of a concept category to multiple regions. Natural transformations represent coherent interconnections between hierarchy representations

4.6 From Points to Models

In the topological-lattice account of [12, 29, 30], the instances of propositions are referred to as *points*. This is suggestive, for the semantics of propositional geometric logic is analyzed in terms of topological systems (similar to the spaces of point-set topology) and continuous functions. However, as mentioned in the aforementioned papers by Vickers, category theory is required for expressing the semantics of the geometric predicate calculus. In this context, the collection of points associated with a predicate formula is a space of models, or possible worlds in which the predicate has a valid interpretation. Category theory is required to express the internal structure of theories, and their model spaces are also categories. The semantic theory presented here replaces formulas with whole theories, with functors associating theories and their morphism with objects and morphisms of a neural category . Here, the models are possible worlds or situations in which a theory is valid.

Each concept morphism $s\colon T \longrightarrow T'$ has an associated *model-space morphism*, a functor $\mathsf{Mod}(s)\colon \mathsf{Mod}(T') \longrightarrow \mathsf{Mod}(T)$. Here, $\mathsf{Mod}(T)$ and $\mathsf{Mod}(T')$ are the model categories for T and T', respectively. Since $\mathsf{Mod}(s)$ reverses the direction of s, each instance of T' has a corresponding instance of T. *This fact has great significance for knowledge representation.* In particular, it suggests a design principle for neural networks, as follows: Suppose that neural category objects $M(T)$ and $M(T')$ are the images of objects (concepts) T and T' under a functor $M\colon \mathbf{Concept} \longrightarrow \mathbf{N}_{A,w}$, and that $M(s)\colon M(T) \longrightarrow M(T')$ is the image of a concept morphism $s\colon T \longrightarrow T'$. We associate the activating inputs for the objects $M(T)$ and $M(T')$ with objects in the model categories $\mathsf{Mod}(T)$ and $\mathsf{Mod}(T')$, respectively. Given this association, *every input that activates $M(T')$ must also activate $M(T)$*, a consequence of the existence of the model-space morphism $\mathsf{Mod}(s)\colon \mathsf{Mod}(T') \longrightarrow$

$\mathrm{Mod}(T)$. This provides a mathematical justification—in fact, an imperative—for the presence of feedback in neural networks. Further, this principle applies in some appropriate form to the design of any knowledge representation system.

The model-space morphism principle has important implications for limit and colimit representations. For example, let T be the apical concept of a limit cone for a diagram Δ in **Concept** and let $\ell\colon T \longrightarrow T'$ be one of the leg morphisms for the limit cone, where T' is an object in the base diagram Δ. Then, $M(T')$ is an object in the image diagram $M(\Delta)$, and the model-space morphism principle dictates that the image $M(T)$ of the limit apical object *must be activated through $M(\ell)$ whenever $M(T')$ is active.* The reverse is true for concept colimits: *Every instance of the functorial image of a colimit apical object must also be an instance of the objects in the image of its base diagram.* One consequence is that the analyst can detect limit and colimit representations in a given neural network, and distinguish between them.

4.7 Conclusion

We express the semantics of a system in terms of a distributed system of concepts. The system is an ontology for the system's environment as experienced through the system inputs, both external to the system and possibly internally-generated as well. The ontology is expressed mathematically as a category **Concept** whose objects are symbolic descriptions (concepts) of a domain of items (sensed entities, events, situations, and their parts). We have discussed a sense in which this category is a kind of generalized lattice, differing from a lattice in having morphisms, which are more expressive than edges, with a consequent multiplicity of relationships between objects and with diagrammatic constructions which provide automated concept derivations useful in expressing learning. Two modes of learning can be expressed, specialization and abstraction. In a system such as a neural network with incremental knowledge gain, these two modes serve to "fill out" a representation of an ontology for the environment and functionality of the system, beginning with basic knowledge about inputs. Mathematical rigor is maintained through structure-preserving mappings from **Concept** to a category representing the system's computational structure, thus showing how concepts, their morphisms, and the consequences of composition including commutative concept diagrams are represented in the system computations. Natural transformations formalize knowledge coherence, the unified operation of the separate knowledge representations in the different subnetworks of a multi-regional architecture. Another system of structure-preserving mappings formalizes the relationship between possible worlds for the concept representations. The semantic theory thus offers a comprehensive mathematical theory to support investigations in concept learning in adaptive, distributed systems.

References

1. Adamek J, Herrlich H, Strecker G (1990) Abstract and Concrete Categories. Cambridge University Press, Cambridge New York
2. Andrews R, Diederich J, Tickle A (1995) Survey and critique of techniques for extracting rules from trained artificial neural networks. Knowledge-Based Systems 8:373–389
3. Arbib M (1987) Brains, Machines, and Mathematics. Springer, Berlin Heidelberg New York
4. Barsalou L (1999) Perceptual symbol systems. Behavioral and Brain Sciences 22:577–660
5. Bennet M (1995) Affine and Projective Geometry. John Wiley and Sons, New York
6. Craven M, Shavlik J (1993) Learning symbolic rules using artificial neural networks. In: Proc 10th Intl Machine Learning Conference pp 73–80
7. Crole R (1993) Categories for Types. Cambridge University Press, Cambridge
8. Damasio A (1989) Time-locked multiregional retroactivation: a systems-level proposal for the neural substrates of recall and recognition. Cognition 33:25–62
9. Damasio A (1999) Decartes' Error. Putnam, New York
10. Eichenbaum H (2004) Hippocampus: cognitive processes and neural representations that underlie declarative memory. Neuron 44:109–120
11. Goguen J, Burstall R (1992) Institutions: abstract model theory for specification and programming. J Assoc Computing Machinery 39:95–146
12. Healy M (1999) A topological semantics for rule extraction with neural networks. Connection Science 11:91–113
13. Healy M, Caudell T (1997) Acquiring rule sets as a product of learning in a logical neural architecture. IEEE Trans Neural Networks 8:461–474
14. Healy M, Caudell T (2003) From categorical semantics to neural network design. In: Proc Intl Joint Conf Neural Networks pp 1981–1986
15. Healy M, Olinger R, Young R, Caudell T, Larson K (2005) Modification of the ART-1 architecture based on category theoretic design principles. In: Proc Intl Joint Conf Neural Networks pp 457–462
16. Healy M, Caudell T (2006) Ontologies and worlds in category theory: implications for neural systems. Axiomathes 16:165–214
17. Heileman G, Georgiopoulos M, Healy M, Verzi S (1997) The generalization capabilities of ARTMAP. In: Proc Intl Conf Neural Networks (ICNN)
18. Kaburlasos V, Petridis V (2000) Fuzzy lattice neurocomputing (FLN) models. Neural Networks 13:1145–1170
19. Kasabov N (1996) Adaptable neuro production systems. Neurocomputing 13:95–117
20. Lawvere F, Schanuel S (1997) Conceptual Mathematics: A First Introduction to Categories. Cambridge University Press, Cambridge
21. Mac Lane S (1971) Categories for the Working Mathematician. Springer, Berlin Heidelberg New York
22. McClelland J, McNaughton B, O'Reilly R (1995) Why there are complementary learning systems in the hippocampus and neocortex: insights from the successes and failures of connectionist models of learning and memory. Psychological Review 102:419–457
23. Meseguer J (1989) General logics. In: Ebbinghaus H-D, et al (eds) Logic Colloquium '87. Science Publishers B V, North-Holland

24. Mitra S, Pal S (1995) Fuzzy multi-layer perceptron, inferencing and rule generation. IEEE Trans Neural Networks 6:51–63
25. Pierce B (1991) Basic Category Theory for Computer Scientists. MIT Press, Cambridge, Mass London
26. Pinkas G (1995) Reasoning, nonmonotonicity and learning in connectionist networks that capture propositional knowledge. Artificial Intelligence 77:203–247
27. Sima J (1995) Neural expert systems. Neural Networks 8:261–271
28. Sussner P (2003) Generalizing operations of binary autoassociative morphological memories using fuzzy set theory. J Math Imaging Vision 19:81–93
29. Vickers S (1992) Geometric theories and databases. In: Fourman M, et al (eds) Applications of Categories in Computer Science, Proc LMS Symp Lec Note Ser 177. Cambridge University Press, Cambridge
30. Vickers S (1993) Topology via Logic. Cambridge University Press, Cambridge
31. Wickelgren I (1997) Getting a grasp on working memory. Science 275:1580–1582
32. Williamson K, Healy M, Barker R (2001) Industrial applications of software synthesis via category theory-case studies using specware. Automated Software Eng 8:7–30

Part II

Mathematical Morphology Applications

Noise Masking for Pattern Recall Using a Single Lattice Matrix Associative Memory

Gonzalo Urcid[1] and Gerhard X. Ritter[2]

[1] National Institute of Astrophysics, Optics & Electronics, Department of Optics
Tonanzintla, Puebla 72000, Mexico
`gurcid@inaoep.mx`

[2] University of Florida, Dept. of Computer & Information Science & Engineering
Gainesville, Florida 32611-6120, USA
`ritter@cise.ufl.edu`

Summary. Lattice matrix associative memories have been developed as an alternative way to work with a set of associated pattern pairs for which the storage and retrieval stages are based in the theory of minimax algebra. Several methods have been proposed to cope with the problem of binary or real valued pattern recall from corrupted inputs and recent results on fixed point sets of matrix lattice transforms have provided for an algebraic characterization as well as a geometrical description of the canonical lattice min and max auto-associative memories. Compared to other correlation type associative memory models, the lattice associative memory schemes have shown better performance for both storage and recall capability; however, the computational techniques devised to achieve that purpose are still cumbersome when inputs have undetermined noise bounds. The procedures explained in this chapter makes use of noise masking to boost the recall performance of either the min or max morphological auto-associative memories. Examples using image patterns show the enhanced recovery of almost correct associations from noisy inputs by a single lattice matrix associative memory.

5.1 Introduction

The neural networks known as *morphological associative memories* (MAMs) were introduced a decade ago by Ritter and coworkers [8, 9, 10] as a new paradigm for the storage and recall of pattern associations. These memories are feedforward, fully connected neural networks, in which the interconnection weights between input and output neurons follow a properly defined hebbian type learning rule. In most correlation type associative memory models, the storage and recall stages use conventional algebra whereas computation in MAMs are based on the mathematical theory of minimax algebra developed by Cuninghame-Green [1, 2] and applied by Ritter [7]; since the standard minimax matrix algebra is a specific instance of a lattice algebraic structure, MAMs

G. Urcid and G.X. Ritter: *Noise Masking for Pattern Recall Using a Single Lattice Matrix Associative Memory*, Studies in Computational Intelligence (SCI) **67**, 81–100 (2007)
`www.springerlink.com`

are also named *lattice matrix associative memories*. Once a set of exemplar pattern pairs is imprinted in the neural network, then it is expected that the memory device will exhibit a certain capability to recall the correct association when presented with an exemplar pattern (perfect input) or with a corrupted version of it (non-perfect input). It is important to remark that both input cases pose a difficult robustness problem for any associative memory model. For perfect input, Ritter and Sussner [8, 10, 16] have proved that the canonical *auto-associative morphological memories* (AMMs) have unlimited storage capacity, give perfect recall for all exemplar patterns, and are robust for exclusively erosive or dilative noise; based on stronger assumptions, similar results were established for hetero-associative morphological memories (HMMs). To deal with inputs distorted by mixed random noise, i.e. both erosive and dilative noise combined, the kernel method that chains the max and min AMMs in a two-stage memory scheme was developed for binary [10, 16] and real valued patterns [11, 12]; in addition, different computational approaches or model extensions were devised to increase its recall capability or applicability in pattern recognition problems. A fast technique for finding kernels and the use of multiple kernels of binary patterns were proposed respectively by Hattori [6] and Hashiguchi [5]; variations on the kernel, the dual kernel methods and the use of fuzzy set theory applied to AMMs were established by Sussner [18, 19] for the binary case; Urcid [22] gave an algorithm based on induced morphological strong independence over a source set of real valued exemplar patterns, and recently, Wang [23] has proposed a fuzzy AMM that uses an empirical map that has also been tested with real valued patterns. In the last two techniques, inputs corrupted by mixed random noise were considered. For real world applications, Graña and coworkers [3, 4] have successfully applied the AMMs and the morphological independence criterion of a pattern set to find endmembers in hyperspectral imagery. With a more theoretical algebraic and geometrical orientation and parallel to technical developments and applications, Sussner [17, 20], Ritter and Gader [15] have establish a deeper insight that provides useful results on the fixed point set of AMMs and therefore, a complete characterization of their response for arbitrary inputs. From this point of view, numerical examples and a practical discussion are given in [15], and a comparison of AMMs with other enhanced associative memory neural networks for gray-scale pattern retrieval appears in [20] although only for erosive or dilative noise. More recently, Sussner and Valle [21] have developed gray-scale AMMs that use an appropriate additive constant to achieve considerable amounts of error correction for inputs degraded by random noise.

We point out the relevant fact that the strengths and weaknesses of the min and max auto-associative memories have driven the different developments of all previous work done on MAMs as briefly described above. The recall failure of the min AMM for inputs degraded by dilative noise and similarly, the poor performance of the max AMM for inputs corrupted with erosive noise, and consequently their inapplicability to deal with mixed random noise has been exposed repeatedly since their introduction [10, 16], in the most recent theoretical results discovered about them [15, 20, 21], and in the development of new lattice based associative dendritic or fuzzy memory models [13, 14, 23]. In these pages we give a complete pragmatical solution to pattern recall from inputs corrupted by mixed random noise using only one of the canonical AMMs by masking the noise contained in the corrupted pattern. Therefore, in spite of the aforementioned limitations of AMMs we will use their robustness to either erosive or dilative noise to regain their inherent computational advantage by extending their functionality to cope with any type and considerable amounts of random noise.

This chapter is organized as follows: Sect. 5.2 gives some mathematical background on minimax algebra covering the fundamental lattice matrix operations used to define and apply the fundamental AMMs, together with the fundamental theoretical results in regards to their recall performance; Sect. 5.3 explains the mechanism behind noise masking that will be used for pattern recall; in Sect. 5.4 we provide the basic algorithms that employ a single AMM capable to recall binary or real valued patterns from inputs contaminated with mixed random noise, and a few concluding remarks to the material exposed here are given in Sect. 5.5. The chapter ends with a list of essential references to which the reader may turn for supplementary information.

5.2 Mathematical Background

5.2.1 Some Minimax Matrix Algebra

The basic numerical operations of taking the maximum or minimum of two numbers usually denoted as functions $\max(x, y)$ and $\min(x, y)$ will be written as binary operators using the "join" and "meet" symbols employed in lattice theory, i.e., $x \vee y = \max(x, y)$ and $x \wedge y = \min(x, y)$. It was shown in [1, 7] that the algebraic systems $(\mathbb{R}_{-\infty}, \vee, +)$ and $(\mathbb{R}_{+\infty}, \wedge, +')$ are *semirings* for the corresponding max-$\vee$, min-$\wedge$ operations and their respective additions, $+$ and $+'$ over the sets $\mathbb{R}_{-\infty} = \mathbb{R} \cup \{-\infty\}$ and $\mathbb{R}_{\infty} = \mathbb{R} \cup \{+\infty\}$. In what follows we consider only *finite* values of x and y, for which, $x +' y = x + y$.

We use lattice matrix operations that are defined componentwise using the underlying structure of $\mathbb{R}_{-\infty}$ or $\mathbb{R}_{\infty}$ as semirings. For example, the maximum and minimum of two matrices X, Y of the same size $m \times n$ is defined as shown in (5.1), i.e., for all $i = 1, \ldots, m$ and $j = 1, \ldots, n$:

$$(X \vee Y)_{ij} = x_{ij} \vee y_{ij} \quad ; \quad (X \wedge Y)_{ij} = x_{ij} \wedge y_{ij} . \tag{5.1}$$

Inequalities between matrices are also verified elementwise, e.g., $X \leq Y$ if and only if $x_{ij} \leq y_{ij}$. On the other hand, the *conjugate matrix* X^* is defined as $-X^{\mathrm{T}}$ where X^{T} denotes usual matrix transposition. In addition, given a transform $T : \mathbb{R}^n \to \mathbb{R}^n$, then $\mathbf{x} \in \mathbb{R}^n$ is called a *fixed point* of T if and only if $T(\mathbf{x}) = \mathbf{x}$. Also, we will need to use the arithmetic mean $\mu(\mathbf{x})$ of a vector $\mathbf{x} \in \mathbb{R}^n$, where $\mu(\mathbf{x}) = \mu(x_1, \ldots, x_n) = \sum_{i=1}^{n} x_i/n$.

Definition 5.1 *The max-product, $X \boxvee Y$ and the min-product, $X \boxwedge Y$, of matrix X of size $m \times p$ with matrix Y of size $p \times n$, for all $i = 1, \ldots, m$ and $j = 1, \ldots, n$ are given by*

$$(X \boxvee Y)_{ij} = \bigvee_{k=1}^{p} (x_{ik} + y_{kj}), \tag{5.2}$$

$$(X \boxwedge Y)_{ij} = \bigwedge_{k=1}^{p} (x_{ik} + y_{kj}). \tag{5.3}$$

Definition 5.2 *The lattice product of two vectors $\boldsymbol{x} = (x_1, \ldots, x_n)^{\mathrm{T}} \in \mathbb{R}^n$ and $\boldsymbol{y} = (y_1, \ldots, y_m)^{\mathrm{T}} \in \mathbb{R}^m$, referred here as the* minimax outer product, *is given by the $m \times n$ matrix*

$$\boldsymbol{y} \times \boldsymbol{x}^{\mathrm{T}} = \begin{pmatrix} y_1 + x_1 & \cdots & y_1 + x_n \\ \vdots & \ddots & \vdots \\ y_m + x_1 & \cdots & y_m + x_n \end{pmatrix}. \tag{5.4}$$

It is worthwhile to note that $\boldsymbol{y} \times \boldsymbol{x}^{\mathrm{T}} = \boldsymbol{y} \boxvee \boldsymbol{x}^{\mathrm{T}} = \boldsymbol{y} \boxwedge \boldsymbol{x}^{\mathrm{T}}.$

Definition 5.3 *If $X = \{\boldsymbol{x}^1, \ldots, \boldsymbol{x}^k\} \subset \mathbb{R}^n$, then a* lattice polynomial *or* linear minimax combination *of vectors from X, denoted by $\mathcal{L}_{\mathrm{P}}(X)$, is any vector $\mathbf{x} \in \mathbb{R}^n$ of the form*

$$\boldsymbol{x} = \mathcal{L}_{\mathrm{P}}(X) = \bigvee_{i \in I} \bigwedge_{\xi=1}^{k} (\boldsymbol{x}^\xi + \alpha_i^\xi), \tag{5.5}$$

where I is a finite set of indices and $\alpha_i^\xi \in \mathbb{R}$ for all i, ξ. The set of all such lattice polynomials is called the real linear minimax span *of X and is denoted by $LMS_{\mathbb{R}}(X)$.*

The previous definitions and equations are the necessary tools to discuss the theoretical background of morphological associative memories as explained in the next subsection. In relation to MAMs, detailed treatments of lattice and minimax algebra as well as additional mathematical results can be found in [7, 15, 21]. See also in this book, Chap. 2 by Ritter on *Learning in Lattice Neural Networks that Employ Dendritic Computing* and Chap. 8 by Sussner and Valle on *Morphological and Certain Fuzzy Morphological Associative Memories for Classification and Prediction* for complementary comprehensive surveys and additional insight on the same background material.

5.2.2 Lattice Matrix Associative Memories

Henceforth, let $(\boldsymbol{x}^1, \boldsymbol{y}^1), \ldots, (\boldsymbol{x}^k, \boldsymbol{y}^k)$ be k vector pairs with $\boldsymbol{x}^\xi = (x_1^\xi, \ldots, x_n^\xi)^{\mathrm{T}}$ and $\boldsymbol{y}^\xi = (y_1^\xi, \ldots, y_m^\xi)^{\mathrm{T}}$ for $\xi = 1, \ldots, k$ where $\boldsymbol{x}^\xi \in \mathbb{R}^n$ and $\boldsymbol{y}^\xi \in \mathbb{R}^m$.

For a given set of pattern associations $\{(\boldsymbol{x}^\xi, \boldsymbol{y}^\xi) : \xi = 1, \ldots, k\}$ we define a pair of associated pattern matrices (X, Y), where $X = (\boldsymbol{x}^1, \ldots, \boldsymbol{x}^k)$ and $Y = (\boldsymbol{y}^1, \ldots, \boldsymbol{y}^k)$. Thus, X is of dimension $n \times k$ with i, jth entry x_i^j and Y is of dimension $m \times k$ with i, jth entry y_i^j. To store k vector pairs $(\boldsymbol{x}^1, \boldsymbol{y}^1), \ldots, (\boldsymbol{x}^k, \boldsymbol{y}^k)$ in an $m \times n$ *lattice matrix associative memory* we use a similar approach for encoding associations in a linear or correlation memory but instead of using the linear outer product, the minimax outer product is used as follows [8, 9, 10].

Definition 5.4 *The* min-memory W_{XY} *and the* max-memory M_{XY}, *both of size* $m \times n$, *that store a set of pattern associations* (X, Y) *are given, respectively, by the matrix and componentwise expressions*

$$W_{\mathrm{XY}} = \bigwedge_{\xi=1}^{k} [\boldsymbol{y}^\xi \times (-\boldsymbol{x}^\xi)^{\mathrm{T}}] \quad ; \quad w_{ij} = \bigwedge_{\xi=1}^{k} (y_i^\xi - x_j^\xi), \tag{5.6}$$

$$M_{\mathrm{XY}} = \bigvee_{\xi=1}^{k} [\boldsymbol{y}^\xi \times (-\boldsymbol{x}^\xi)^{\mathrm{T}}] \quad ; \quad m_{ij} = \bigvee_{\xi=1}^{k} (y_i^\xi - x_j^\xi). \tag{5.7}$$

We speak of a hetero-associative *morphological memory (HMM) if* $X \neq Y$ *and an* auto-associative *morphological memory (AMM) if* $X = Y$.

The expressions to the left of each equation are in matrix form and the right expression is the ijth entry of the corresponding memory. Note that according to (5.4), for each ξ, $\boldsymbol{y}^\xi \times (-\boldsymbol{x}^\xi)^{\mathrm{T}}$ is a matrix E^ξ of size $m \times n$ that memorizes the association pair $(\boldsymbol{x}^\xi, \boldsymbol{y}^\xi)$ hence $W_{\mathrm{XY}} = \bigwedge_{\xi=1}^{k} E^\xi$ and $M_{\mathrm{XY}} = \bigvee_{\xi=1}^{k} E^\xi$, which suggests the given names. The retrieval of pattern $\boldsymbol{y}^\xi$ from pattern $\boldsymbol{x}^\xi$ can be expressed using the following one pass memory scheme,

$$\boldsymbol{x}^\xi \rightarrow \{W_{\mathrm{XY}} \mid M_{\mathrm{XY}}\} \rightarrow \boldsymbol{y}^\xi, \tag{5.8}$$

where, the vertical bar means that, either one of W_{XY} or M_{XY} may be used. In this exposition we restrict our discussion to AMMs, i.e., to W_{XX} or M_{XX} of size $n \times n$; also, for application purposes we will assume that pattern entries are *non-negative* and *finite*, i.e., $0 \leq x_i^\xi < \infty$ for all i, ξ.

5.2.3 Main Theoretical Results

The theorems listed here give a quick overview of the theoretical foundations of AMMs that are of practical significance in applications; their proofs can be found in [10, 16, 17] or [15]. To safe space, statements in the theorems refer only to the min-memory W_{XX} since analogous properties for the max-memory M_{XX} are readily obtained using the *duality principle*. AMMs have unlimited storage capacity, the pattern domain can be binary or real valued, give *perfect recall* for perfect input, and computation is performed in one step, free of any convergence problems [8, 10]. In mathematical form:

Theorem 5.1 $W_{XX} \boxtimes X = X$, *for any matrix X of size $n \times k$ whose columns are the exemplar pattern vectors $x^1, \ldots, x^k \in \mathbb{R}^n$.*

Theorem 5.2 *If $W_{XX} \boxtimes x = x'$, then $W_{XX} \boxtimes x' = x'$, where $x \in \mathbb{R}^n$ is any input vector and x' is the recalled pattern.*

Notice that the max-product of W_{XX} with vector x is an example of a *lattice matrix transform* defined by $T_W(x) = W_{XX} \boxtimes x$; therefore, Theorem 5.1 means that any exemplar pattern x^ξ for $\xi = 1, \ldots, k$ is a fixed point of T_W (complete perfect recall). From Theorem 5.2 it is clear that the recalled pattern x' obtained in the first step is a fixed point for the second application of W_{XX} and therefore one step convergence is guaranteed.

Definition 5.5 *The* fixed point set *of the min-memory W is given by*

$$F(X) = \{ x \in \mathbb{R}^n : W_{XX} \boxtimes x = x \}. \tag{5.9}$$

The next theorem gives a complete algebraic characterization of the recall response of the lattice matrix auto-associative memories in terms of lattice polynomials.

Theorem 5.3 *The fixed point set $F(X)$ of the AMMs, W_{XX} and M_{XX} coincides, in either case it is given by $F(X) = LMS_{\mathbb{R}}(X)$.*

Consequently, from (5.9) and (5.5), the result established in Theorem 5.2 means that, for any input vector $x \in \mathbb{R}^n$ the recalled pattern x' belongs to $F(X)$. Hence, if $\beta_i^\xi \in \mathbb{R}$ for all $i \in I, \xi \in \{1, \ldots, k\}$, the following equation is valid

$$W_{XX} \boxtimes x = x' = \mathcal{L}_P(X) = \bigvee_{i \in I} \bigwedge_{\xi=1}^{k} (x^\xi + \beta_i^\xi). \tag{5.10}$$

Observe that the input vector x appearing in (5.10) is a lattice polynomial in X that can be an exemplar pattern x^ξ, a noisy version of it, usually symbolized by $\widetilde{x}^\xi$, or even a non-corrupted non-exemplar pattern. Any way, the righthand side of (5.10) means also that W_{XX} has many *spurious states* that are stable. This equation is a deep and nice theoretical result but still does not provide a practical criterion to enhanced the recall performance of the canonical AMMs for inputs corrupted by noise. An alternative approach to lattice associative matrix memories is based on morphological neurons endowed with dendrites whose description the reader will find in Chap. 2 of the present book, *Learning in Lattice Neural Networks that Employ Dendritic Computing* by Ritter. Application of the canonical AMMs to hyperspectral imaging and pattern recognition are surveyed in Chap. 6 on *Convex Coordinates from Lattice Independent Sets for Visual Pattern Recognition* by Graña, Villaverde, Moreno, and Albizuri.

Missing parts, occlusions or corruption of exemplar patterns can be considered as "noise". Particularly, if alterations in pattern entries follow a probability law, then the pattern is said to be contaminated with *random noise.* If pattern entries change in a deterministic way, then the pattern is said to be distorted with *structured noise.* In order to specify the recall capabilities of AMMs when presented with non-perfect inputs, noise can be classified in three types relative to the total ordered structure of a given numerical scale such as the real line.

Definition 5.6 *Let* $I = \{1, \ldots, n\}$ *then, a distorted version* $\widetilde{x}$ *of pattern* x *has undergone an* erosive *change whenever* $\widetilde{x} \leq x$ *or equivalently if* $\forall i \in I$, $\widetilde{x}_i \leq x_i$. *A* dilative *change occurs whenever* $\widetilde{x} \geq x$ *or equivalently if* $\forall i \in I$, $\widetilde{x}_i \geq x_i$. *Let* $L, G \subset I$ *be two non-empty disjoint sets of indexes. If* $\forall i \in L$, $\widetilde{x}_i < x_i$ *and* $\forall i \in G$, $\widetilde{x}_i > x_i$, *then the distorted pattern* $\widetilde{x}$ *is said to contain* mixed *noise (random or structured).*

Theorem 5.4 *Let* $\widetilde{x}^\gamma$ *be an eroded version of pattern* x^γ, *then the equation* $W_{XX} \boxdot \widetilde{x}^\gamma = x^\gamma$ *holds if and only if for each row index* $i \in \{1, \ldots, n\}$ *there exists a column index* $j(i) \in \{1, \ldots, n\}$ *such that*

$$\widetilde{x}^\gamma_{j(i)} = max\left(x^\gamma_{j(i)}, \bigvee_{\xi \neq \gamma} \left[x^\gamma_i - x^\xi_i + x^\xi_{j(i)} \right] \right). \tag{5.11}$$

Corollary 5.1 *If* $W_{XX} \boxdot \widetilde{x}^\gamma = x^\gamma$ *then* $\widetilde{x}^\gamma$ *must be an eroded version of* x^γ, *i.e.,* $\widetilde{x}^\gamma \leq x^\gamma$.

Theorem 5.4 and Corollary 5.1 still remain the fundamental results that give the conditions to guarantee *perfect recall* from a distorted exemplar corrupted by erosive noise using the min-memory W_{XX}. Similar conditions provide for perfect recall applying the max-memory M_{XX} to an exemplar pattern that has been changed by dilative noise. Additional theoretical discussions [16, 20], useful commentaries and explanations using numerical examples [10, 15], as well as a wide spectrum of computational experiments with different sets of binary or gray-scale images [10, 11, 12, 16, 20], have demonstrated the surprising fact that W_{XX} is *very robust* to erosive noise and dually, that M_{XX} is very robust to dilative noise. However, both AMMs fail completely for inputs corrupted with mixed noise, a handicap that has been emphasized several times in all enhanced models developed so far, such as, the kernel based methods, the extended fuzzy models, and the lattice dendrite associative memory models.

A reasonable, simple question pops up in our minds. If the canonical AMMs have a firm mathematical foundation, give results of far reaching applications, and are quite robust to erosive or dilative distortions – closely related to the more general situation of mixed noise – is it possible to implement a mechanism that takes all their known advantages and therefore regain their status as good associative memory models for perfect recall in the presence of mixed noise? The answer is affirmative and we explain it in detail in the following two sections.

5.3 Noise Masking

The only roadblock that prevents the use of the min-memory W_{XX} as a direct memory scheme for arbitrary inputs, is the condition that it can work very well only with exemplar patterns corrupted by erosive noise. We propose to use *noise masking* to changed an input vector degraded by mixed noise into a vector corrupted only by erosive or dilative noise.

Definition 5.7 *Let $X = \{x^1, \ldots, x^k\} \subset \mathbb{R}^n$ be a finite set of exemplar patterns and let $\widetilde{x}$ be a mixed noisy version of $x \in X$, then the* min-masked *pattern of $\widetilde{x}$ is given by $\widetilde{x} \wedge x = \widetilde{x}_e$ and the* max-masked *pattern of $\widetilde{x}$ is given by $\widetilde{x} \vee x = \widetilde{x}_d$.*

It turns out, using the axioms of a lattice structure, that $\widetilde{x}_e \leq x$, hence $\widetilde{x}_e$ is an eroded version of x that can be used as input to W_{XX}; dually, $\widetilde{x}_d \geq x$, i.e., $\widetilde{x}_d$ is a dilated version of x useful for M_{XX}. We point out the important fact that in any associative memory model, the set X of fundamental memories or exemplars constitutes available "known information". Therefore, determination of the min- and max-masked input patterns makes sense and both are readily obtained since computation with lattice vector operations is fast.

Example 5.1 *In binary image processing it is usual to consider two possible encodings of objects. The W/B encoding assigns a value of 1 (white) for the object or foreground, and 0 (black) for its background. Alternatively, the encoding B/W considers a value of 0 (black) for the object and 1 (white) for its background. Figure 5.1 illustrates binary masking for the uppercase letter 'G' with the W/B encoding and the uppercase letter 'R' with the B/W encoding. For each exemplar letter four samples of noisy versions and their corresponding masked results are displayed.*

In the third column of the left panel in Fig. 5.1, noise min-masking with the $\wedge$-operator is equivalent to a pointwise logical AND operation and gives the eroded versions of the 'G' letter that can be used as inputs to W_{XX}. Similarly, in the third column of the right panel, noise max-masking using the $\vee$-operator is equivalent to a pointwise logical OR operation and produces the dilated versions of the complement of the 'R' letter which can be used as inputs to M_{XX}. Note that, in the case of binary contaminated image patterns, mixed random noise is synonymous with impulsive noise or pixel value reversal.

If random noise occurs with probability $p \in (0, 1)$, then the *masking rules* that generate an adequate noisy version of an exemplar binary pattern $x \in \{0, 1\}^n$, are specified as follows:

Fig. 5.1. Binary masking: in each image panel, the 1st entry in the 1st column shows an uppercase letter with different encoding as an exemplar pattern, 2nd column of both panels shows corrupted versions with $0.25, 0.40, 0.60, 0.75$ random noise probability, and the 3rd column displays the results of noise masking

- eroded noisy patterns obtained by min-masking,

$$\textbf{if } p \in (0, 0.5) \textbf{ then } \widetilde{\boldsymbol{x}}_{+e} = \boldsymbol{x} \wedge \widetilde{\boldsymbol{x}} \tag{5.12}$$

$$\textbf{else } \widetilde{\boldsymbol{x}}_{-e} = \boldsymbol{x} \wedge (1 - \widetilde{\boldsymbol{x}}), \tag{5.13}$$

- dilated noisy patterns obtained by max-masking,

$$\textbf{if } p \in (0, 0.5) \textbf{ then } \widetilde{\boldsymbol{x}}_{+d} = \boldsymbol{x} \vee \widetilde{\boldsymbol{x}} \tag{5.14}$$

$$\textbf{else } \widetilde{\boldsymbol{x}}_{-d} = \boldsymbol{x} \vee (1 - \widetilde{\boldsymbol{x}}). \tag{5.15}$$

In (5.12) and (5.14) we denote positive masking with a plus sign $(+)$ as subindex and similarly, in (5.13) and (5.15) negative masking is distinguished by a minus sign $(-)$. In this last case, use is made of the *inverted* or *negative* of the noisy input which is given by $1 - \widetilde{\boldsymbol{x}}$ where $\boldsymbol{1} = (1, 1, \ldots, 1) \in \mathbb{R}^n$.

Example 5.2 *Figure 5.2 illustrates the advantage of using (5.13) or (5.15) to properly mask the corrupted input pattern when it comes with high levels of noise. Under this situation, more and more pixels are inverted and the corrupted patterns begins to change its initial encoding, hence the need of negative masking is justified. Encoding interchange for the exemplar uppercase letters 'U' and 'S' is clearly seen against their 85% noisy versions shown in the 4th row and 2nd column of their corresponding image panels.*

It is important to remark, that visual object recognition as illustrated here with binary image patterns is a function of the spatial organization and structure of the object as well as the noise probability distorting it. In the examples given in Fig. 5.1 and Fig. 5.2, it is evident that low-high levels of noise with probability p in the range $(0, 0.35] \cup [0.65, 1)$ still allows for visual identification of the selected letters or their negatives; however, if $p \in (0.35, 0.65)$, then the corrupted letter is no longer visually recognizable. The extreme values

Fig. 5.2. Binary masking: in each image panel, 1st entry in 1st column shows an uppercase letter with different encoding as an exemplar pattern, 2nd column of both panels shows highly corrupted versions of letters "U" and "S" with $0.55, 0.65, 0.75, 0.85$ random noise probability, 3rd column displays the results obtained with positive masking, and the 4th column displays better results produced by negative masking

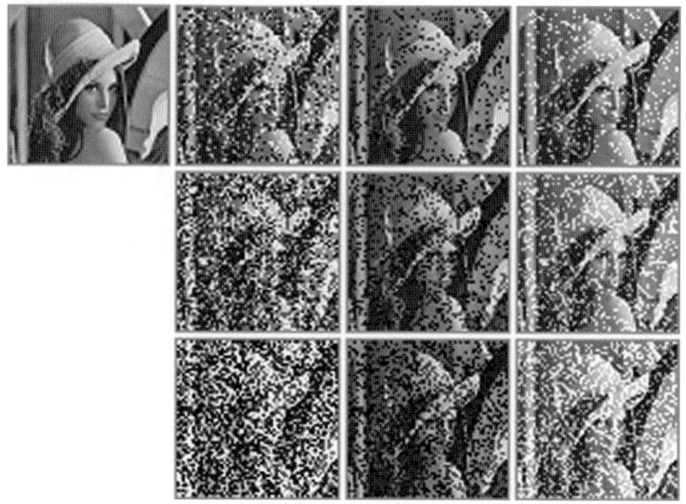

Fig. 5.3. Gray-scale masking: 1st entry of 1st column shows the exemplar pattern, 2nd column shows corrupted versions of the "Lena" image with $0.25, 0.50, 0.75$ random noise probability, 3rd column displays the eroded results obtained by min masking, and the 4th column displays the dilated results produced by max masking

given in these intervals do not define crisp boundaries due to psychophysical aspects of human visual perception. As a general comment, the technique of noise masking combined with subtraction is equivalent to *unmix* the erosive and dilative components that are present in the input pattern.

Example 5.3 *We end this section with Fig. 5.3 that illustrates noise masking for gray-scale image patterns, in this case the "Lena" image has been corrupted* *by additive random salt and pepper noise with a gray reference value of 128. The eroded noisy versions of the "Lena" image displayed in the 3rd column can be used as inputs to W_{XX} and the dilated corrupted images of the 4th column may serve as inputs to M_{XX}. An interesting fact is that for any amount of noise added to an L-gray-scale image pattern $\mathbf{x} \in [0, L-1]^n$, only the masking rules given by (5.12) and (5.14) are used, respectively, to generate eroded or dilated noisy versions. Therefore, negative masking is not required for sets of real-valued patterns.*

5.4 Recall from Noisy Inputs

This section contains the computational procedures based on noise masking, to change a corrupted exemplar input pattern into a similar pattern that has only erosive or dilative noise. Since the conditions for perfect recall with AMMs given in Theorem 5.4 are not always satisfied for arbitrary finite sets of exemplar patterns, it is necessary to measure the difference between the noisy input and the output recalled. For convenience of exposition we treat the binary and real valued cases separately.

5.4.1 Binary Patterns

Let $X = \{\boldsymbol{x}^1, \ldots, \boldsymbol{x}^k\} \subset \{0,1\}^n$ be a finite set of k exemplar binary patterns that are stored in W_{XX} following (5.6) by taking $Y = X$. If $\widetilde{\boldsymbol{x}}$ denotes a noisy version of the exemplar $\boldsymbol{x} = \boldsymbol{x}^\xi$ for some $\xi \in \{1, \ldots, k\}$, we will use the *normalized Hamming distance* (NHD) between both patterns as a measure of their proximity; it is defined by any of the two expressions in (5.16), i.e.

$$h(\boldsymbol{x}, \widetilde{\boldsymbol{x}}) = \frac{1}{n} \sum_{i=1}^{n} \mid x_i - \widetilde{x}_i \mid = \frac{1}{n} \sum_{i=1}^{n} \mathtt{mod}(x_i + \widetilde{x}_i, 2) . \tag{5.16}$$

Observe that, $0 \leq h(\boldsymbol{x}, \widetilde{\boldsymbol{x}}) \leq 1$. If the vectors compared match each other in all entries, then $h(\boldsymbol{x}, \widetilde{\boldsymbol{x}}) = 0$, if there is a single mismatch then $h(\boldsymbol{x}, \widetilde{\boldsymbol{x}}) = 1/n$, and if all entries differ in value then $h(\boldsymbol{x}, \widetilde{\boldsymbol{x}}) = 1$. With the help of this measure we can formalize the notion of "almost perfect recall" for the binary case in the following sense:

Definition 5.8 *The lattice matrix autoassociative memory W_{XX} is an* almost perfect recall *memory for a set X of binary patterns if and only if there exists a small rational number $\varepsilon > 0$ close to zero, such that $h(W_{XX} \boxtimes \widetilde{\boldsymbol{x}}, \boldsymbol{x}) \leq \varepsilon$ for all $\boldsymbol{x} \in X$ with respect to finite sets of noisy versions $\widetilde{\boldsymbol{x}}$ of $\boldsymbol{x}$. Under similar conditions, M_{XX} is an almost perfect recall memory whenever $h(M_{XX} \boxtimes \widetilde{\boldsymbol{x}}, \boldsymbol{x}) \leq \varepsilon$.*

The next algorithm, expressed in mathematical notation, establishes the technique of noise masking for binary patterns as a mechanism to provide an eroded noisy input pattern to the min memory W_{XX}. Numbered steps are prefixed by S and comments are provided within brackets.

Algorithm 5.1. Binary Pattern Recall with W_{XX}

S1. Input $\widetilde{x} \in \{0, 1\}^n$
[Assume $\widetilde{x}$ is an unknown noisy exemplar corrupted with mixed noise.]

S2. For $\xi = 1$ **to** k **do**
$$\widetilde{x}^\xi_{+e} = x^\xi \wedge \widetilde{x}$$
$$\widetilde{x}^\xi_{-e} = x^\xi \wedge (1 - \widetilde{x})$$
[Min-positive/negative masked vectors of k memorized patterns.]

S3. For $\xi = 1$ **to** k **do**
$$D^\xi_{+e} = \mu[h(\widetilde{x}^\xi_{+e}, x^\xi), h(\widetilde{x}^\xi_{+e}, \widetilde{x})]$$
$$D^\xi_{-e} = \mu[h(\widetilde{x}^\xi_{-e}, x^\xi), h(\widetilde{x}^\xi_{-e}, 1 - \widetilde{x})]$$
$$D_\xi = D^\xi_{+e} \wedge D^\xi_{-e}$$
[Mean positive, mean negative and minimum mean distance vectors.]

S4. Find $\gamma \in \{1, \ldots, k\}$ such that $D_\gamma = \bigwedge_{\xi=1}^{k} D_\xi$.
[Min-masked candidate vectors, eroded only versions of $\widetilde{x}$ are $\widetilde{x}^\gamma_{+e}$ and $\widetilde{x}^\gamma_{-e}$.]

S5. Let $x^\gamma_+ = W_{XX} \boxtimes \widetilde{x}^\gamma_{+e}$ and **let** $x^\gamma_- = W_{XX} \boxtimes \widetilde{x}^\gamma_{-e}$
[The masked γ-vectors in S4 are inputs to W_{XX} to get output candidate patterns.]

S6. Let $\delta_+ = h(x^\gamma_+, x^\gamma)$ and **let** $\delta_- = h(x^\gamma_-, x^\gamma)$
[Positive and negative Hamming distances of recalled candidate patterns.]

S7. If $\delta_+ < \delta_-$ **then** $x^\gamma = x^\gamma_+$ **else** $x^\gamma = x^\gamma_-$
[Select the exemplar pattern that is most correctly associated by W_{XX} with the noisy input $\widetilde{x}$. For another input return to S1.]

Observe that, after step 7 the following relationship holds

$$\delta_+ \wedge \delta_- = \bigwedge_{\xi=1}^{k} h(\widetilde{x}, x^\xi). \tag{5.17}$$

The operations needed in steps 2, 3, 5 and 6 of Algorithm 5.1 can be performed in *parallel* and do not pose a computational burden in possible applications that may use the single autoassociative memory W_{XX} since all arithmetical operations are very simple. Equivalently, the search operation needed in step 4 may be replaced by a fast ascending sort. Algorithm 5.1 can be readily modified to be used with the dual max memory M_{XX}. Specifically, step 2 will use the $\vee$-operator to generate dilated versions ($+d, -d$ subindexes), step 5 will compute two min-products using M_{XX}, and the other steps remain exactly the same.

Example 5.4 *In this computational experiment, the set of exemplar patterns consisted of the 26 uppercase letters of the English alphabet, each letter coded as a binary image in a matrix of size 32×32 pixels; here, the smallest non-zero value of* NHD $= 1/1024 \approx 0.001$. *Table 5.1 gives the numerical mean values of the normalized Hamming distances related to the W/B encoding over 100 eroded corrupted versions of the binary letters 'G' and 'U' that where used as noisy inputs for the min-memory W_{XX}. In a similar fashion, Table 5.2 lists the numerical results for the dual encoding of the letters R' and 'S' of which 100 complemented dilative versions were fed into the max-memory M_{XX}. A single trial has been shown in Fig. 5.1 and Fig. 5.2; specifically, letter 'G,U' in the left image panel and letters 'R,S' in the right image panel.*

From the last column in Tables 5.1 and 5.2, it can be observed that the difference between the recalled pattern and the stored pattern is small, since *all* the numerical values listed for $\mu[h_i(\widetilde{\boldsymbol{x}}^{\gamma}, \boldsymbol{x})]$ for $i = 1, \ldots, 100$, are at most an order of magnitude above the minimum difference 0.001 given earlier.

Example 5.5 *Further computational experiments reveal that, for several runs of 100 noisy versions of each uppercase letter, the mean of the distances D_{γ} computed in step 4 of Algorithm 5.1, calculated over a run, is a piecewise linear function of noise probability, i.e.,*

$$\mu(D_{\gamma}^1, \ldots, D_{\gamma}^{100})(p) \simeq 0.25(1 - 2|p - 0.5|)\,. \tag{5.18}$$

In (5.18), the value of γ gives the correct pattern index for $p \in [0, 0.45] \cup [0.55, 1]$ for each letter, even for similar letters such as 'D' and 'O' ($\gamma = 4, 15$), or 'I' and 'J' ($\gamma = 9, 10$). For noise probability levels in the range $[0.46, 0.48] \cup [0.52, 0.54]$, the value of γ may correspond to similar letters; for

Table 5.1. Mean normalized Hamming distances for mixed noise, eroded versions, and recall performance of memory W_{XX}

Letter	Noise p	Mean $h(\widetilde{\boldsymbol{x}}, \boldsymbol{x})$	Mean $h(\widetilde{\boldsymbol{x}}_e, \boldsymbol{x})$	Mean $h(\widetilde{\boldsymbol{x}}^{\gamma}, \boldsymbol{x})$
'G'	0.25	0.250	0.053 (+e)	0.007
'G'	0.45	0.452	0.095 (+e)	0.014
'U'	0.55	0.549	0.080 (−e)	0.021
'U'	0.75	0.748	0.045 (−e)	0.010

Table 5.2. Mean normalized Hamming distances for mixed noise, dilated versions, and recall performance of memory M_{XX}

Letter	Noise p	Mean $h(\widetilde{\boldsymbol{x}}, \boldsymbol{x})$	Mean $h(\widetilde{\boldsymbol{x}}_d, \mathbf{x})$	Mean $h(\widetilde{\boldsymbol{x}}^{\gamma}, \boldsymbol{x})$
'R'	0.25	0.250	0.051 (+d)	0.015
'R'	0.45	0.452	0.092 (+d)	0.032
'S'	0.55	0.554	0.086 (−d)	0.025
'S'	0.75	0.752	0.047 (−d)	0.011

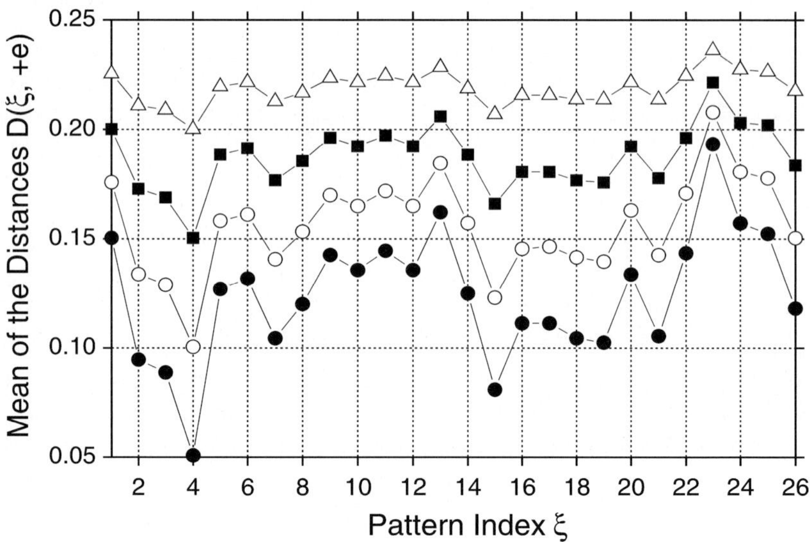

Fig. 5.4. Binary positive masking: mean of the distances D_{+e}^{ξ} computed over a run of 100 random corrupted versions of the letter 'D' against each letter with $p = 0.1(\bullet), 0.2(\circ), 0.3(\blacksquare), 0.4(\triangle)$. Vertical scale is normalized and $D(\xi, +e) = D_{+e}^{\xi}$

$p \in (0.48, 0.52)$, *the mean distance values are almost the same for all letters and there is no way to tell what letter is truly associated with such a noisy input. Figure 5.4 shown below, displays the value of the mean of the distances D_{+e}^{ξ} computed over a run of 100 random corrupted versions of 'D' against each letter with probabilities equal to 0.1, 0.2, 0.3, 0.4 which are marked respectively, with a disk (bottom curve), a circle, a solid square, and a triangle (top curve). Observe that in each curve, the pattern index $\gamma = 4$ obtained in step 4 of Algorithm 5.1 gives the correct letter; in this case, since $p < 0.5$, it follows that $D_4 = D_{+e}^{4} \wedge D_{-e}^{4} = D_{+e}^{4}$.*

Similarly, Fig. 5.5 shows the value of the mean of the distances D_{-e}^{ξ} computed over a run of 100 random corrupted versions of 'I' against each letter with probabilities equal to 0.9, 0.8, 0.7, 0.6 which are marked respectively, with a disk (bottom curve), a circle, a solid square, and a triangle (top curve). Observe that in each curve, the pattern index $\gamma = 9$ obtained in Step 4 of Algorithm 5.1 gives the correct letter; in this case, since $p > 0.5$, it follows that $D_9 = D_{+e}^{9} \wedge D_{-e}^{9} = D_{-e}^{9}$. These curves also illustrate that for noise probability values near 0.5, letter 'I' will be confused more times with 'T' ($\gamma = 20$) than with 'J' ($\gamma = 10$).

Thus, aided by noise masking, each AMM, W_{XX} or M_{XX} used as a single *robust* associative memory device, gives almost perfect recall for binary inputs that have been corrupted with random noise for probability values below $0.5 - \delta$ or above $0.5 + \delta$ where $\delta > 0$ is a small number which is problem dependent; in example 5.5, $\delta = 0.04$.

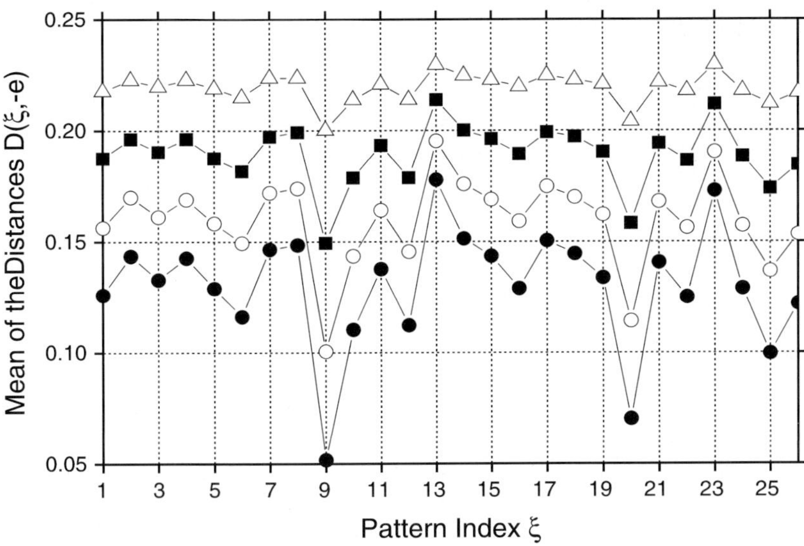

Fig. 5.5. Binary negative masking: mean of the distances D^ξ_{-e} computed over a run of 100 random corrupted versions of the letter 'I' against each letter with $p = 0.9(\bullet), 0.8(\circ), 0.7(\blacksquare), 0.6(\triangle)$. Vertical scale is normalized and $D(\xi, -e) = D^\xi_{-e}$

5.4.2 Real Valued Patterns

Let $X = \{\boldsymbol{x}^1, \ldots, \boldsymbol{x}^k\} \subset \mathbb{R}^n$ be a finite set of k exemplar real valued patterns that are stored in W_{XX} following (5.6) by taking $Y = X$. If $\widetilde{\boldsymbol{x}}$ denotes a noisy version of the exemplar $\boldsymbol{x} = \boldsymbol{x}^\xi$ for some $\xi \in \{1, \ldots, k\}$, we can use the *normalized absolute error* (NAE) or the *normalized mean squared error* between both patterns to measure their proximity. They are defined by

$$\alpha(\boldsymbol{x}, \widetilde{\boldsymbol{x}}) = \frac{1}{n} \sum_{i=1}^{n} \mid x_i - \widetilde{x}_i \mid, \tag{5.19}$$

$$\sigma(\boldsymbol{x}, \widetilde{\boldsymbol{x}}) = \sum_{i=1}^{n} (x_i - \widetilde{x}_i)^2 / \sum_{i=1}^{n} (x_i^\xi)^2. \tag{5.20}$$

Although NAE is the natural extension to real vectors of NHD for binary vectors, it is common practice to compute distances between pattern vectors using the NMSE, since it gives numerical values that agree more objectively with the perceived visual differences when comparing two gray-scale image patterns. With the aid of this metric we formalize again the idea of "almost perfect recall" in the real valued case as follows:

Definition 5.9 *The lattice matrix autoassociative memory W_{XX} is an almost perfect recall memory for a set X of real valued patterns if and only if there exists a small real number $\varepsilon > 0$ close to zero, such that $\sigma(W_{XX} \boxtimes \widetilde{\boldsymbol{x}}, \boldsymbol{x}) \leq \varepsilon$ for all $\boldsymbol{x} \in X$ with respect to finite sets of noisy versions $\widetilde{\boldsymbol{x}}$ of $\boldsymbol{x}$. Under similar conditions, M_{XX} is an almost perfect recall memory whenever $\sigma(M_{XX} \boxtimes \widetilde{\boldsymbol{x}}, \boldsymbol{x}) \leq \varepsilon$.*

The next algorithm, expressed in mathematical notation, establishes the technique of noise masking for real valued patterns as a mechanism to provide an eroded noisy input pattern to the min memory W_{XX}. Numbered steps are prefixed by S and comments are provided within brackets.

Algorithm 5.2. REAL VALUED PATTERN RECALL WITH W_{XX}

S1. Input $\widetilde{x} \in \mathbb{R}^n$
[Assume $\widetilde{x}$ is an unknown noisy exemplar corrupted with mixed noise.]

S2. For $\xi = 1$ **to** k **do**
$$\widetilde{x}_e^\xi = x^\xi \wedge \widetilde{x}$$
[Min-masked vectors of k memorized patterns.]

S3. For $\xi = 1$ **to** k **do**
$$D_\xi = \mu[\sigma(\widetilde{x}_e^\xi, x^\xi), \sigma(\widetilde{x}_e^\xi, \widetilde{x})]$$
[Mean distance vector.]

S4. Find $\gamma \in \{1, \ldots, k\}$ such that $D_\gamma = \bigwedge_{\xi=1}^{k} D_\xi$.
[Min-masked candidate vector, eroded only version of $\widetilde{x}$ is $\widetilde{x}_e^\gamma$.]

S5. Let $x_+^\gamma = W_{XX} \boxdot \widetilde{x}_e^\gamma$
[The masked γ-vector in S4 is an input to W_{XX} to get output candidate pattern.]

Note that, after step 5, the exemplar pattern that is most correctly associated by W_{XX} with the noisy input $\widetilde{x}$ is given by x^γ. Therefore, the following relationship holds

$$\sigma(\widetilde{x}, x^\gamma) = \bigwedge_{\xi=1}^{k} \sigma(\widetilde{x}, x^\xi). \tag{5.21}$$

The same remarks made for Algorithm 5.1 apply to Algorithm 5.2, except that in the real valued case, the algorithm is much simpler and requires a less number of arithmetical operations. Algorithm 5.2 can be readily modified to be used with the dual max memory M_{XX}. Specifically, step 2 will use the $\vee$-operator to generate dilated versions ($+d$ subindex) and step 5 will compute one min-product using M_{XX}. In the next two examples, the set of exemplar patterns consisted of 10 grayscale public domain images shown in Fig. 5.6, where each image is registered in a matrix of size 64×64 pixels.

Example 5.6 *Table 5.3 gives several NMSE mean values over 25 noisy versions of the gray-scale "Lena" image, whose eroded versions were used as inputs to the min-memory W_{XX}. The same Table lists similar numerical values for the same face image of which 25 dilated corrupted versions were fed into the max-memory M_{XX}. Note that in the 4th (W-Mean) and 7th (M-Mean) columns, the difference between the recalled pattern and the exemplar is small for any $p \leq 0.85$, since $\mu[\sigma_i(\widetilde{x}^\gamma, x)] \leq 5 \times 10^{-3}$ for $i = 1, \ldots, 25$.*

Fig. 5.6. Exemplar gray-scale images. 1st row, $\xi = 1, \ldots, 5$; 2nd row, $\xi = 6, \ldots, 10$

Table 5.3. Mean NMSE's for mixed noise, eroded versions, recall performance of W_{XX}; mixed noise, dilated versions, and recall performance of M_{XX}

Noise p	Mean $\sigma(\widetilde{x}, x)$	Mean $\sigma(\widetilde{x}_{+e}, x)$	W-Mean $\sigma(\widetilde{x}^{\gamma}, x)$	Mean $\sigma(\widetilde{x}, x)$	Mean $\sigma(\widetilde{x}_{+d}, x)$	M-Mean $\sigma(\widetilde{x}^{\gamma}, x)$
0.15	0.119	0.057	0.00011	0.124	0.075	0.00026
0.25	0.188	0.102	0.00016	0.188	0.106	0.00033
0.35	0.250	0.154	0.00011	0.242	0.135	0.00064
0.45	0.296	0.215	0.00047	0.284	0.153	0.00075
0.55	0.340	0.270	0.00030	0.341	0.178	0.00089
0.65	0.385	0.356	0.00094	0.362	0.190	0.00100
0.75	0.418	0.436	0.00100	0.403	0.208	0.00200
0.85	0.436	0.499	0.00098	0.440	0.221	0.00300

Example 5.7 *In several runs of 25 noisy versions of each gray-scale image, the mean of the distances D_γ computed in step 4 of Algorithm 5.2 gives the correct pattern index γ for any $p \in [0, 0.65]$. For noise probability levels in the range $(0.66, 0.70]$, the value of γ may correspond to similar noisy images, and for $p > 0.70$, the mean distance values for at least two images are very close and there is no way to recall the correct exemplar. Figure 5.7 displays the mean of the distances $D(\xi, e) = D_e^{\xi}$ computed over a run of 25 random corrupted versions of four selected images against each other image with $p = 0.35$, marked respectively, with a disk (Baboon), a circle (Lena), a solid square (Bird), and a triangle (Fruits). The lowest point in each curve gives the correct value of γ. In similar fashion, Fig. 5.8 shows an analogous behavior for $p = 0.65$; again, the lowest point in each curve gives the correct pattern index.*

The kind of computer experiments described in Examples 5.6–5.7, show that grayscale noise masking as a preprocessing step, turns each AMM, W_{XX} or M_{XX} as a single *robust* associative memory device, into an almost perfect recall associative memory for noisy real valued inputs.

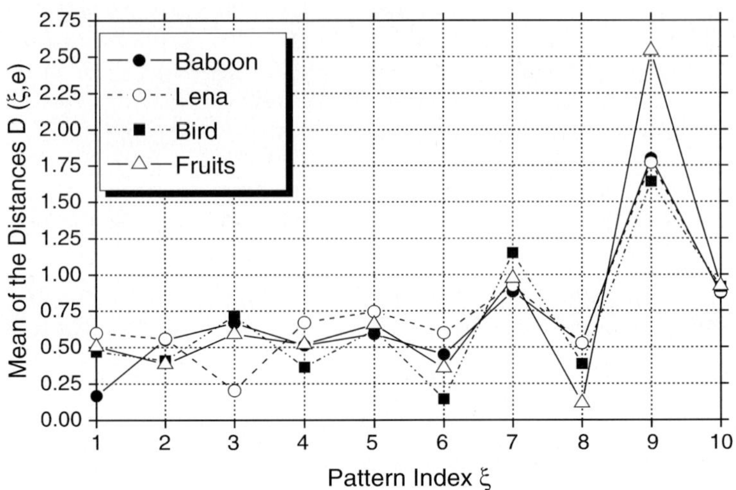

Fig. 5.7. Gray-scale masking: mean of the distances D_e^ξ computed over a run of 25 random corrupted versions of 4 selected images vs. each other image with $p = 0.35$

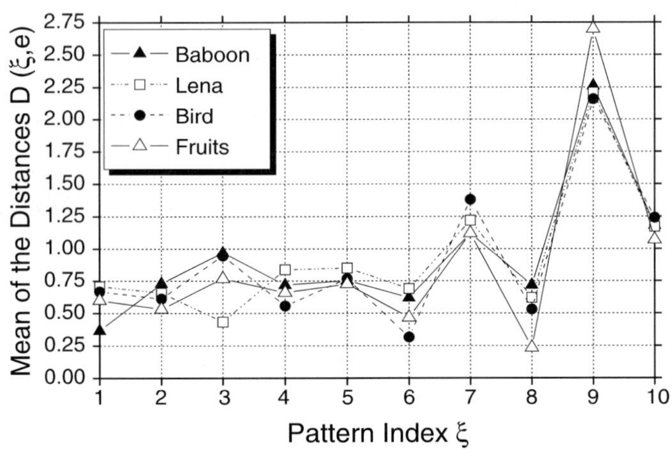

Fig. 5.8. Gray-scale masking: mean of the distances D_e^ξ computed over a run of 25 random corrupted versions of 4 selected images vs. each other image with $p = 0.65$

5.5 Conclusions

This chapter exposes a noise masking technique, based on simple lattice algebra operations, to unmix the erosive and dilative noise components of mixed noise present in a corrupted input pattern. In this way, any input corrupted with any kind and a wide range of noise levels can be recognized by a single lattice matrix associative memory, either the min-memory W_{XX} that has been thoroughly tested with erosive noise, or the max-memory M_{XX} that is also very robust in handling dilative noise. The algorithms given here, for binary or real valued patterns, can be optimized by parallelizing several of their steps and are good candidates to be implemented as neuromorphic or FPGA systems for real-time applications. Noise masking provides a practical and complete solution to the problem of pattern recall from noisy binary or real-valued inputs, that is highly competitive against other associative memory models described in detail in [20, 21].

References

1. Cuninghame-Green R (1979) Minimax Algebra. Lectures Notes in Economics and Mathematical Systems 166. Springer, New York
2. Cuninghame-Green R (1995) Minimax algebra and applications. In: Hawkes P (ed) Advances in Imaging and Electron Physics 90:1–121. Academic Press, New York
3. Graña M, Sussner P, Ritter GX (2003) Innovative applications of associative morphological memories for image processing and pattern recognition. Mathware & Soft Computing 10:155–168
4. Graña M, Maldonado O, Vicente D (2005) Morphological independence and hyperspectral image indexing. In: Astola JT, Tabus I, Barrera J (eds) Proc SPIE Math Methods in Pattern and Image Analysis 5916 pp 59160L:1–10
5. Hashiguchi H, Hattori M (2003) Robust morphological associative memory using multiple kernel patterns. In: Proc 10th Int Conf on Neural Information Processing
6. Hattori M, Fukui A, Ito H (2002) A fast method of constructing kernel patterns for morphological associative memories. In: Proc 9th Int Conf on Neural Information Processing pp 1058–1063
7. Ritter GX (1999) Image Algebra. Center for Computer Vision and Visualization, CISE Dept, Univ of Florida, Gainesville FL. Unpublished manuscript pp 121–135; `ftp://ftp.cise.ufl.edu/pub/scr/ia/documents`
8. Ritter GX, Sussner P (1996) An introduction to morphological neural networks. In: Proc 13th Int Conf on Pattern Recognition pp 709–717
9. Ritter GX, Sussner P (1997) Associative memories based on lattice algebra. In: IEEE Int Conf on Systems, Man, and Cybernetics pp 3570–3575
10. Ritter GX, Sussner S, Diaz de Leon JL (1998) Morphological associative memories. IEEE Trans Neural Networks 9(2):281–293
11. Ritter GX, Urcid G (2002) Minimal representations and morphological associative memories for pattern recall. In: Proc 7th Iberoamerican Congress in Pattern Recognition pp 27–38
12. Ritter GX, Urcid G, Iancu L (2003) Reconstruction of noisy patterns using morphological associative memories. J Math Imaging and Vision 19(5): 95–111
13. Ritter GX, Iancu L, Schmalz MS (2004) A new auto-associative memory based on lattice algebra. In: Sanfeliu A, Trinidad JF, Ochoa JA (eds) Progress in Pattern Recognition and Image Analysis, LNCS 3287:148–155. Springer, Berlin, Germany
14. Ritter GX, Urcid G, Schmalz MS (2005) Lattice associative memories that are robust in the presence of noise. In: Astola JT, Tabus I, Barrera J (eds) Proc SPIE Math Methods in Pattern and Image Analysis 5916 pp 59160Q:1–6

15. Ritter GX, Gader PD (2006) Fixed points of lattice transforms and lattice associative memories. In: Hawkes P (ed) Advances in Imaging and Electron Physics 144:165–242. Elsevier, Amsterdam, The Netherlands
16. Sussner P (2000) Observations on morphological associative memories and the kernel method. Neurocomputing 31:167–183
17. Sussner P (2000) Fixed points of autoassociative morphological memories. In: Proc Int Joint Conf on Neural Networks
18. Sussner P (2003) Associative morphological memories based on variations of the kernel and dual kernel methods. Neural Networks 16:625–632
19. Sussner P (2003) Generalizing operations of binary morphological autoassociative memories using fuzzy set theory. J of Math Imaging and Vision 19(5):81–93
20. Sussner P (2005) Recall of patterns using binary and gray-scale autoassociative morphological memories. In: Astola JT, Tabus I, Barrera J (eds) Proc SPIE Math Methods in Pattern and Image Analysis 5916 pp 59160M:1–10
21. Sussner P, Valle ME (2006) Gray-scale morphological associative memories. IEEE Trans Neural Networks 17(3):559–570
22. Urcid G, Ritter GX (2003) Kernel computation in morphological associative memories for grayscale image recollection. In: Proc 5th IASTED Int Conf on Signal and Image Processing pp 450–455
23. Wang M, Chen S (2005) Enhanced fuzzy morphological auto-associative memory based on empirical kernel map. IEEE Trans on Neural Networks 16(3):557–564

Convex Coordinates From Lattice Independent Sets for Visual Pattern Recognition

Manuel Graña, Ivan Villaverde, Ramon Moreno, and Francisco X. Albizuri

Computational Intelligence Group
Dept. CCIA, UPV/EHU, Apdo. 649, 20080 San Sebastian, Spain
ccpgrrom@si.ehu.es

Summary. One of the key processes in nowadays intelligent systems is feature extraction. It pervades applications from computer vision to bioinformatics and data mining. The purpose of this chapter is to introduce a new feature extraction process based on the detection of extremal points on the cloud of points that represent the high dimensional data sample. These extremal points are assumed to define an approximation to the convex hull covering the data sample points. The features extracted are the coordinates of the data points relative to the extremal points, the convex coordinates. We have experimented this approach in several applications that will be summarized in the chapter.

6.1 Introduction

Feature extraction is the process of extracting a vector of (real) values from a sample data item. The pioneer works on the application of Principal Component Analysis (PCA) [10] to this task, have been followed by a cohort of methods, most of them linear (i.e. Independent Component Analysis (ICA) [18, 19]). Feature extraction implies dimensionality reduction, because the extracted feature vector is usually of much lower dimension than the original data items. This dimensionality reduction can be guided by minimum error criterion (i.e. PCA), maximum independence (i.e. ICA), maximum discrimination power, etc. Depending on the application and type of the data, the diverse methods can be of different usefulness.

The method proposed in this chapter has the following outline:

1. The extremal points of the data sample are extracted. They provide an approximation to the vertices of the minimal convex set that covers the data sample.
2. The data points in the sample are represented as linear combination of the extremal points: the convex coordinates are the extracted features.

This simple outline has several difficulties. First, in high dimensional spaces, computing the convex hull of a set points is not trivial matter. Second, the

M. Graña et al.: *Convex Coordinates From Lattice Independent Sets for Visual Pattern Recognition*, Studies in Computational Intelligence (SCI) **67**, 101–128 (2007)
www.springerlink.com

algorithms give approximations to the correct convex hull vertices, hence there is no guarantee that the simplex defined by them encloses all the data. The convex coordinates are expected to be positive, but this is seldom true for the approximations to the convex hull obtained.

We have tried our approach on a variety of applications:

Hyperspectral image unsupervised segmentation The starting point of our research [11, 12]. The motivation is to obtain a fast segmentation algorithm for hyperspectral images, that could serve as a first hand analysis. Unsupervised means that no a priori classes are give, so that the results do not identify materials but regions of high abundance.

Hyperspectral image supervised segmentation The motivation for unsupervised segmentation is the scarcity of the ground truth information. However, when trying supervised classifier design, the process becomes a true feature extraction process. We did some experiences with available ground truth for hyperspectral data that encouraged the work reviewed in this chapter [13, 14].

Content Based Image Retrieval The issue here is to be able to extract image features that work as indices for image database search. We have worked with both hyperspectral images and shape features from conventional images.

Robot Visual Localization From our early attempts [30, 31] at brute force visual navigation with morphological neural networks, we have changed the approach to that of using them for feature extraction. The problem of robot self-localization is stated as a classification problem [48, 49] based on the features obtained from the endmember extraction algorithm.

We started our works based on the works on Morphological Neural Networks from Ritter, Sussner and other people [34, 35, 43, 44]. The introduction in [36] of the idea of morphological independence has been very influential in our work. This idea has been elaborated in depth in [38] relating strong lattice independence with linear independence, a necessary condition for the vertices of the convex hull of the data sample points. Although we do not ensure strong lattice independence, our procedure obtains almost always linearly independent sets of vectors that can be used for the computation of convex coordinates.

We will start the summary description of our approach revising the definition of the convex coordinates in Sect. 6.2. Next we introduce some methods for linear feature extraction in Sect. 6.3 and the algorihtms for the induction of the endmembers from the data in Sect. 6.4, including our proposed approach. The works on hyperspectral images are related in Sects. 6.5 and 6.6. The application to shape CBIR to a database of mushrooms is presented in Sect. 6.7. The application to robot navigation is described in Sect. 6.8. We gather our conclusions in Sect. 6.9.

6.2 Convex Coordinates

The linear model:

$$\mathbf{x} = \sum_{i=1}^{M} a_i \mathbf{s}_i + \mathbf{w} = \mathbf{Sa} + \mathbf{w}, \tag{6.1}$$

where $\mathbf{x}$ is the d-dimensional explained data point, $\mathbf{S}$ is the $d \times M$ mixture matrix whose columns are the d-dimensional basis vectors $\mathbf{s}_i, i = 1, .., M$, $\mathbf{a}$ is the M-dimension coordinate vector, and $\mathbf{w}$ is the d-dimension additive observation noise vector. If the coordinates fulfill the following conditions:

Nonnegative $a_i \geq 0, i = 1, .., M$,
Additive normalization $\sum_{i=1}^{M} a_i = 1$,

then the basis vectors can be assumed to be the vertices of a convex polytope enclosing the data points $\mathbf{x}$. When $M \ll d$ the computation of the convex coordinates can be interpreted as a dimension reduction process, or a feature extraction process.

Depending on the application domain, the convex coordinates are interpreted differently. For instance, if the input data are hyperspectral remote sensing images, the linear model in (6.1) is called a linear mixing model [21], the basis vectors $\mathbf{s}_i$ are the endmember' spectra, $\mathbf{a}$ is the fractional abundance vector, and $\mathbf{x}$ is the sensed pixel spectrum. The convexity conditions amount to the physical feasibility of the linear model, because observed spectra are made up from the positive contributions of the spectra of materials in the physical world, and the fractional contributions of the endmembers always sum up to one.

Once the convex polytope vertices $\mathbf{S}$ have been determined, computing the convex coordinates amounts to solve the linear system of equations:

$$\mathbf{x} = \mathbf{Sa}. \tag{6.2}$$

The simplest approach is the unconstrained least squared error estimation given by:

$$\widehat{\mathbf{a}} = \left(\mathbf{S}^T \mathbf{S}\right)^{-1} \mathbf{S}^T \mathbf{x}. \tag{6.3}$$

The key problem in this simple process is that of finding out the vertices of a convex polytope that covers the data sample points. In the hyperspectral image processing these convex polytope vertices are called endmembers. The first significant work on the automated induction of endmember spectra from image data is [7], whose starting observation is that the scatter plots of remotely sensed data are tear shaped or pyramidal, if two or three spectral bands are considered. The apex lies in the so-called dark point. The endmember detection becomes the search for non-orthogonal planes that enclose the data defining a minimum volume simplex, hence the name of the method. The method is computationally expensive and requires the prior specification of the number of endmembers. A recent approach to the automatic endmember detection is the Convex Cone Analysis (CCA) method proposed in [20], applied to target detection. The CCA selects the greatest eigenvalue eigenvectors, as many as the specified number of endmembers. These eigenvectors define the basis of the convex cone that covers the image data. The vertices of the convex cone correspond to spectra with as many zero elements as the number of eigenvectors minus one. The search for the convex cone vertices involves the exploration of the combination of bands and the solution of a linear system for each combination. The complexity of the search for these vertices is $O(b^c)$ where b is the number of bands and c the number of eigenvectors. At present we use a raw random search to obtain the experimental results reported below. Another approach to endmember induction from the data from a morphological background is presented in [29]. The authors generalize the erosion and dilation morphological operators based on the distances between pixels in a neighborhood. They introduce a measure of eccentricity as a measure of the variance in the neighborhood that allows to decide on the validity of the neighborhood to extract an endmember. The method uses neighborhoods located on region boundaries and discards smooth regions. The next section describes some endmember induction algorithms.

6.3 Linear Feature Extraction: ICA and PCA

The Independent Component Analysis (ICA) [19] assumes that the data is a linear combination of nongaussian, mutually independent latent variables with an unknown mixing matrix. The ICA reveals the hidden independent sources and the mixing matrix. That is, given a set of observations represented by a d dimensional vector $\mathbf{x}$, ICA assumes a generative model $\mathbf{x} = \mathbf{As}$, where $\mathbf{s}$ is the M dimensional vector of independent sources and $\mathbf{A}$ is the $d \times M$ unknown basis matrix. The ICA searches for the linear transformation of the data $\mathbf{W}$, such that the projected variables $\mathbf{Wx} = \mathbf{s}$ are as independent as possible. It has been shown that the model is completely identifiable if the sources are statistically independent and at least $M - 1$ of them are non gaussian. If the sources are gaussian the ICA transformation could be estimated up to an orthogonal transformation. Estimation of mixing and unmixing matrices can be done maximizing diverse objective functions, among them the non gaussianity of the sources and the likelihood of the sample. We have used the FastICA [18] algorithm available at `http://www.cis.hut/projects/ica/fastica`. The Principal Component Analysis (PCA) [10] is a well-known linear dimension reduction procedure that is optimal in the sense of the mean squared error. It consists in the selection of the largest eigenvalue eigenvectors of the data covariance matrix. These eigenvectors constitute the transformation matrix. The selection of the number of eigenvectors or the independent components can be made attending to some quantitative criteria, but in our experiment below we selected the number of components in the ground truth image.

6.4 Endmember Induction

The main problem in the approach of computing the convex coordinates as features extracted from the data, lies in the need to obtain the vertices of the convex polytope from the data. We start this section recalling a fast review of Morphological Associative Memories (MAM). Next, we introduce our method based on lattice computing, which in fact was derived from properties of the MAMs. Finally we include the definition of an algorithm originally intended for the extraction of endmembers in hyperspectral images, that we take as an alternative method for the extraction of the convex polytope vertices, which we continue to call endmembers following the conventions of hyperspectral image processing.

6.4.1 Morphological associative memories

The work on Morphological Associative Memories stems from the consideration of an algebraic lattice structure $(\mathbb{R}, \vee, \wedge, +)$ as the alternative to the algebraic $(\mathbb{R}, +, \cdot)$ framework for the definition of Neural Networks computation [34, 35]. The operators $\vee$ and $\wedge$ denote, respectively, the discrete `max` and `min` operators (resp. `sup` and `inf` in a continuous setting). The approach is a highly successful instance of lattice computing with several evolutions and upgradings being published in this volume. The name "morphological" came from the common grounds between this approach and the field of mathematical morphology in image processing. Given a set of input/output pairs of pattern $(X, Y) = \left\{ \left(\mathbf{x}^\xi, \mathbf{y}^\xi \right) ; \xi = 1, .., k \right\}$, an heteroassociative neural network based on the pattern's cross correlation [17] is built up as $W = \sum_\xi \mathbf{y}^\xi \cdot \left(\mathbf{x}^\xi \right)'$. Mimicking this construction procedure [34, 35] propose the following constructions of Heteroassociative Morphological Memories (HMM's):

$$W_{XY} = \bigwedge_{\xi=1}^{k} \left[\mathbf{y}^\xi \times \left(-\mathbf{x}^\xi \right)' \right] \text{ and } M_{XY} = \bigvee_{\xi=1}^{k} \left[\mathbf{y}^\xi \times \left(-\mathbf{x}^\xi \right)' \right], \tag{6.4}$$

where $\times$ is any of the $\boxtimes$ or $\boxdot$ operators. Here $\boxtimes$ and $\boxdot$ denote the max and min matrix product, respectively defined as follows:

$$C = A \boxtimes B = [c_{ij}] \Leftrightarrow c_{ij} = \bigvee_{k=1..n} \{a_{ik} + b_{kj}\}, \tag{6.5}$$

$$C = A \boxdot B = [c_{ij}] \Leftrightarrow c_{ij} = \bigwedge_{k=1..n} \{a_{ik} + b_{kj}\}. \tag{6.6}$$

If $X = Y$ then the HMM memories are Autoassociative Morphological Memories (AMM). Conditions of perfect recall by the HMM's and AMM's of the stored patterns are proved in [34, 35]. In the continuous case, the AMM's are able to store and recall any set of patterns:

$$W_{XX} \boxvoid X = X = M_{XX} \boxslash X, \tag{6.7}$$

for any X.

These results hold when we try to recover the output patterns from the noise-free input pattern. Let it be $\widetilde{\mathbf{x}}^\gamma$ a noisy version of $\mathbf{x}^\gamma$. If $\widetilde{\mathbf{x}}^\gamma \le \mathbf{x}^\gamma$ then $\widetilde{\mathbf{x}}^\gamma$ is an eroded version of $\mathbf{x}^\gamma$, alternatively we say that $\widetilde{\mathbf{x}}^\gamma$ is corrupted by erosive noise. If $\widetilde{\mathbf{x}}^\gamma \ge \mathbf{x}^\gamma$ then $\widetilde{\mathbf{x}}^\gamma$ is a dilated version of $\mathbf{x}^\gamma$, alternatively we say that $\widetilde{\mathbf{x}}^\gamma$ is corrupted by dilative noise.

Morphological memories are selectively sensitive to these kinds of noise. The conditions of *robust* perfect recall are proven in [34, 35]. Here we will remember them for the sake of the reader, because they are on the basis of the proposed algorithm. Given patterns X, the equality

$$W_{XX} \boxvoid \widetilde{\mathbf{x}}^\gamma = \mathbf{x}^\gamma \tag{6.8}$$

holds when the noise affecting the pattern is erosive $\widetilde{\mathbf{x}}^\gamma \le \mathbf{x}^\gamma$ and the following relation holds:

$$\forall i \exists j_i; \widetilde{x}_{j_i}^\gamma = x_{j_i}^\gamma \vee \left(\bigvee_{\xi \ne \gamma} \left(x_i^\gamma - x_i^\xi + x_{j_i}^\xi \right) \right). \tag{6.9}$$

Similarly, the equality

$$M_{XY} \boxslash \widetilde{\mathbf{x}}^\gamma = \mathbf{x}^\gamma \tag{6.10}$$

holds when the noise affecting the pattern is dilative $\widetilde{\mathbf{x}}^\gamma \ge \mathbf{x}^\gamma$ and the following relation holds:

$$\forall i \exists j_i; \widetilde{x}_{j_i}^\gamma = x_{j_i}^\gamma \wedge \left(\bigwedge_{\xi \ne \gamma} \left(x_i^\gamma - x_i^\xi + x_{j_i}^\xi \right) \right). \tag{6.11}$$

Therefore, the AMM will fail to recall the pattern if the noise is a mixture of erosive and dilative noise. In [32] we have proposed a morphological scale space method to increase the robustness of AMM.

An approach to obtain general noise robustness is based on the so-called kernel patterns [35, 36, 43]. Related to the construction of the kernels, [36] introduced the notion of *morphological independence*, distinguishing erosive and dilative versions of its definition: Given a set of pattern vectors $X = (\mathbf{x}^1, ..., \mathbf{x}^k)$, a pattern vector $\mathbf{y}$ is said to be morphologically independent of X in the erosive sense if $\mathbf{y} \nleq \mathbf{x}^\gamma; \gamma = \{1, .., k\}$, and morphologically independent of X in the dilative sense if $\mathbf{y} \ngeq \mathbf{x}^\gamma; \gamma = \{1, .., k\}$. The set of pattern vectors X is said to be morphologically independent in either sense when all the patterns are morphologically independent of the remaining patterns in the set. For the current application we want to use AMM as detectors of the set extreme points, to obtain a rough approximation of the minimal simplex that covers the data points. We note that given a set of pattern vectors $X = \left(\mathbf{x}^1, ..., \mathbf{x}^k\right)$, and the erosive W_{XX} and dilative M_{XX} memories constructed from it, and a test pattern $\mathbf{y} \notin X$, if $\mathbf{y}$ is morphologically independent of X in the erosive sense, then $W_{XX} \boxtimes \mathbf{y} \notin X$. Also, if $\mathbf{y}$ is morphologically independent of X in the dilative sense, then $M_{XX} \boxtimes \mathbf{y} \notin X$. Therefore the AMM's can be used as detectors of morphological independence. The works in [38] have generalized the notion of morphological independence to that of lattice independence, showing that *strong lattice independence* implies linear independence. Therefore a set of strongly lattice independent points can be considered as the vertices of a convex polytope. In our approach we find lattice independent vectors from the data looking for morphologically independent vectors. Although we do not test strong lattice independence, we have found from experience that it holds almost always when the number of induced endmembers is much lower than the data dimensionality. That is, almost always the induced endmembers are linearly independent.

A final remark before the presentation of the algorithm for endmember induction from the data. The set of vector patterns that we are searching for are morphologically independent vectors both in the erosive and dilative senses, and they enclose the remaining vectors. Working with integer valued vectors, given a set of pattern vectors $X = \left(\mathbf{x}^1, ..., \mathbf{x}^k\right)$ and the erosive W_{XX} and dilative M_{XX} memories constructed from it, if a test pattern $\mathbf{y} < \mathbf{x}^\gamma$ for some $\gamma \in \{1, .., k\}$ then $W_{XX} \boxtimes \mathbf{y} \notin X$. Also, if the test pattern $\mathbf{y} > \mathbf{x}^\gamma$ for some $\gamma \in \{1, .., k\}$ then $M_{XX} \boxtimes \mathbf{y} \notin X$. Therefore, working with integer valued patterns the AMM will be useless for the detection of morphologically independent patterns. However, if we consider the binary vectors obtained as the sign of the vector components, then morphological independence would be detected as suggested above: The already detected endmembers are used to build the erosive and dilative AMM. If the output recalled by a new pattern does not coincide with any of the endmembers, then the new pattern is a new endmember.

6.4.2 The selection of endmembers from the data

The region of the space enclosed by a set of vectors which are morphologically independent in both erosive and dilative senses simultaneously is a high dimensional box. Here we try that this box approaches as much as possible the minimal simplex enclosing the data points. Let us denote $\left\{\mathbf{f}\left(i\right) \in \mathbb{R}^d; i = 1, .., n\right\}$ the high dimensional data that may be the pixels in a multispectral or hyperspectral image, or selected points in shape representation, μ and σ are, respectively, the mean vector and the vector of standard deviations computed over the data sample, α the noise correction factor and E the set of already discovered vertices. The noise amplitude of the additive noise in (1) is σ, the patterns are corrected by the addition and subtraction of $\alpha\sigma$, before being presented to the AMM's. The gain parameter α controls the amount of flexibility in the discovering of new endmembers. Let us denote by the expression $\mathbf{x} > \mathbf{0}$ the construction of the binary vector $(\{b_i = 1 \text{ if } x_i > 0; b_i = 0 \text{ if } x_i \leq 0\}; i = 1, .., n)$.

The steps in the procedure are the following:

1. Shift the data sample to zero mean
   $\{\mathbf{f}^c(i) = \mathbf{f}(i) - \mu; i = 1, .., n\}$.
2. Initialize the set of vertices $E = \{\mathbf{e}_1\}$ with a randomly picked sample. Initialize the set of morphologically independent binary signatures $X = \{\mathbf{x}_1\} = \{(e_k^1 > 0; k = 1, .., d)\}$
3. Construct the AMM's based on the morphologically independent binary signatures: M_{XX} and W_{XX}.
4. For each pixel $\mathbf{f}^c(i)$

   a) compute the noise corrections sign vectors $\mathbf{f}^+(i) = (\mathbf{f}^c(i) + \alpha\sigma > \mathbf{0})$ and $\mathbf{f}^-(i) = (\mathbf{f}^c(i) - \alpha\sigma > \mathbf{0})$
   b) compute $y^+ = M_{XX} \boxtimes \mathbf{f}^+(i)$
   c) compute $y^- = W_{XX} \boxtimes \mathbf{f}^-(i)$
   d) if $y^+ \notin X$ or $y^- \notin X$ then $\mathbf{f}^c(i)$ is a new vertex to be added to E, execute once 3 with the new E and resume the exploration of the data sample.
   e) if $y^+ \in X$ and $\mathbf{f}^c(i) > \mathbf{e}_{y+}$ the pixel spectral signature is more extreme than the stored vertex, then substitute $\mathbf{e}_{y+}$ with $\mathbf{f}^c(i)$.
   f) if $y^- \in X$ and $\mathbf{f}^c(i) < \mathbf{e}_{y-}$ the new data point is more extreme than the stored vertex, then substitute $\mathbf{e}_{y-}$ with $\mathbf{f}^c(i)$.

5. The final set of endmembers is the set of original data vectors $\mathbf{f}(i)$ corresponding to the sign vectors selected as members of E.

6.4.3 The convex cone analysis (CCA)

The CCA was proposed by [20]. The basic idea is that after PCA of the spectral correlation matrix, the data falls in a cone shaped region in the positive subspace centered in the first eigenvector. Given the $N \times M \times d$ hyperspectral image, it is reorganized as a $NM \times d$ matrix $\mathbf{X}$. The spectral correlation matrix is computed as $\mathbf{C} = \mathbf{X}^T\mathbf{X}$. Let it be $\mathbf{C} = \mathbf{PLP}^T$ the PCA decomposition of the correlation matrix, select the first c eigenvectors $[\mathbf{p}_1, .., \mathbf{p}_c] = \mathbf{P}_c$ and search for the boundaries of the convex region characterized by $\mathbf{x} = \mathbf{p}_1 + a_1\mathbf{p}_2 + .. + a_{c-1}\mathbf{p}_c \geq \mathbf{0}$. The vertices of this region are the points with exactly $c-1$ zero components. The CCA algorithm searches among all the $\binom{b}{c-1}$ possible combinations of eigenvectors performing the following test. Let it be $[\mathbf{p}(\gamma_1), .., \mathbf{p}(\gamma_{c-1})] = \mathbf{P}'$ the selected set of eigenvectors. Solve the set of equations $\mathbf{P}'\mathbf{a} = \mathbf{0}$ and compute $\mathbf{x} = \mathbf{P}_c\mathbf{a}$. If $\mathbf{x}$ has exactly $c-1$ zero components it is a vertex of the convex region data. In practice, each component is tested against a threshold. However, as the combinatorial space grows the problem becomes intractable. We implemented an straightforward random search. Application of more sophisticated random search algorithms like genetic algorithms may be of interest for large problems. The CCA algorithm has been used in some works as a competing algorithm to ours.

6.5 Hyperspectral Image CBIR

In Content Based Image Retrieval (CBIR) [42] systems, the images stored in the database are labeled by feature vectors, which are extracted from the images by means of computer vision and digital image processing techniques. In CBIR systems, the query to a database is specified by an image. The query's feature vector is computed and the closest items in the database, according to a similarity measure defined in feature space, are returned as the answers to the query. Of course, there is no universal definition of the feature vectors because of the diversity of image kinds.

Hyperspectral images are a special kind of images, in which each pixel contains a fine sampling of the visible and near infrared spectrum, represented by a high dimensional vector. Hyperspectral sensing in hundreds of spectral bands allows the recognition of physical materials in image pixels, and the decomposition of mixed pixel spectrum into their constituent material spectra by *spectral unmixing* [21]. There is a growing need for the maintenance of large collections of hyperspectral images, and for the automated search within these collections. The attempts to define CBIR systems for them are scarce and partial. The only clear example of an attempt to build up such a system found in the literature searched is [1]. Even there, the use of the spectral information is rather marginal. It is our main concern that the spectral information must be used as much as possible in the definition of search strategies in large collections of hyperspectral images. That is, the indexing of those images must be based, at least partiall, in the spectral information extracted from the image itself.

We have proposed the characterization of the hyperspectral images by their endmembers. Endmembers may be defined by the domain experts (geologists, biologists, etc.) selecting them from available spectral libraries, or induced from the hyperspectral image data using machine learning techniques. It is our working hypothesis that much of the spectral information of the image is summarized in the convex region representation provided by the induced endmembers. Also, the abundance images produced by the spectral unmixing may be used as dimensionally reduced images for searches based on spatial features.

A big drawback for the definition and validation of the kind of systems that we are discussing is the absence of publicly available large collections of experimental images. While there are some collections for the "conventional" CBIR community, like the COIL dataset [5], in the case of hyperspectral images there are some free images provided by the AVIRIS site, and some other miscellaneous that come from the works of Landgrebe. Most of them do not have ground truth information, or it is rather inexact. As the experimental setup to demonstrate our ideas, we have constructed a small database of synthetic images. We start defining the similarity between images based on the image induced endmembers.

6.5.1 Similarity between images

Let it be $S_k = \left[s_1^k, ..., s_{n_k}^k\right]$ the set of endmembers, obtained as described before from the k-th image $\mathbf{f}_k\,(i,j)$ in the database, where n_k is the number of endmembers detected in this image. Given two images $\mathbf{f}_k\,(i,j)$ and $\mathbf{f}_l\,(i,j)$, we compute the following matrix whose elements are the Euclidean distances between the endmembers of each image:

$$D_{k.l} = [d_{i,j}; i = 1, .., n_k; j = 1, ..., n_l] \tag{6.12}$$

where

$$d_{i,j} = \left\| s_i^k - s_j^l \right\|. \tag{6.13}$$

We compute the vectors of the minimal values by rows and columns,

$$\mathbf{m}_k = \left[m_i^k = \min_j \left\{ d_{ij} \right\} \right] \tag{6.14}$$

and

$$\mathbf{m}_l = \left[m_j^l = \min_i \left\{ d_{ij} \right\} \right] \tag{6.15}$$

respectively. Then the similarity between the images is given by the following expression:

$$d\left(\mathbf{f}_k, \mathbf{f}_l\right) = \left(\|\mathbf{m}_k\| + \|\mathbf{m}_l\|\right)\left(|n_k - n_l| + 1\right). \tag{6.16}$$

The endmember induction procedure may give different number of endmembers and endmember features for two hyperspectral images. The similarity measure of (6.16) is a composition of two asymmetrical views: each vector of minimal distances measures how close are the endmembers of one image to some endmember of the other image. Suppose that all the endmembers S_k of an image are close to a subset of the endmembers S_l of the other image. Then the vector of minimal distances $\mathbf{m}_k$ will be very small, not taking into account the unlike endmembers in the second image. However, the vector of minimal distances $\mathbf{m}_l$ will be larger than $\mathbf{m}_k$ because it will take into account the distances of endmembers in S_l which are unlike to those in S_k. Thus the similarity measure of (6.16) can cope with the asymmetry of the situation. It avoids the combinatorial problem of trying to decide which endmembers can be matched and what to do in case that the number of endmembers is different from one image to the other. The difference in the number of endmembers is introduced as an penalty factor. The measure is independent of image size and, as the endmember induction algorithm is very fast, it can be computed in acceptable time. Also the endmember set poses no storage problem.

6.5.2 Experimental results and discussion

The validation of CBIR approaches is a subtle issue, because it is not possible in general to determine the response to a query that best fits the user expectations or needs. Our approach to demonstrate the usefulness of the proposed similarity measure is to built up a database of simulated hyperspectral images. The hyperspectral images are generated as linear mixtures of a set of spectra (the ground truth endmembers) with synthesized abundance images. The ground truth endmembers were randomly selected from a subset of the USGS spectral libraries corresponding to the AVIRIS flights. The synthetic ground truth abundance images were generated in a two step procedure, first we simulate each as an gaussian random field with Matern correlation function of parameters varying between 2 and 20. We applied the procedures proposed by [22] for the efficient generation of big domain gaussian random fields. Second to ensure that there are regions of almost pure endmembers we selected for each pixel the abundance coefficient with the greater value and we normalize the remaining to ensure that the abundance coefficients in this pixel sum up to one. It can be appreciated on the abundance images that each endmember has several region of almost pure pixels, viewed as brighter regions in the images. Image size is 256×256 pixels of 224 spectral bands each. We have generated collections of 100 images with a given number of ground truth endmember/abundances, this number varying from 2 to 5 for total number of 400 images.

The experiment performed on these images consists on the following steps:

1. Compute the distances between the images in the database, on the basis of (6.16), using the ground truth endmembers. The distances are computed between images with the same number of ground truth endmembers, and with all the remaining images.
2. Extract the endmembers from the images using the approach described in Sect. 6.4.2.
3. Compute the distances between the images in the database, on the basis of (6.16), using the morphologically independent induced endmembers. The distances are computed between images with the same number of ground truth endmembers, and with all the remaining images.
4. We consider the R closer images to each image in each case (ground truth and morphologically independent induced endmembers) as the responses to a potential query represented by the image.
5. The images that appear in both responses (based on the ground truth and the morphologically independent induced endmembers) are considered as relevant images, or correct responses.

In Table 6.1 we present the results from the experiment with the 400 images, in terms of the average number of correct responses. First row presents the results when we pool together all the images, regardless of the number of ground truth endmembers. The next rows present the results when we

Table 6.1. Average number of relevant images per query

	R=1	R=3	R=5	R=10
All images	0.94	1.21	1.61	2.96
2 endmembers	0.81	1.55	2.27	4.67
3 endmembers	0.98	1.44	2.21	4.9
4 endmembers	0.99	1.53	2.36	4.81
5 endmembers	1.00	1.57	2.37	4.74

only try to search in the subcollection of images with the same number of endmembers as the query image. Each row corresponds to a different number of images in the response to the query. The value of the noise gain was set to $\alpha = 0.5$. In Table 6.1 it can be appreciated that the consideration of all the images as responses to the query introduces some confusion and reduce the average number of correct images obtained in the query. This effect can be due to the fact that the morphological independence algorithm can find a number of endmembers different from the ground truth making it possible for the image to match with images outside its natural collection of images. Then images with different ground truth numbers of endmembers may become similar enough to enter in their respective response sets.

When we restrict the search to the collections with identical number of ground truth endmembers, all the results improve, except when $R = 1$. We have that near 50% of the responses are significative when $R > 1$. The case R=1 can be interpreted as the probability of obtaining the closest image in the database according to the distance defined in (6.16), or the probability of success. It can be seen that it is very high, close to 1 for all search instances, except for the case of 2 ground truth endmembers.

In Fig. 6.1 we show an interrogation to the database of simulated images. The left part of the image presents the ground truth endmembers of a query to the database (above) and the ground truth abundance images (below). The right part of the image presents the closest image in the database represented by endmembers extracted from the image by the AMM based algorithm (above) and the abundance images computed based on these endmembers (below).

6.6 Supervised Classification

Construction of supervised classifiers often employs some feature extraction algorithms, which are data dimension reduction procedures applied to the experimental data prior to training or operation of the classifier. The goals of feature extraction are both computational efficiency and enhanced discrimination of the data classes. Linear feature extraction algorithms, like Principal Component Analysis (PCA) [10], Linear Discriminant Analysis (LDA) [10],

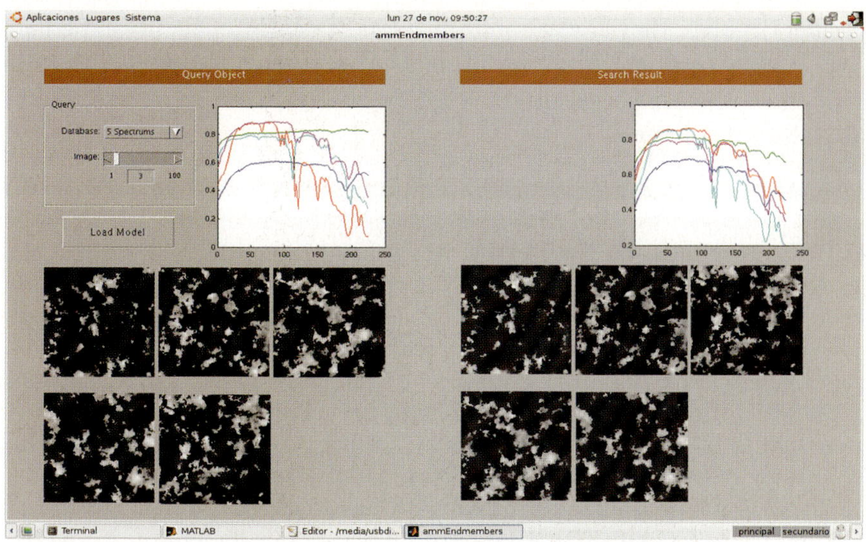

Fig. 6.1. An instance of a single response query to the database of simulated images. See text for explanations

Independent Component Analysis (ICA) [19] are defined as linear transformations that minimize some criterion function, like the mean square error (PCA), a class separability criterion (LDA) or an independence criterion (ICA). In this section we reproduce some results on the construction of supervised classifiers of hyperspectral images that compare some linear feature extraction methods (PCA and ICA) and the convex coordinates computed on the endmembers obtained from the CCA and the AMM algorithms described in Sect. 6.4. The classifiers consists of two steps:

1. An unsupervised feature extraction algorithm that reduces the data dimensionality. This role is performe by the PCA and the ICA algorithms, as linear feature extraction methods, and by the convex coordinate computed from sets of endmembers obtained from the CCA and AMM algorithms.
2. A supervised classifier, constructed using the given ground truth information and the features obtained by the algorithms tested in the previous step. The role of the supervised classifiers is neutral. They are not specially tuned to any feature extraction method.

The supervised classifiers employed were the Nearest Neighbor (1-NN), the Gaussian Classifier (GC) using the Euclidean distance, and the Support Vector Machines (SVM) [46] with a Radial Basis Function (RBF) kernel of identical unit variances, using the implementation by Anton Schwaighofer available at `http://www.cis.tugraz.at/igi/aschwaig/software.html`. No attempt has been made to fine tune the SVM. The motivation for this selection of classifiers is that they do not introduce additional bias in the experiment, which is aimed to show the value of the LSU as feature extraction algorithm.

6.6.1 The data sets

The data used in the experiments are two well known real hyperspectral images, obtained by the Airborne Visible/Infrared Imaging Spectrometer (AVIRIS) developed by NASA JPL which has 224 contiguous spectral channels covering a spectral region from 0.4 to 2.5 mm in 10 nm steps. The first hyperspectral image used for this work correspond to the Indian Pines 1992 image. It is a 145 by 145 pixel image with 220 spectral bands. The available image ground truth designates 16 mutually exclusive classes of land cover [45].

The second is real hyperspectral data collected by the AVIRIS imaging spectrometer in 1998 over Salinas Valley, California. The full scene consists of 512 lines by 217 samples with 224 spectral bands with a spatial pixel resolution of 3.7m×3.7m. The available ground truth has 15 classes. When applying the PCA, ICA and CCA methods we have set the target dimension to the exact number of ground truth components. Our AMM approach needed the setting of the noise gain parameter α. Setting $\alpha = 2$ we obtained 12 endmembers on the Salinas image, whereas setting $\alpha = 3$ we obtained 6 endmembers.

6.6.2 The experimental results

The experiment consisted in 30 repetitions of the construction and validation of the clasifiers over 50% random partitions of the data, which preserve the a priori distributions of the classes. We did not perform any band selection or smoothing of the pixel spectra in the experimental results presented here. The results of the experiment are presented in Tables 6.2 and 6.3. They consist of the average accuracy of the classifiers.

Table 6.2. Correct recognition over the Indian Pines data

	Nearest Neigh.	Gaussian C.	SVM
raw	0.19	0.08	0.23
PCA	0.16	0.05	0.24
ICA	0.14	0.03	0.35
CCA	0.33	0.20	0.48
AMM	0.34	0.25	0.64

Table 6.3. Correct recognition over the Salinas data

	Nearest Neigh.	Gaussian C.	SVM
raw	0.08	0.11	0.21
PCA	0.33	0.17	0.36
ICA	0.13	0.07	0.38
CCA	0.57	0.41	0.75
AMM	0.61	0.50	0.89

The SVM improves greatly over the other classifiers, as may be expected from the results in the literature. However, we are more interested in analyzing the results by rows. The results on the raw data are very bad, but, surprisingly, PCA and ICA do not improve very much over them most of the times. Finally, both methods based on convex coordinates, CCA and AMM improve substantially over the linear projection methods. The SVM with the convex coordinates obtained from AMM gives an almost state of the art result.

6.7 Shape CBIR

In Sect. 6.5 we have introduced the ideas of CBIR, describing an experiment in the domain of hyperspectral images. In this section the application of interest is the CBIR of shape images [47]. The CBIR problem is that of finding within a set of images the ones more similar to a given one. The role of the feature extraction algorithm is to map the images to low dimensional vectors whose topology reflects the visual similarity between shapes. The shape of the objects is described by the spatial distribution of a set of significative shape points [6]. The data we are interested in consists of the shapes of Mushroom images in a custom image database that we are building for the design of remote mushroom identification through mobile communications. Although the problem is one of database access, the validation of the approach becomes a classification problem, once the validation data set is fixed. As in Sect. 6.6, we consider the classifier composed of two stages: the unsupervised feature extraction and the supervised classifier. Again we compare the approach to feature extraction based on convex coordinates, with endmembers computed either with AMM or CCA algorithms, and the ICA. As the quality of the original shape contours is critical, we review the approach based on active contours that we have applied to obtain them from the mushroom images.

6.7.1 Active contours

In order to obtain smooth contours, we have applied an active contour technique [3, 51], based on the minimization of the following energy functional:

$$E = \sum_s (\alpha E_c(s) + \beta E_g(s) + \gamma E_e(s) + \delta E_p(s)). \tag{6.17}$$

Where the contour is given by an implicit function $c(s)$. In practice the contour is given by a discrete set of control points. The first two terms of the energy functional refer to the internal energy of the contour, while the last two terms model the external influences on the contour shape. The first term models the contour continuity, that is the spread of the points of the contour:

$$E_c(s) = (\bar{d} - \|c(s) - c(s-1)\|)^2. \tag{6.18}$$

Where $\bar{d}$ is the average distance between points in the contour. The second term models the contour smoothness via an approximation of the second derivative:

$$E_s(s) = \|c(s-1) - 2c(s) + c(s+1)\|^2. \qquad (6.19)$$

The third term models the attraction of the contour to the edges in the image

$$E_e(s) = -\|\nabla I\|^2. \qquad (6.20)$$

Where ∇I is the spatial gradient in the image, which can be computed by conventional operators, such as the Sobel operator, or by morphological gradient operators. The use of the gradient information to push the contour to the objects in the image can lead to frozen dynamics in flat regions of the image. To avoid this situation we introduce a potential term, which models the attraction to the image objects in regions with null spatial gradient.

$$E_p(s) = K \sum_{r \neq s} \frac{I(r)}{d_{r,s}}. \qquad (6.21)$$

Where $I(r)$ is the image intensity at pixel r and $d_{r,s}$ is the distance between pixel sites r and s. This definition is quite similar to the physical one of electrical potential. The constant factor is set to $K = -1$.

In our experiments we did use 200 control points to represent the shape contour. The active contour algorithm was applied to them, with a distance renormalization interleaved step that redistributed the control points over the contour at equal distances. This step avoided excessive concentration of the control points on highly noise regions of the contour. When the active contour algorithm reaches its stable final state, we shift the centroid of the shape to the origin and we normalize the coordinates to a maximum value of 1. (Not a norm normalization). We compute the polar coordinates of the normalized shapes and take as our data sample vectors the polar coordinate magnitude starting with the lowest point in the normalized shape. It is important to note that we did not need to register the shapes to have a standard orientation, because we assume the image capture was performed with enough care. We keep the magnitude of the polar coordinates as the shape description. That way, each shape is represented by a 1D discrete function, a high dimensional vector, and all shapes have the same dimensionality.

6.7.2 Experimental results

From these 1D representations we extracted the vertices of a convex set covering most of the data applying the procedure described in Sect. 6.4.2 setting the noise filtering parameter to the standard value $\alpha = 2$. In Fig. 6.2 we present the extreme shapes found by one execution of the algorithm. Repetitions of the algorithm with the same control parameter give similar number of extreme shapes, although the precise shapes may vary between executions.

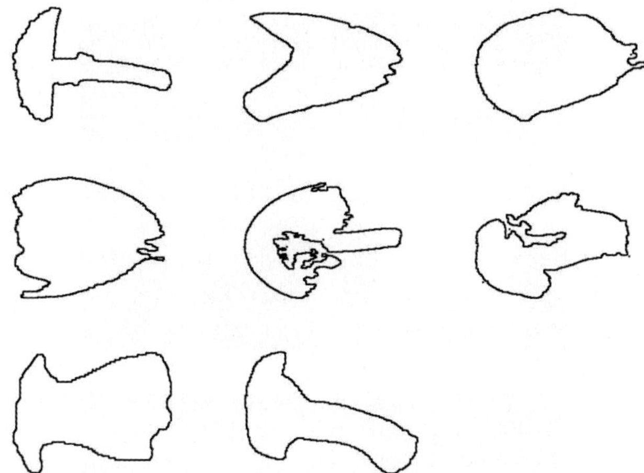

Fig. 6.2. Lattice independent shapes found by our algorithm

As can be appreciated there is a great variability between shapes, partly due to the capture conditions, the surface texture and reflectance of the mushrooms, partly due to the intraclass and interclass variability of the mushroom shapes. For instance, the shape of the Amanita Muscaria shown in Fig. 6.3(a) is quite different from the exemplar of Boletus Granulatus in Fig. 6.3(b) or the exemplar of Lepista Inversa in Fig. 6.3(d). This interclass variability, which can be very positive for classification purposes is compensated negatively by a high intraclass variability. The exemplars of Boletus Granulatus shown in Fig. 6.3(b) and Fig. 6.3(c) correspond to mushrooms of different stages of development and they may be closer to other species in analogous stages of development than to the ones of its own species. Besides the diverse sizes produce contours of diverse size (in number of pixels). The effect of these size changes have not been taken into account.

To test of the power of the representation obtained, we computed the convex coordinates of each shape in the database on the basis of selected extreme shapes, applying the unmixing algorithm in (2). Then we perform the search of the most similar shapes to a given one according to the Euclidean distance between their feature vectors (the convex coordinates). Figure 6.4 shows one query applied to the database. Figure 6.5 shows the four most similar shapes found in the database according to the convex coordinates computed relative to the shapes shown in Fig. 6.2. These kind of responses encourage the use of our approach for CBIR. However, to obtain some kind of quantitative backup of our proposition we have performed a recognition experiment using a Nearest Neighbor algorithm over the database of shapes described by their convex coordinates. The validation strategy is one-leave-out. Besides we computed the same experiment over the results of ICA and the convex coordinates over the vertices obtained by the CCA algorithm.

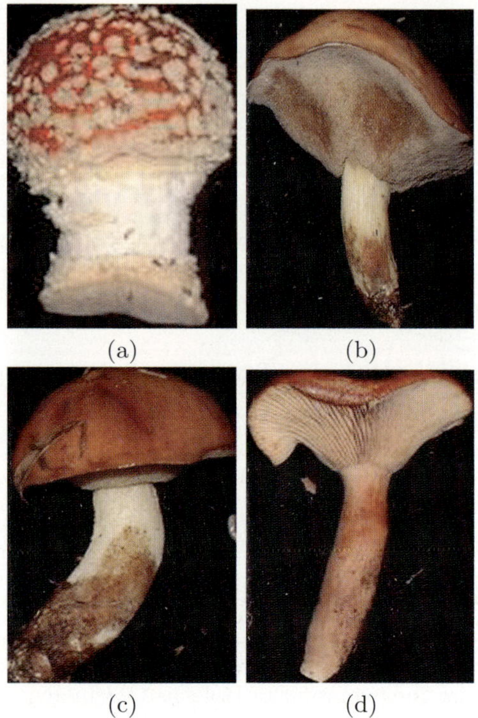

Fig. 6.3. Some mushrooms in our database (a) Amanita Muscaria, (b) and (c) Boletus Granulatus, (d) Lepista Inversa

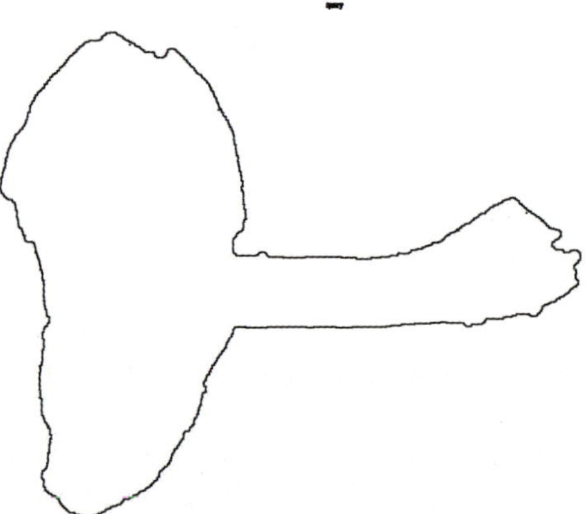

Fig. 6.4. A test shape to query the database of shapes

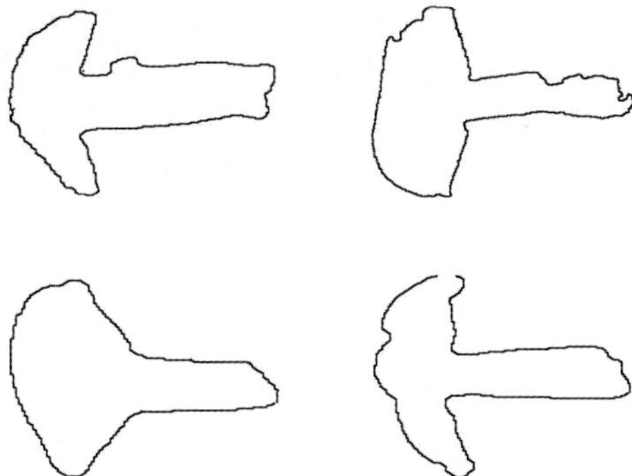

Fig. 6.5. Most similar shapes found in the database of shapes

Table 6.4. Recognition ratio over the Mushroom data set with diverse number of classes

#Clases	#Extrema	AMM	ICA	CCA
32	16	0.3557	0.0772	0.2852
32	15	0.3356	0.0872	0.1913
32	14	0.2987	0.0369	0.2114
15	12	0.4772	0.1066	0.3706
15	11	0.4365	0.0761	0.4112
15	10	0.3957	0.1015	0.2386
10	8	0.5319	0.1489	0.4326
10	7	0.5177	0.1384	0.4894
10	6	0.4823	0.1844	0.4823
5	9	0.7512	0.1375	0.7125
5	8	0.7625	0.2250	0.7750
5	7	0.7375	0.1875	0.6625
3	5	0.9333	0.4889	0.8889
3	4	0.9556	0.5556	0.7111
3	3	0.9333	0.4889	0.8000

In Table 6.4 we show the recognition results (ratio of correct classification) for the AMM, ICA and CCA approaches using a 1-NN classification algorithm over the different subsets of mushroom classes, using a one-leave-out validation methodology. The criteria to select the classes has been the number of individual images gathered for each class. The complete database contains 39 classes. The lowest number of classes considered (the ones with most individuals) is 3. To set the number of endmembers for CCA and the dimensional reduction of ICA, we compute first the AMM algorithm and then we set both CCA and ICA to the number of endmembers found by the AMM algorithm. The inspection of the results show that both convex coordinate approaches (AMM and CCA) improve over the ICA transformation, and that AMM improves almost always over CCA, although, both approaches being stochastic in some way, there may be instances of repetitions of the CCA which improve other repetitions of the AMM approach. Discarding the classes with the lowest number of items we obtain databases with 15, 10, 5 and 3 classes. Reducing the number of classes reduces confusion among classes and improves the recognition ratio up to levels of significant performance using a simple 1-NN classifier.

6.8 Indoor Robot Navigation

Navigation is the ability of an agent to move around its environment with a specific purpose [4]. It implies some knowledge of the environment, be it topological or not. A needed ability for navigation is self-localization: the ability of the robot to ascertain, more or less accurately, where it is from the information provided by its sensors. This knowledge makes possible other navigation related tasks like planning. The basic self-localization procedure is odometry: self-sensing and keeping track of motion commands. However the uncertainties related to the environment and the robot internal status call for sensor based external confirmation of the position internal estimation. Self-localization based on low range external sensors has been formulated in a probabilistic framework [9]. Vision based [8] and mixed [27, 40] systems are proposed to increase the sensing range and self-localization robustness. Visual self-localization methods usually are based on landmark recognition [2, 15, 23, 24, 26, 28, 33, 39, 41]. For instance, the system described in [2] computes for each stored view a graph model representation of salient points in the image as measured by the information content of a neighborhood of the point in a gradient image. To recognize the view, the salient points in the current image are compared to the stored models. Model matching as performed in [2] is not invariant or robust against translations and rotations. Therefore each view is only recognized when the robot is within a small neighborhood of the physical position and orientation where the landmark was detected originally. Recognition in this case is not continuous, unless the stored views built up a dense map. Our approach falls within the class of holistic approaches. The stated goal is to recognize, with some degree of robustness, several predetermined robot placements and orientations based on the recognition of the visual information captured by the robot. We can associate to each position an area of the physical environment where the robot recognizes this position, like in [2, 24]. The robot is supposed to wander looking forward, taking images at a steady rate. Each image corresponds to a view of the world, characterized by a physical position and orientation. Images are analyzed continuously and when a scene is recognized a certain spatial position is assumed for the robot. Robustness implies that the recognition must cope with some variations in lighting and small rotations and translations of the camera due to the uncertainty of the robot position, which, in its turn, is due to the uncertainties in the motion of the robot. This work described here departs from previous work [30, 31, 48] where we first proposed MAMs to solve the self-localization problem, although we continue working on the same platform: a Pioneer 2DX (ActivMedia, Ma.). The approach we follow is to characterize the data by a convex region that encloses them or most of them. The features extracted are the relative coordinates of the data points in this region, the convex coordinates, as discussed in Sect. 6.2, when the vertices of the convex region are approximated by those found by the procedure described in Sect. 6.4.2.

6.8.1 Experimental results

The experimental setup is as follows. First a path was defined from our laboratory to the stairs hall on 3d floor of our building. The mobile robot platform was guided manually seven times following this path. In each of those trips, the odometry was recorded and the images taken from the camera, at an average of 10 frames per second, were also recorded.

In the computational experiments that follow, the first trip was used to train the system parameters, and the six other trips were used as test sequences, simulating a real trip. The task to perform is to recognize a given set of map positions from their landmark views. The positions were selected on the floor plane, selecting places of practical relevance, like doors to other laboratories, and the corresponding landmark views were selected from the first trip image sequence based on odometry readings. Figure 6.6 shows the map position landmark views as extracted from the image sequence.

Fig. 6.6. The landmark views corresponding to the positions selected to build up the map

Classes are composed for each of the selected landmark position, assigning the images in the sequences to the closest map position according to its odometry reading. Therefore the task becomes the classification of the images into one of the map classes, where each class is composed of images taken in robot path positions before and after the map position for which it is the closest map position. The clasification was done using k-NN. Each of the images of the training trip was assigned to a class, using them as a cluster of images representing each selected position. For the test trips, each image was classified on those classes using 3-NN.

From the image sequence of the first trip a PCA transformation consisting of 230 eigenvectors was computed. All the following computations were done on the PCA coefficients of the images. We apply the algorithm described in 6.4.2 several times to the different image sequences, after their transformation by the 230 eigenvector PCA. The noise tolerance parameter with best results was set, after some tuning, to $\alpha = 5$ An instance of the extrema selected with this value is given in Fig. 6.7. Despite their similarity to the collection of images shown in Fig. 6.6, these are obtained from a completely unsupervised process, while those in Fig. 6.6 correspond to a human made selection. An interesting question is whether our approach could be used as an automatic landmark selection algorithm for simultaneous map building and localization.

As the algorithm described in Sect. 6.4.2 has a random start, different runs of it may give different results. In Tables 6.6 and 6.7 we present the classification success ratios for each of the sequences on the convex coordinates computed from the extrema found by the algorithm after each run, with different values of α. The last column shows the number of extrema found at

Fig. 6.7. The views corresponding to the extrema selected in one instance of the execution of the algorithm described in [49]

Table 6.5. Landmark recognition success rate based on the PCA representation of the navigation images for several sets of eigenvectors selected, using a 3-NN classifier

Set	Train	P1	P2	P3	P4	P5	P6	Av.
PCA 197	0.95	0.95	0.84	0.76	0.65	0.79	0.77	0.82
PCA 150	0.95	0.95	0.84	0.76	0.65	0.78	0.76	0.81
PCA 100	0.95	0.95	0.85	0.76	0.65	0.79	0.76	0.82
PCA 50	0.95	0.94	0.85	0.76	0.65	0.78	0.78	0.81
PCA 30	0.96	0.94	0.87	0.77	0.64	0.78	0.78	0.82
PCA 10	0.96	0.94	0.86	0.78	0.66	0.76	0.73	0.81
Av.	0.96	0.94	0.85	0.76	0.65	0.78	0.76	0.82

Table 6.6. Landmark recognition success rate based on the convex coordinates representation of the navigation images for several runs of the extrema extraction algorithm with $\alpha = 6$ and using 3-NN

Run	Train	P1	P2	P3	P4	P5	P6	Av.	#e
1	0.92	0.91	0.78	0.74	0.66	0.68	0.62	0.76	8
2	0.95	0.93	0.77	0.73	0.74	0.72	0.62	0.78	7
3	0.95	0.93	0.83	0.73	0.67	0.72	0.64	0.78	8
4	0.94	0.93	0.78	0.71	0.67	0.69	0.64	0.77	7
5	0.93	0.90	0.76	0.71	0.65	0.67	0.62	0.75	7
6	0.93	0.91	0.77	0.72	0.69	0.68	0.57	0.75	8
7	0.95	0.93	0.78	0.70	0.61	0.66	0.62	0.75	8
8	0.95	0.93	0.78	0.69	0.58	0.69	0.62	0.75	8
9	0.93	0.92	0.80	0.73	0.70	0.73	0.63	0.78	8
10	0.94	0.94	0.80	0.73	0.65	0.69	0.67	0.77	9
Av.	0.94	0.92	0.79	0.72	0.66	0.69	0.62	0.76	

each run. Notice that the number of extrema is in the order of 10 for the best results, meaning a dimension reduction from 230 to 10. The Av. row shows the average success for each image sequence, including the training image sequence. For comparison we perform some further dimension reduction on the PCA coefficient vectors selecting the most significant eigenvectors for the transformation. The average results of the image classification are given in Table 6.5, the average being computed as in the last column of the other Tables.

In all the tables the success ratio decreases as the image sequence is farther in time from the initial one used for training. The comparison of the results in Tables 6.5, 6.6, and 6.7 shows that there is at least an instance of the convex coordinates features which improves the best results of the PCA, and that the convex coordinates are comparable or improve the results of the PCA with a the same or much stronger dimensionality reduction. These results are much less favorable to our approach than the ones reported in previous sections. The cause of the performance loss may be in the nature of the images

Table 6.7. Landmark recognition success rate based on the convex coordinates representation of the navigation images for several runs of the extrema extraction algorithm with $\alpha = 5$ and using 3-NN

Run	Train	P1	P2	P3	P4	P5	P6	Av.	#e
1	0.94	0.93	0.81	0.76	0.72	0.73	0.67	0.79	13
2	0.94	0.93	0.85	0.77	0.69	0.78	0.71	0.81	14
3	0.94	0.93	0.84	0.75	0.70	0.75	0.74	0.81	13
4	0.94	0.93	0.83	0.71	0.63	0.73	0.67	0.78	14
5	0.94	0.93	0.85	0.79	0.69	0.78	0.72	0.81	12
6	0.93	0.93	0.80	0.70	0.67	0.69	0.70	0.77	12
7	0.94	0.93	0.83	0.71	0.59	0.70	0.66	0.77	12
8	0.93	0.93	0.82	0.76	0.69	0.74	0.66	0.79	12
9	0.94	0.93	0.79	0.73	0.64	0.70	0.63	0.77	14
10	0.92	0.92	0.79	0.70	0.63	0.65	0.60	0.75	12
Av.	0.94	0.93	0.82	0.74	0.67	0.73	0.68	0.78	

used for the experiment. The effect of translations of the robot is a radial displacement of the pixels in the image, with some new values appearing at the image center. This is an effect that is highly non linear and the most likely conclusion is that both PCA and our approach have comparable difficulties to cope with this kind of data. The effect of rotations is a displacement of the pixels along the lines of the image, with new values appearing at the border. Again, the appearance of new values is a highly nonlinear phenomena. When using the images selected from a previous walk, we find that the robot minor path deviations are translated into image translations or changes in image conditions. These factors are at the root of the behavior degradation in time.

6.9 Conclusions

We propose the use of convex coordinates as a feature extraction method. Given a convex polytope defined by a set of linearly independent endmbembers, the computation of such coordinates is done by a simple matrix inversion. The key problem is that of inducing the endmembers from the data. We have proposed an algorithm for the induction from the data of a set of endmembers or vertices of a convex polytope approaching the minimum one covering the data. The definition of the algorithm was based on lattice independence, or as known at the time of its earlier proposals, morphological independence. In this chapter we have gathered several applications of this approach.

In Sect. 6.5 we have proposed an approach to CBIR in homogeneous databases of hyperspectral images based on the collection of endmembers induced by our algorithm. The image characterization is independent of image size and spatial undesired variations, like distortions due to variations in the flying path. We have performed an experiment of exhaustive search on a collection of simulated hyperespectral images. The results are encouraging: almost

100% success in providing the closest image in terms of the ground truth endmembers. It is possible to define other distances based on the endmembers extracted by the AMM (or any alternative algorithm). For example, the Euclidean distance between individual endmembers may be substituted by max/min distances. The whole set of endmbembers may be evaluated according to the Haussdorf distance. There are also diverse ways to introduce penalization terms to evaluate the diverse number of endmembers found in the images. Another case of use of our approach in CBIR applications is shown in Sect. 6.7 for shape based search on a database of mushroom images. The visual results of the search for the most similar images are rather consistent with the desired behavior of the system.

In Sect. 6.6 we have explored the use of the convex coordinates as feature extraction methods for supervised construction classification methods. We have performed an experiment on two well known hyperspectral images. Both convex coordinate instances were superior to the linear feature extraction algorithms. These results are pointing to one line of research little explored. Instead of trying to build classifiers based on lattice computing, using lattice computing to obtain good data transformations that allow to construct better classifiers using conventional techniques. Besides, this experiment confirms also that our approach to the induction of endmembers from the data, the AMM approach is, at least, comparable to other well-established methods, like CCA. The quantitative validation of the shape CBIR system described in Sect. 6.7 are also an instance of supervised classifier construction and confirm the results obtained on the hyperspectral image pixel classification.

The less optimistic results were obtained on the robot navigation experiment described in Sect. 6.8. Mobile robot self-localization is stated as a classification of images taken from the camera. We use the PCA and the convex coordinates, based on the MAM detected endmembers, as the features for classification with a 1-NN classifier. We found that our approach improves the PCA approach in some runs, while on average performs similar than most PCA eigenvector selections tried. Both approaches degrade their performance with time. We attribute those results to the highly non-linear image transformations introduced by the robot motion.

Acknowledgments

The authors received support from the Ministerio de Ciencia y Tecnologia through grant VIMS-2003-20088-c04-04

References

1. Alber IE, Xiong Z, Yeager N, Farber M, Pottenger W M (2001) Fast retrieval of multi- and hyperspectral images using relevance feedback. In: Proc Geosci Rem Sens Symp (IGARSS) 3:1149–1151

2. Balkenius C, Kopp L (1997) Robust self-localization using elastic templates. In: Lindberg T (ed) Proc Swedish Symposium on Image Analysis
3. Blake A, Isard M (1998) Active Contours. Springer-Verlag, New York
4. Chatila R (1995) Deliberation and reactivity in autonomous mobile robots. Robotics and Autonomous Systems 16:197–211
5. http://www1.cs.columbia.edu/CAVE/research/softlib/coil-100.html
6. Cootes TF, Taylor CJ, Cooper DH, Graham J (1995) Active shape models- their training and application. Computer Vision and Image Understanding 61:38–59
7. Craig M (1994) Minimum volume transformations for remotely sensed data. IEEE Trans Geoscience and Remote Sensing 32(3):542–552
8. DeSouza GN, Kak AC (2002). Vision for mobile robot navigation: a survey. IEEE Trans Pattern Analysis and Machine Intelligence 24(2):237–267
9. Fox D (1998) Markov Localization: A Probabilistic Framework for Mobile Robot Localization and Navigation. PhD Thesis, University of Bonn, Germany
10. Fukunaga K (1990) Introduction to Statistical Pattern Recognition. Academic Press, Boston, MA
11. Graña M, Gallego J (2003) Associative mophological memories for endmember induction. In: Proc IGARSS
12. Graña M, Sussner P, Ritter GX (2003) Associative morphological memories for endmember determination in spectral unmixing. In: Proc FUZZ-IEEE
13. Graña M, d'Anjou A (2004) Feature extraction by linear spectral unmixing. In: Negoita M, Howlett RJ, Jain LC (eds) Knowledge-Based Intelligent Information and Engineering Systems — I. Springer-Verlag, LNAI 3213:692–697
14. Graña M, d'Anjou A, Albizuri FX (2005) Morphological memories for feature extraction in hyperspectral images. In: Verleysen M (ed) Proc ESANN pp 497–502
15. Gross HM, Koening A, Boehme HJ, Schroeter Ch (2002) Vision-based Monte Carlo self-localization for a mobile service robot acting as shopping assistant in a home store. In: Proc IEEE Intl Conf on Intelligent Robots and Systems
16. Gualtieri JA, Chettri S (2000) Support vector machines for classification of hyperspectral data. In: Proc Geosci Rem Sens Symp (IGARSS) 2:813–815
17. Hopfield JJ (1982) Neural networks and physical systems with emergent collective computational abilities. In: Proc Nat Acad Sciences 79:2554–2558
18. Hyvarynën A, Oja E (1999) A fast fixed-point algorithm for independen component analysis. Neural Computation 9:1483–1492
19. Hyvarynën A, Karhunen J, Oja E (2001) Independent Component Analysis. John Wiley & Sons, New York.
20. Ifarraguerri A, Chang CI (1999) Multispectral and hyperspectral image analysis with convex cones. IEEE Trans Geoscience and Remote Sensing 37(2):756–770
21. Keshava N, Mustard JF (2002) Spectral unimixing. IEEE Signal Processing Magazine 19(1):44–57
22. Kozintsev B (1999) Computations With Gaussian Random Fields. PhD Thesis, ISR99-3 University of Maryland
23. Livatino S, Madsen C (1999) Optimization of robot self-localization accuracy by automatic visual-landmark selection. In: Proc 11th Scandinavian Conference on Image Analysis (SCIA) pp 501–506
24. Livatino S, Madsen C (1999) Autonomous robot navigation with automatic learning of visual landmarks. In: International Symposium of Intelligent Robotic Systems (SIRS)

25. Manolakis D, Shaw G (2002) Detection algorithms for hyperspectral imaging applications. IEEE Signal Processing Magazine 19(1):29–43
26. Marando F, Piaggio M, Scalzo A (2001) Real time self localization using a single frontal camera. In: Intl Symposium of Intelligent Robotic Systems (SIRS)
27. Ohya A, Kosaka A, Kak AC (1998) Vision-based navigation by a mobile robot with obstacle avoidance using single-camera vision and ultrasonic sensing. IEEE Journal of Robotics and Automation 14(6):969–978
28. Olson CF (2000) Landmark selection for terrain matching. In: Proc ICRA
29. Plaza A, Martinez P, Perez R, Plaza J (2002) Spatial spatial/spectral end-member extraction by multidimensional morphological operations. IEEE Trans Geoscience Remote Sensing 40(9):2025–2041
30. Raducanu B, Graña M, Sussner P (2001) Morphological neural networks for vision based self-localization. In: Proc ICRA
31. Raducanu B, Graña M, Sussner P (2002) Steps towards one-shot vision-based self-localization. In: Duro R, Santos J, Graña M (eds) Biologically Inspired Robot Behavior Engineering pp 265–294. Springer-Verlag
32. Raducanu B, Graña M, Albizuri FX (2003) Morphological scale spaces and associative morphological memories: results on robustness and practical applications. J Mathematical Imaging and Vision 19(2):113–122
33. Reuter J (2000) Mobile robot self-localization using PDAB. In: Proc ICRA
34. Ritter GX, Sussner P, Diaz-de-Leon JL (1998) Morphological associative memories. IEEE Trans on Neural Networks 9(2):281–292
35. Ritter GX, Diaz-de-Leon JL, Sussner P (1999) Morphological bidirectional associative memories. Neural Networks 12:851–867
36. Ritter GX, Urcid G, Iancu L (2003) Reconstruction of patterns from moisy inputs using morphological associative memories. J Mathematical Imaging and Vision 19(2):95–112
37. Ritter GX, Urcid G (2003) Lattice algebra approach to single-neuron computation. IEEE Trans Neural Networks 14(2): 282–295
38. Ritter GX, Gader PD (2006) Fixed points of lattice transforms and lattice associative memories. In: Hawkes P (ed) Advances in Imaging and Electron Physics 144:165–242. Elsevier, Amsterdam, The Netherlands
39. Rizzi A, Duina D, Inelli S, Cassinis R (2002) Unsupervised matching of visual landmarks for robotic homing using Fourier-Mellin transform. Robotics and Autonomous Systems 40:131–138
40. Saffiotti A, Wesley LP (1996) Perception-based self-localization using fuzzy location. In: Dorst L, Van Lambalgen M, Voorbraak F (eds) Springer-Verlag, LNAI 1093:368–385
41. Sekimori D, Usui T, Masutani Y, Miyazaki F (2002) High-speed obstacle avoidance and self-localization for mobile robots based on omni-directional imaging of floor region. IPSJ Transactions on Computer Vision and Image Media 42 NoSIG13-012
42. Smeulders AWM, Worring M, Santini S, Gupta A, Jain R (2000) Content-based image retrieval at the end of the early years. IEEE Trans Patttern Analysis Machine Intelligence 22(12):1349–1380
43. Sussner P (2001) Observations on morphological associative memories and the kernel method. In: Proc IJCNN
44. Sussner P (2003) Generalizing operations of binary autoassociative morphological memories using fuzzy set theory. J Mathematical Imaging and Vision 19(2):81–94

45. Tadjudin S, Landgrebe D (1999) Covariance estimation with limited training samples. IEEE Trans Geoscience Remote Sensing 37(4):2113–2118
46. Vapnik VN (1995) The Nature of Statistical Learning Theory. Springer-Verlag, New York
47. Veltkamp RC, Hagedoorn M (2001) State of the art in shape matching. In: Lew MS (ed) Principles of Visual Information Retrieval. Springer-Verlag, Berlin
48. Villaverde I, Ibañez S, Albizuri FX, Graña M (2005) Morphological neural networks for real-time vision based self-localization. In: Abraham A, Dote Y, Furuhashi T, Köpen M, Ohuchi A, Ohsawa Y (eds) Proc Soft Computing as Transdisciplinary Science and Techonology (WSTST) pp 70–79. Springer-Verlag
49. Villaverde I, Graña M, D'Anjou A (2006) Morphological neural networks for localization and mapping. In: Proc IEEE Intl Conf Computational Intelligence for Measurement Systems and Applications (CIMSA)
50. Villman T, Merenyi E, Hammer B (2003) Neural maps in remote sensing image analysis. Neural Networks 16:389–403
51. Xu C, Prince L (1998) Snakes, shapes, and gradient vector flow. IEEE Trans Image Processing 7(3):359–369

A Lattice-Based Approach to Mathematical Morphology for Greyscale and Colour Images

Valérie De Witte, Stefan Schulte, Mike Nachtegael, Tom Mélange,
and Etienne E. Kerre

Ghent University, Department of Applied Mathematics and Computer Science,
Krijgslaan 281 (Building S9), B-9000 Gent, Belgium
Valerie.DeWitte@UGent.be, Stefan.Schulte@UGent.be,
Mike.Nachtegael@UGent.be, Tom.Melange@UGent.be, Etienne.Kerre@UGent.be

Summary. Mathematical morphology (MM) is a theory for the analysis of spatial structures, based on set-theoretical notions and on the concept of translation. MM has many applications in image analysis such as edge detection, noise removal, object recognition, pattern recognition and image segmentation in a.o. geosciences, materials science, the biological and medical world [13, 15]. MM was originally developed for binary images only. The basic tools of MM are the morphological operators, which transform an image A we want to analyse, using a structuring element B into a new image $P(A, B)$ in order to obtain additional information about the objects in A like shape, size, orientation, image measurements. Apart from the threshold and umbra approach, binary morphology can be extended to morphology for greyscale images using fuzzy set theory, called fuzzy morphology. In this work we will present a new vector-based approach for the extension of MM for greyscale images to colour morphology. We will extend the basic morphological operators dilation and erosion based on the threshold and fuzzy set approach to colour images. Finally in the last section we illustrate an image denoising method using MM to reduce stripes' artefacts in satellite images.

7.1 Basic Notions

7.1.1 Modelling of images

Digital images are often represented by a 2-dimensional array, where a pair (i, j) denotes the position of a pixel of the image. Binary images assume two possible pixel values, e.g. 0 and 1, corresponding to black and white respectively. White represents the objects in an image, whereas black represents the background. Mathematically, a 2-dimensional binary image can be represented as a mapping f from a universe X of pixels (usually X is a finite subset of $\mathbb{R}^2$, in practice, it will even be a subset of $\mathbb{Z}^2$) into $\{0, 1\}$, which is completely determined by $f^{-1}(\{1\})$, i.e., the set of white pixels, so that f can be identified with the set $f^{-1}(\{1\})$, a subset of X, the so-called domain of the image.

V.D. Witte et al.: *A Lattice-Based Approach to Mathematical Morphology for Greyscale and Colour Images*, Studies in Computational Intelligence (SCI) **67**, 129–148 (2007)
www.springerlink.com

A 2-dimensional greyscale image can be represented as a mapping from X to the universe of grey values $[0, 1]$, where 0 corresponds to black, 1 to white and in between we have all shades of grey. Colour images are then represented as mappings from X to a 'colour interval' that can be for example the product interval $[0, 1] \times [0, 1] \times [0, 1]$ (the RGB colour model). So a digital colour image in RGB is stored as a two-dimensional array of (three-dimensional) vectors that defines the red, green and blue colour component for each pixel. Colour can be modelled in different colour models; more information about colour models can be found in [12, 14].

7.1.2 Fuzzy sets

A fuzzy set F in a universe X is characterised by a $X - [0, 1]$ mapping, the so-called membership function, where for all x in X, $F(x)$ denotes the degree in which x belongs to the fuzzy set F. An extensive study of fuzzy sets can be found in [7]. Further on we need the extension of the binary logical operators negation ($\neg$), conjunction ($\wedge$) and implication ($\Rightarrow$) to fuzzy logic, where these operators are called negators, conjunctors and implicators. The most popular conjunctors $\mathcal{C}$ on $[0, 1]$ are the triangular norms minimum $\mathcal{T}_M$, algebraic product $\mathcal{T}_P$ and Lukasiewicz triangular norm $\mathcal{T}_W$; the most popular implicators $\mathcal{I}$ on $[0, 1]$ are the Kleene-Dienes implicator $\mathcal{I}_{KD}$, the Reichenbach implicator $\mathcal{I}_R$ and the Lukasiewicz implicator $\mathcal{I}_W$ given by

conjunctor	implicator
$\mathcal{T}_M(a, b) = \min(a, b)$	$\mathcal{I}_{KD}(a, b) = \max(1 - a, b)$
$\mathcal{T}_P(a, b) = a \cdot b$	$\mathcal{I}_R(a, b) = 1 - a + a \cdot b$
$\mathcal{T}_W(a, b) = \max(0, a + b - 1)$	$\mathcal{I}_W(a, b) = \min(1, 1 - a + b)$

The standard negator $\mathcal{N}_s$ on $[0, 1]$ is defined as $\mathcal{N}_s(a) = 1 - a$ for all a in $[0, 1]$.

7.2 Binary Morphology

Consider a binary image A and a binary structuring element B. The translation $T_y(B)$ of B by a vector $y \in \mathbb{R}^2$ is defined as $T_y(B) = \{x \in \mathbb{R}^2 \mid x - y \in B\}$; the reflection of B is defined as $-B = \{-x \in \mathbb{R}^2 \mid x \in B\}$.

Definition 7.1 *Let A be a binary image and B a binary structuring element. The binary dilation $D(A, B)$ of A by B is the binary image given by*

$$D(A, B) = \{y \in \mathbb{R}^2 \mid T_y(B) \cap A \neq \emptyset\}.$$

The binary erosion $E(A, B)$ of A by B is defined as

$$E(A, B) = \{y \in \mathbb{R}^2 \mid T_y(B) \subseteq A\}.$$

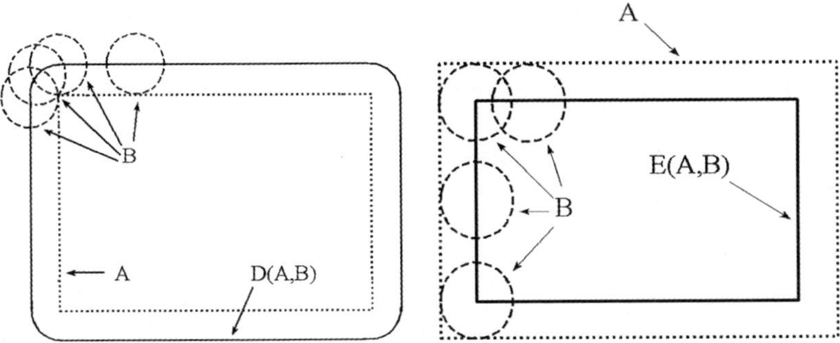

Fig. 7.1. Geometrical interpretation of the binary dilation (left) and the binary erosion (right). The centre of the structuring element B coincides with the origin of the coordinate system

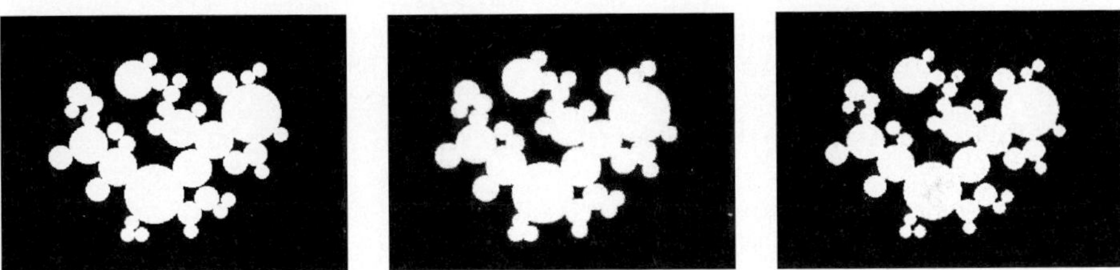

Fig. 7.2. The original binary image A (left), the binary dilation $D(A, B)$ (middle) and the binary erosion $E(A, B)$ (right). You see that the dilation enlarges the objects in an image, while the erosion reduces them

The binary dilation and erosion have a beautiful geometrical interpretation, see Fig. 7.1. The dilation $D(A, B)$ contains all points y in $\mathbb{R}^2$ for which the translation $T_y(B)$ of B has a non-empty intersection with the image A. A point y belongs to the dilation $D(A, B)$ if and only if the translation $T_y(B)$ and A hit each other. The binary erosion $E(A, B)$ consists of all points $y \in \mathbb{R}^2$ for which the translation $T_y(B)$ of B is contained in A. A point y belongs to the erosion $E(A, B)$ if and only if the translation $T_y(B)$ and coA don't hit. When we use the following structuring element B given by (the underlined element corresponds to the origin of coordinates)

$$B = \begin{pmatrix} 1\ 1\ 1 \\ 1\ \underline{1}\ 1 \\ 1\ 1\ 1 \end{pmatrix},$$

we get the results shown in Fig. 7.2 for the binary dilation and erosion.

7.3 Greyscale Morphology

7.3.1 Greyscale morphology based on the threshold approach

Consider a greyscale image A represented as a $\mathbb{R}^2 - [0,1]$ mapping and a binary structuring element B modelled as a crisp subset of $\mathbb{R}^2$. The support d_A of A is defined as the set $d_A = \{x \in \mathbb{R}^2 \mid A(x) > 0\}$; the reflection of A is the $\mathbb{R}^2 - [0,1]$ mapping $-A$ characterised by $(-A)(x) = A(-x)$, for all x in $\mathbb{R}^2$.

Definition 7.2 *Let A be a greyscale image and B a binary structuring element. The t-dilation $D_t(A,B)$ and the t-erosion $E_t(A,B)$ are the greyscale images given by*

$$D_t(A,B)(y) = \sup_{x \in T_y(B) \cap d_A} A(x) \ \text{ and } E_t(A,B)(y) = \inf_{x \in T_y(B)} A(x), \ \text{for } y \in \mathbb{R}^2.$$

Figure 7.3 gives an example of the t-morphological dilation and erosion.

7.3.2 Fuzzy mathematical morphology

Since greyscale images can be modelled as $\mathbb{R}^2 - [0,1]$ mappings, we can identify greyscale images with fuzzy sets and extend binary morphology to greyscale morphology using fuzzy set theory.

Definition 7.3 *Let A be a greyscale image and B a greyscale structuring element (both seen as fuzzy sets), $\mathcal{C}$ a conjunctor on $[0,1]$ and $\mathcal{I}$ an implicator on $[0,1]$. The fuzzy dilation $D_{\mathcal{C}}(A,B)$ and the fuzzy erosion $E_{\mathcal{I}}(A,B)$ are the fuzzy sets defined as*

$$D_{\mathcal{C}}(A,B)(y) = \sup_{x \in T_y(d_B) \cap d_A} \mathcal{C}(B(x-y), A(x)) \ \text{ for } y \in \mathbb{R}^2,$$

$$E_{\mathcal{I}}(A,B)(y) = \inf_{x \in T_y(d_B)} \mathcal{I}(B(x-y), A(x)) \ \text{ for } y \in \mathbb{R}^2.$$

Fig. 7.3. The original greyscale image A (left), the t-dilation $D_t(A,B)$ (middle) and the t-erosion $E_t(A,B)$ (right)

Proposition 7.1 *Because* $\mathcal{T}_M \geq \mathcal{T}_P \geq \mathcal{T}_W$ *and* $\mathcal{I}_{KD} \leq \mathcal{I}_R \leq \mathcal{I}_W$, *we obtain*

$$D_{\mathcal{T}_M}(A,B) \supseteq D_{\mathcal{T}_P}(A,B) \supseteq D_{\mathcal{T}_W}(A,B) \quad \text{and}$$
$$E_{\mathcal{I}_{KD}}(A,B) \subseteq E_{\mathcal{I}_R}(A,B) \subseteq E_{\mathcal{I}_W}(A,B),$$

for every greyscale image A and greyscale structuring element B.

Figure 7.4 illustrates the fuzzy dilation and fuzzy erosion using the following greyscale structuring element

$$B(:,:,1) = B(:,:,2) = B(:,:,3) = \frac{1}{255}\begin{pmatrix} 200 & 220 & 200 \\ 220 & \underline{255} & 220 \\ 200 & 220 & 200 \end{pmatrix}.$$

For a detailed study of binary and greyscale morphology we refer to [5, 6, 11]. More information about fuzzy MM can be found in [1, 3, 4, 10]. A study about the aspects of the algebraic theory of MM from the viewpoints of minimax algebra and translation-invariant systems and an extension to a more general algebraic structure that includes generalized Minkowski operators and fuzzy image operators can be found in [9].

Fig. 7.4. Above: the fuzzy dilation $D_{\mathcal{T}_M}(A,B)$ (left), the fuzzy dilation $D_{\mathcal{T}_P}(A,B)$ (middle) and the fuzzy dilation $D_{\mathcal{T}_W}(A,B)$ (right), below: the fuzzy erosion $E_{\mathcal{I}_{KD}}(A,B)$ (left), the fuzzy erosion $E_{\mathcal{I}_R}(A,B)$ (middle) and the fuzzy erosion $E_{\mathcal{I}_W}(A,B)$ (right)

7.4 Colour Morphology

Colour images can be represented as $\mathbb{R}^2 - [0,1] \times [0,1] \times [0,1]$ mappings. A first way to extend MM for greyscale images to colour images is the component-based approach. MM can be naturally extended to colour morphology by processing the morphological operators on each of the colour components separately. A major disadvantage of this approach is that the existing correlations between the different colour components are not taken into account and this often leads to disturbing artefacts. Another approach is to treat the colour at each pixel as a vector. Since we need the concept of a supremum and infimum to define morphological operators, we first have to define an ordering between these colour vectors. We have considered the three colour models RGB, HSV and L*a*b*.

7.4.1 New colour vector ordering

In the RGB colour model

A colour in the RGB colour model is obtained by adding the three colours red, green and blue in different combinations. Therefore a colour can be defined as a vector in a 3-D space that can be represented as a unit cube using a Cartesian coordinate scheme. This way every point in the cube represents a vector (colour). The greyscale spectrum is characterised by the line between the black top Bl with coordinates $(0,0,0)$ and the white top Wh $(1,1,1)$.

On the RGB cube in Fig. 7.5 [2] we can observe the colour hue red for example (we can also choose green or blue). If we start at the white top (with coordinates $(1,1,1)$) and go along the diagonal to the red top $(1,0,0)$ and from there on along the edge to the black top $(0,0,0)$, we see that we go from 'light' red over the most 'bright' colour red to 'dark' red. Inspired by this observation we will sort the colours in the RGB colour model from 'dark'

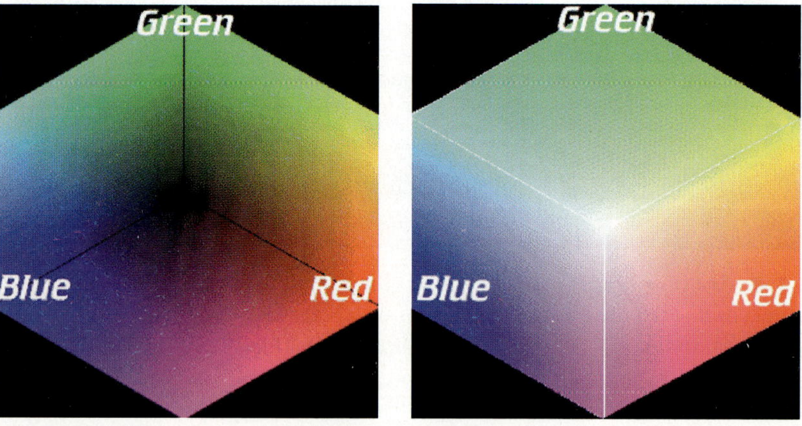

Fig. 7.5. Representation of the RGB colour model

colours (close to black) to 'light' colours (close to white), with respect to their distance to black and white. So we can define three relations $R_<$, $R_>$ and $R_=$ on RGB, given for all colours $c(r_c, g_c, b_c)$ and $c'(r_{c'}, g_{c'}, b_{c'})$ in RGB by

$$(c, c') \in R_< \Leftrightarrow d(c, Bl) < d(c', Bl) \text{ or}$$
$$(d(c, Bl) = d(c', Bl) \text{ and } d(c, Wh) > d(c', Wh))$$
$$\Leftrightarrow c \text{ lies strict closer to black than } c' \text{ or } c \text{ lies as far}$$
$$\text{from black as } c' \text{ and } c \text{ lies strict farther from white than } c'$$
$$(c, c') \in R_> \Leftrightarrow d(c, Wh) < d(c', Wh) \text{ or}$$
$$(d(c, Wh) = d(c', Wh) \text{ and } d(c, Bl) > d(c', Bl))$$
$$\Leftrightarrow c \text{ lies strict closer to white than } c' \text{ or } c \text{ lies as far}$$
$$\text{from white as } c' \text{ and } c \text{ lies strict farther from black than } c'$$
$$(c, c') \in R_= \Leftrightarrow (d(c, Bl) = d(c', Bl)) \text{ and } (d(c, Wh) = d(c', Wh)).$$

1. With the relation $R_<$ colours are first ordered from vectors with smallest distance to black to vectors with largest distance to black. The smaller the distance to black, the lower the colour is ranked. This way the RGB cube is sliced into different parts of spheres around the black top. Colours that are part of the same sphere (around the black top) are then ordered according to their distance with respect to white, from colours with largest distance to white to colours with smallest distance to white. So we will 'cut' the spheres around the black top with spheres with the white top as centre. Those colours closest to white are farthest away from black and vice versa.

2. With the relation $R_>$ we look at the distance with respect to white to know which one of two colours is ranked highest in the RGB colour model. The colour with the smallest distance to white is ordered higher than the other colour. If the distance to white is equal, i.e., if both colours lie on the same sphere around the white top, we select that colour lying farthest from black. Again, the RGB cube is sliced into parts of spheres, but now first towards the white top and then towards the black top.

3. Finally in the relation $R_=$ we combine both relations $R_<$ and $R_>$ to say that colours that have the same distance to the black top and the same distance to the white top, and thus lie on a circle (as profile of two spheres) in the RGB cube, are ranked equally.

Inspired by our idea to look at the black top to determine the smallest colour and to look at the white top to determine the largest colour, we have investigated the cases HSV and L*a*b*.

In the HSV colour model

In the HSV colour model a colour is characterised by the three quantities hue, saturation and value. Because of the opposite colour theory [12] all colour hues can be arranged in an opponent colour wheel along two axes (red-green and blue-yellow) that begins and ends by the same colour. So we can range the hue component in a circle from $0°$ to $360°$, which begins and ends by red. Values for the saturation component range from 0% if the colour isn't saturated (grey values) to 100% if the colour is completely saturated (pure colours). The value component V in the HSV colour model varies from 0 (black) to 1 (white), where the colours become increasingly brighter. While adding black to a certain colour, the value of the colour will decrease. The value axis begins by black, ends by white and in between we get all shades of grey. So that we get a cone for the representation of the three-dimensional colour model HSV. We denote a colour c in the HSV colour model as $c(h_c, s_c, v_c)$.

We want to order the colour vectors in the HSV colour model with respect to black and white, so we get:

1. Because the value component V of each colour in HSV gives us the 'grey level' of that colour, we can first order colours by looking at their V-value. A large V-value for a colour means that the colour lies closer to white than to black, while a colour with a small V-value lies closer to black than to white. The smaller the value component, the smaller the colour is seen.
2. If the V component is equal, we look at the saturation component S of both colours. An S-value of 100% indicates that the colour is completely saturated and contains no white light, the colour is pure. An S-value of 0% indicates that the colour is a grey value. That's why we sort the colours from colours with greatest S-value to colours with smallest S-value.
3. Finally, if two colours have the same V-value and the same S-value, we consider the hue component to order these colours. We have made the choice that the smaller the H-value, the lower the colour is ranked. We want to notice that the ordering we obtained here is equal to the one introduced by G. Louverdis, M.I. Vardavoulia, I. Andreadis and Ph. Tsalides in [8]. Their idea for ordering the colours in HSV this way is that the larger the mixture of a colour with black, the smaller the colour is considered, regardless of its saturation and its hue. In case two colours have been mixed with the same amount of black, the smaller is the one with the less mixture of white. If two colours have been mixed with the same amount of black and white, the smaller is the one with the smaller hue value.

This gives us an ordering $\leq_{HSV}$ of colour vectors in the HSV colour model, defined for two colours $c(h_c, s_c, v_c)$ and $c'(h_{c'}, s_{c'}, v_{c'})$ as:

$$
\begin{aligned}
c <_{HSV} c' &\Leftrightarrow v_c < v_{c'} \text{ or}\\
&\quad (v_c = v_{c'} \text{ and } s_c > s_{c'}) \text{ or}\\
&\quad (v_c = v_{c'} \text{ and } s_c = s_{c'} \text{ and } h_c < h_{c'})\\
c >_{HSV} c' &\Leftrightarrow c' <_{HSV} c\\
c =_{HSV} c' &\Leftrightarrow (v_c = v_{c'} \text{ and } s_c = s_{c'} \text{ and } h_c = h_{c'}).
\end{aligned}
$$

In the L*a*b* and L*u*v* colour model

The colour models L*a*b* and L*u*v* use a common lightness scale L*. The vertical axis L* in the centre of both colour models represents the lightness/brightness of a colour where the value L^* range from 0 (black) to 100 (white), with in between grey values. Both colour models use different uniform colour axes: the colour axes a* versus b* and u* versus v* (red-green versus yellow-blue) are based on the fact that a colour can't be red and green at the same time or both blue and yellow because these colours are opposite (opposite colour theory) [12]. At every colour axis values go from positive to negative. At the a* and u* axis the positive values give the amount of red and the negative values the amount of green, while at the b* and v* axis yellow is positive and blue negative. For these axes 0 is neutral grey.

Here we will only consider the L*a*b* colour model, but the same reasoning can be done for the L*u*v* colour model. A colour c in the L*a*b* colour model can be presented as $c(L^*, a^*, b^*)$, with $a^*, b^* \in [-1, 1]$. If we order the colours in the L*a*b* colour model by looking at their "distance" to black and white (just as in the RGB colour model), we first have to consider the lightness component L* of the colours, in the same way as described for the value component in the HSV colour model. Secondly, we calculate the hue and chroma of the colours (by converting the rectangular axes a* and b* into polar coordinates)

$$h_{ab}^* = \arctan(b^*/a^*) \quad \text{(hue)} \quad \text{and} \quad C_{ab}^* = \sqrt{(a^{*2} + b^{*2})} \quad \text{(chroma)}.$$

In Fig. 7.6 the hue and chroma is shown in a graphical representation of the L*a*b* colour model. Chroma is defined as the colourfulness of an area judged as a proportion of the brightness of a similarly illuminated reference white [12]. The scales h^* and C^* together with lightness L^* correspond to perceptual colour appearance. So we can order colours with the same L^*-value according to their C^*-component: the greater the C^*-value, the lower the colour is ranked. If the colours have the same L^*-value and also the same C^*-value, then we look for the h^*-value and decide to order colours from smallest h^*-values to greatest h^*-values.

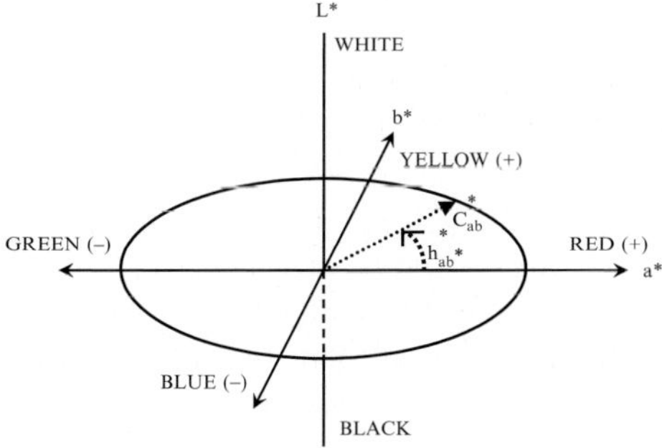

Fig. 7.6. Hue and chroma in a graphical representation of the L*a*b* colour model

Summarised, a new colour ordering $\leq_{Lab}$ in the L*a*b* colour model is defined for two colours $c(L_c^*, a_c^*, b_c^*)$ and $c'(L_{c'}^*, a_{c'}^*, b_{c'}^*)$ as:

$$c <_{Lab} c' \Leftrightarrow L_c^* < L_{c'}^*, \text{ or}$$
$$(L_c^* = L_{c'}^* \text{ and } C_c^* > C_{c'}^*) \text{ or}$$
$$(L_c^* = L_{c'}^* \text{ and } C_c^* = C_{c'}^* \text{ and } h_c^* < h_{c'}^*)$$
$$c >_{Lab} c' \Leftrightarrow c' <_{Lab} c$$
$$c =_{Lab} c' \Leftrightarrow (L_c^* = L_{c'}^* \text{ and } C_c^* = C_{c'}^* \text{ and } h_c^* = h_{c'}^*).$$

We want to notice that $(HSV, \leq_{HSV})$ and $(L*a*b*, \leq_{Lab})$ are both lattices and complete, because we work in 'finite' colour spaces (in practice only a finite number of colours can be obtained in a colour space). The greatest element in $(HSV, \leq_{HSV})$ is $\mathbf{1} = (360, 100, 1)$ and the smallest element is $\mathbf{0} = (0, 0, 0)$; the greatest element in $(L*a*b*, \leq_{Lab})$ is $\mathbf{1} = (100, 1, 1)$ and the smallest element is $\mathbf{0} = (0, 0, 0)$. Notice that the posets $(HSV, \leq_{HSV})$ and $(Lab, \leq_{Lab})$ are totally ordered sets; for every two colours c and c' in HSV and Lab hold, by definition of the order relation $\leq_{HSV}$ and $\leq_{Lab}$, that $c \leq c'$ or $c' \leq c$.

7.4.2 Associated minimum and maximum operators

Based on the vector ordering for colours introduced in the previous section in the HSV and L*a*b* colour model, we now define new minimum and maximum operators.

The HSV colour model

With the ordering defined in Sect. 7.4.1 for colour vectors $c(h_c, s_c, v_c)$ with hue component $h_c \in [0°, 360°]$, saturation component $s_c \in [0, 100]$ and value component v_c varying in $[0, 1]$ in the HSV model, we define the minimum of a given set S of n colours $c_1(h_1, s_1, v_1), \ldots, c_n(h_n, s_n, v_n)$ in HSV as the colour $c_\alpha(h_\alpha, s_\alpha, v_\alpha) \in \{c_1, c_2, \ldots, c_n\}$ satisfying:

if $(\exists! \alpha)(v_\alpha = \min(v_1, \ldots, v_n))$ then $\wedge S = c_\alpha$
else if $(\exists! \alpha)(v_\alpha = \min(v_1, \ldots, v_n)$ and $s_\alpha = \max(s_1, \ldots s_n))$ then $\wedge S = c_\alpha$
else if $(\exists! \alpha)(v_\alpha = \min(v_1, \ldots, v_n)$ and $s_\alpha = \max(s_1, \ldots s_n)$ and $h_\alpha = \min(h_1, \ldots, h_n))$ then $\wedge S = c_\alpha$,   where $1 \leq \alpha \leq n$.
We define the maximum of S as the colour $c_\alpha(h_\alpha, s_\alpha, v_\alpha) \in \{c_1, c_2, \ldots, c_n\}$ wherefore

if $(\exists! \alpha)(v_\alpha = \max(v_1, \ldots, v_n))$ then $\vee S = c_\alpha$
else if $(\exists! \alpha)(v_\alpha = \max(v_1, \ldots, v_n)$ and $s_\alpha = \min(s_1, \ldots s_n))$ then $\vee S = c_\alpha$
else if $(\exists! \alpha)(v_\alpha = \max(v_1, \ldots, v_n)$ and $s_\alpha = \min(s_1, \ldots s_n)$ and $h_\alpha = \max(h_1, \ldots, h_n))$ then $\vee S = c_\alpha$,   where $1 \leq \alpha \leq n$.

The L*a*b* colour model

In Sect. 7.4.1 we have determined an ordering of colour vectors $c_i(L_i^*, a_i^*, b_i^*)$ ($L_i^* \in [0, 100]$, a_i^* and b_i^* in $[-1, 1]$) in the L*a*b* colour model. The minimum of a set S of n colours $c_1(L_1^*, a_1^*, b_1^*), c_2(L_2^*, a_2^*, b_2^*), \ldots, c_n(L_n^*, a_n^*, b_n^*)$ in L*a*b* is the colour $c_\alpha(L_\alpha^*, a_\alpha^*, b_\alpha^*) \in \{c_1, c_2, \ldots, c_n\}$ given by

if $(\exists!\alpha)(L_\alpha^* = \min(L_1^*, \ldots, L_n^*))$ then $\wedge S = c_\alpha$
else if $(\exists!\alpha)(L_\alpha^* = \min(L_1^*, \ldots, L_n^*)$ and $h_\alpha^* = \max(h_1^*, \ldots, h_n^*))$ then $\wedge S = c_\alpha$
else if $(\exists!\alpha)(L_\alpha^* = \min(L_1^*, \ldots, L_n^*)$ and $h_\alpha^* = \max(h_1^*, \ldots, h_n^*)$ and $C_\alpha^* = \min(C_1^*, \ldots, C_n^*))$ then $\wedge S = c_\alpha,$ with $1 \le \alpha \le n.$

The maximum of S is the colour $c_\alpha(L_\alpha^*, a_\alpha^*, b_\alpha^*) \in \{c_1, c_2, \ldots, c_n\}$ with

if $(\exists!\alpha)(L_\alpha^* = \max(L_1^*, \ldots, L_n^*))$ then $\wedge S = c_\alpha$
else if $(\exists!\alpha)(L_\alpha^* = \max(L_1^*, \ldots, L_n^*)$ and $h_\alpha^* = \min(h_1^*, \ldots, h_n^*))$ then $\wedge S = c_\alpha$
else if $(\exists!\alpha)(L_\alpha^* = \max(L_1^*, \ldots, L_n^*)$ and $h_\alpha^* = \min(h_1^*, \ldots, h_n^*)$ and $C_\alpha^* = \max(C_1^*, \ldots, C_n^*))$ then $\wedge S = c_\alpha,$ with $1 \le \alpha \le n.$

7.4.3 New $(+)$, $(-)$ and $(*)$ operations between colours

Apart from a colour ordering, minimum and maximum operators, we also have to define the operations $+$ and $-$ between two colours to extend the fuzzy morphological operators to colour images.

In RGB

If $c(r_c, g_c, b_c)$ and $c'(r_{c'}, g_{c'}, b_{c'})$ are two colours in RGB, then we define the complement $co(c)$, the sum $c + c'$ and the difference $c - c'$ as:

- $(co(c))(r, g, b)$ with $r \overset{def}{=} 1 - r_c, g \overset{def}{=} 1 - g_c, b \overset{def}{=} 1 - b_c;$
- $(c + c')(r, g, b)$ with $r \overset{def}{=} (r_c + r_{c'})/2, g \overset{def}{=} (g_c + g_{c'})/2, b \overset{def}{=} (b_c + b_{c'})/2;$
- $(c - c') \overset{def}{=} c + co(c') = c + (\mathbf{1}_{RGB} - c').$

In HSV

We define the complement co of a colour $c(h_c, s_c, v_c)$ and the operations $+$ and $-$ between two colours $c(h_c, s_c, v_c)$ and $c'(h_{c'}, s_{c'}, v_{c'})$ in the HSV colour model as:

- $(co(c))(h, s, v)$ with
    1. $h \overset{def}{=} h_c, s \overset{def}{=} s_c, v \overset{def}{=} 1 - v_c$, if c is a shade of grey (thus $h_c = s_c = 0$)
    2. $h \overset{def}{=} (h_c + 180°) \bmod 360°, s \overset{def}{=} s_c, v \overset{def}{=} v_c$, otherwise;

- $(c + c')(h, s, v)$ with
  1. $h \overset{def}{=} h_c, s \overset{def}{=} s_c, v \overset{def}{=} (v_c + v_{c'})/2$, if c' is a shade of grey (analogous if c is a shade of grey)
  2. $h \overset{def}{=} (h_c + h_{c'})/2, s \overset{def}{=} (s_c + s_{c'})/2, v \overset{def}{=} (v_c + v_{c'})/2$, otherwise;
- $(c - c') \overset{def}{=} c + co(c') = c + (\mathbf{1}_{HSV} - c')$.

In L*a*b*

If $c(L_c^*, a_c^*, b_c^*)$ and $c'(L_{c'}^*, a_{c'}^*, b_{c'}^*)$ are two colours in the L*a*b* model, then we define the complement co of the colour c and the operations $+$ and $-$ between the two colours c and c' as

- $(co(c))(L^*, a^*, b^*)$ with
  1. $L^* \overset{def}{=} 1 - L_c^*, h^* \overset{def}{=} h_c^*, C^* \overset{def}{=} C_c^*$, if c is a shade of grey (thus $h_c^* = C_c^* = 0$)
  2. $L^* \overset{def}{=} L_c^*, h^* \overset{def}{=} (h_c^* + 180°) \bmod 360°, C^* \overset{def}{=} C_c^*$, otherwise;
- $(c + c')(l^*, a^*, b^*)$ with
  1. $L^* \overset{def}{=} (L_c^* + L_{c'}^*)/2, h^* \overset{def}{=} h_c^*, C^* \overset{def}{=} C_c^*$, if c' is a shade of grey (analogous if c is a shade of grey)
  2. $L^* \overset{def}{=} (L_c^* + L_{c'}^*)/2, h^* \overset{def}{=} (h_c^* + h_{c'}^*)/2, C^* \overset{def}{=} (C_c^* + C_{c'}^*)/2$, otherwise;
- $(c - c') \overset{def}{=} c + co(c') = c + (\mathbf{1}_{Lab} - c')$.

To apply the fuzzy mathematical morphological operators to a colour image in HSV or L*a*b* we also need to define an operation multiplication $*$ between two colours.

In RGB

As 3-dimensional structuring element in RGB we will usually take a symmetric greyscale structuring element because we consider all three colour components R, G and B to be equally important

$$B(:,:,1) = B(:,:,2) = B(:,:,3) = \begin{pmatrix} c_{B_1} & c_{B_{11}} & c_{B_1} \\ c_{B_{11}} & 1 & c_{B_{11}} \\ c_{B_1} & c_{B_{11}} & c_{B_1} \end{pmatrix}$$

to give a certain weight, thus a certain grade of importance, to each observed colour in the window. But we can also choose for example a structuring element of the form

$$B(:,:,1) = \begin{pmatrix} \circ & \circ & \circ \\ \circ & 1 & \circ \\ \circ & \circ & \circ \end{pmatrix}, B(:,:,2) = B(:,:,3) = \begin{pmatrix} 1 & 1 & 1 \\ 1 & 1 & 1 \\ 1 & 1 & 1 \end{pmatrix}$$

to give a weight to the R-component only, so that the G- and B-component remain unchanged. We then define the product $*$ of a colour $c(r_c, g_c, b_c)$ and a colour $c_B(r_{c_B}, g_{c_B}, b_{c_B})$ of the chosen structuring element B component-wisely as

$$(c * c_B)(r, g, b) \text{ with } r \stackrel{def}{=} r_c \cdot r_{c_B}, g \stackrel{def}{=} g_c \cdot g_{c_B}, b \stackrel{def}{=} b_c \cdot b_{c_B}.$$

In HSV

To get a 3-dimensional structuring element in the HSV colour model we can transform the structuring element B chosen in RGB into HSV. Now if we want to define a multiplication $*$ between a colour c in HSV and a colour $c_{B_{HSV}}$ of a structuring element B_{HSV} in HSV, we always have to scale the values of the H-, S- and V-component of the colour $c_{B_{HSV}}$ to the interval $[0, 1]$ (note $c^*_{B_{HSV}}$) to give a weight to the colour c. We can also choose immediately a structuring element B^*_{HSV}, for example,

$$B^*_{HSV}(:,:,1) = B^*_{HSV}(:,:,2) = \begin{pmatrix} 1\ 1\ 1 \\ 1\ 1\ 1 \\ 1\ 1\ 1 \end{pmatrix}, B^*_{HSV}(:,:,3) = \begin{pmatrix} \circ\ \bullet\ \circ \\ \bullet\ 1\ \bullet \\ \circ\ \bullet\ \circ \end{pmatrix},$$

to attach importance to the value component only. So we define the product $*$ of a colour $c(h_c, s_c, v_c)$ and a colour $c^*_{B_{HSV}}(h_{c^*_{B_{HSV}}}, s_{c^*_{B_{HSV}}}, v_{c^*_{B_{HSV}}})$ of the chosen structuring element B^*_{HSV} as

$$(c * c^*_{B_{HSV}})(h, s, v) \text{ with } h \stackrel{def}{=} h_c \cdot h_{c^*_{B_{HSV}}}, s \stackrel{def}{=} s_c \cdot s_{c^*_{B_{HSV}}}, v \stackrel{def}{=} v_c \cdot v_{c^*_{B_{HSV}}}.$$

Some t-norms on the lattice $(HSV, \leq_{HSV})$ are for all γ and δ in HSV given by

$$\mathcal{T}_{\min}(\gamma, \delta) = \min{}_{HSV}(\gamma, \delta) \text{ and } \mathcal{T}_*(\gamma, \delta) = \gamma * \delta.$$

The $\mathcal{S}$-implicators induced by $\mathcal{T}_{\min}$ (and $\mathcal{T}_*$) and the standard negator $\mathcal{N}_s$ on $(HSV, \leq_{HSV})$ are then for all γ and δ in HSV given by

$$\mathcal{I}_{\mathcal{T}_{\min}, \mathcal{N}_s}(\gamma, \delta) = \max{}_{HSV}(1 - \gamma, \delta) \text{ and } \mathcal{I}_{\mathcal{T}_*, \mathcal{N}_s}(\gamma, \delta) = 1 - (\gamma * (1 - \delta)).$$

In L*a*b*

In the L*a*b* colour model we proceed analogously as in the HSV colour model. Two t-norms on the lattice $(L^*a^*b^*, \leq_{Lab})$ are $\mathcal{T}_{\min}(\gamma, \delta) = \min{}_{Lab}(\gamma, \delta)$ and $\mathcal{T}_*(\gamma, \delta) = \gamma * \delta$ defined $\forall \gamma, \delta \in L^*a^*b^*$. The $\mathcal{S}$-implicators induced by $\mathcal{T}_{\min}$ (and $\mathcal{T}_*$) and the standard negator $\mathcal{N}_s$ on $(L^*a^*b^*, \leq_{Lab})$ are $\mathcal{I}_{\mathcal{T}_{\min}, \mathcal{N}_s}(\gamma, \delta) = \max{}_{Lab}(1 - \gamma, \delta)$ and $\mathcal{I}_{\mathcal{T}_*, \mathcal{N}_s}(\gamma, \delta) = 1 - (\gamma * (1 - \delta)), \forall \gamma, \delta \in L^*a^*b^*.$

7.4.4 New vector-based approach to colour morphology

Consider now a colour image C, modelled in the HSV or L*a*b* colour model, and a one- or three-dimensional structuring element B_{HSV} or B_{Lab}. For the extension of the greyscale morphological operators to morphological operators acting on colour images we get

1. The t-morphological operators (threshold approach):
   We calculate the maximum and minimum of the set of colours of the image C contained in an $m \times m$ window (structuring element) around a chosen central colour pixel. The t-dilation and t-erosion are the original colours of the pixels where this maximum, resp. minimum, is obtained.
2. The fuzzy morphological operators (fuzzy logic):
   Again, we have to determine the maximum and minimum of a (new) set of colours (after adding, subtracting or multiplying original colours of C with colours of the structuring element B).

7.4.5 Experimental results

Finally in our experimental results (Fig. 7.7 to Fig. 7.9) we have compared our new approach with the component-based approach. We have used different test images in our experiments (the well-known Tulips, Trees and Lena images). Since the dilation is a supremum operator, this operator will suppress dark colours and intensify light colours: objects/areas in the image that have a dark colour become smaller while objects/areas that have a light colour become larger. The erosion on the other hand is an infimum operator so that light colours are suppressed and dark colours intensified. The choice of the structuring element has of course a great influence on the result and will obviously depend on the application. As 'binary' structuring element we have used

$$B'(:,:,1) = B'(:,:,2) = B'(:,:,3) = \begin{pmatrix} 0 & 1 & 0 \\ 1 & \underline{1} & 1 \\ 0 & 1 & 0 \end{pmatrix},$$

and as greyscale structuring element we have used

$$B''_{RGB}(:,:,1) = B''_{RGB}(:,:,2) = B''_{RGB}(:,:,3) = \frac{1}{255} \begin{pmatrix} 0 & 255 & 0 \\ 255 & \underline{255} & 255 \\ 0 & 255 & 0 \end{pmatrix},$$

or

$$B^{Wh}_{RGB}(:,:,1) = B^{Wh}_{RGB}(:,:,2) = B^{Wh}_{RGB}(:,:,3) = \frac{1}{255} \begin{pmatrix} 255 & 255 & 255 \\ 255 & \underline{255} & 255 \\ 255 & 255 & 255 \end{pmatrix},$$

where the underlined element corresponds to the origin of coordinates. Notice that since in both the HSV and L*a*b* colour model we can separate intensity from chrominance($=$ hue and saturation/chroma), we will obtain the best

Fig. 7.7. T-morphological operators in HSV: at the top: the original image C, the t-dilation $D_t(C, B')$ (left) and the t-erosion $E_t(C, B')$ (right): from top to bottom: the component-based approach and our new vector-based approach

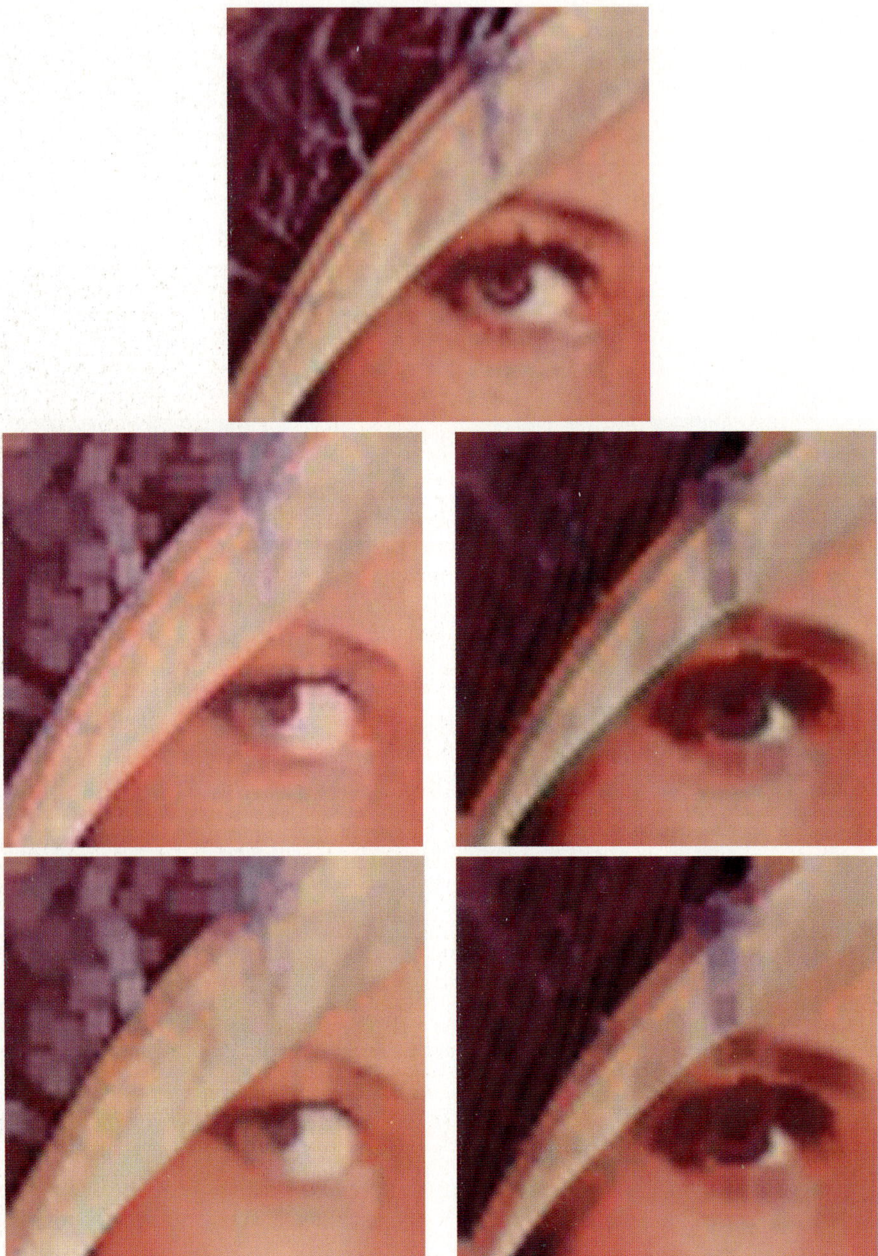

Fig. 7.8. Fuzzy morphological operators for $(\mathcal{C},\mathcal{I}) = (\mathcal{T}_{\min},\mathcal{I}_{\mathcal{T}_{\min},\mathcal{N}_s})$ in L*a*b*: at the top: original image C, left column: the dilation $D_{\mathcal{T}_{\min}}(C, B^{Wh})$ and right column: the erosion $E_{\mathcal{I}_{\mathcal{T}_{\min}},\mathcal{N}_s}(C, B^{Wh})$: from top to bottom: the component-based approach and our new approach

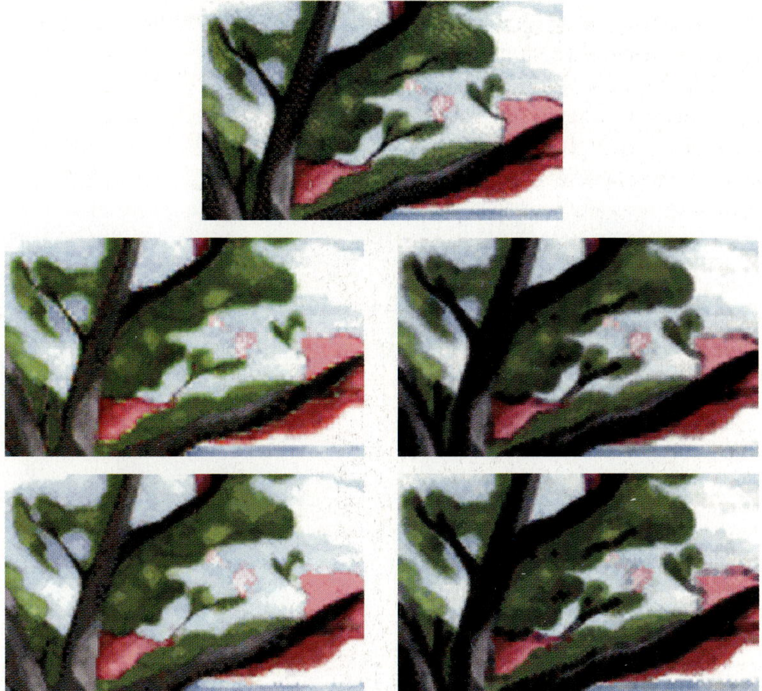

Fig. 7.9. Fuzzy morphological operators for $(\mathcal{C}, \mathcal{I}) = (\mathcal{T}_*, \mathcal{I}_{\mathcal{T}_*, \mathcal{N}_s})$ in HSV: at the top: original image C, left column: the dilation $D_{\mathcal{T}_*}(C, B'')$ and right column: the erosion $E_{\mathcal{I}_{\mathcal{T}_*, \mathcal{N}_s}}(C, B'')$: from top to bottom: the component-based approach and our new approach

results for the component-based approach by applying the greyscale morphological operators on the intensity component only. Then we add the 'new' intensity component to the original chrominance components to get again a colour image in HSV or L*a*b*.

We may conclude that our new method provides an improvement on the component-based approach of morphological operators applied to colour images. Firstly, one great advantage is that the colours are preserved and thus no new colours appear after applying the new vector t- and fuzzy morphological operators for $(\mathcal{C}, \mathcal{I}) = (\mathcal{T}_{\min}, \mathcal{I}_{\mathcal{T}_{\min}, \mathcal{N}_s})$ to colour images. Secondly, more details from the original colour image are preserved, and thus visible.

7.5 Image Denoising using Mathematical Morphology

Especially on satellite images it can happen that systematically some stripes or bands appear. One kind of satellite images where this noise appears are SAR images (Synthetic Aperture Radar images). These images are captured

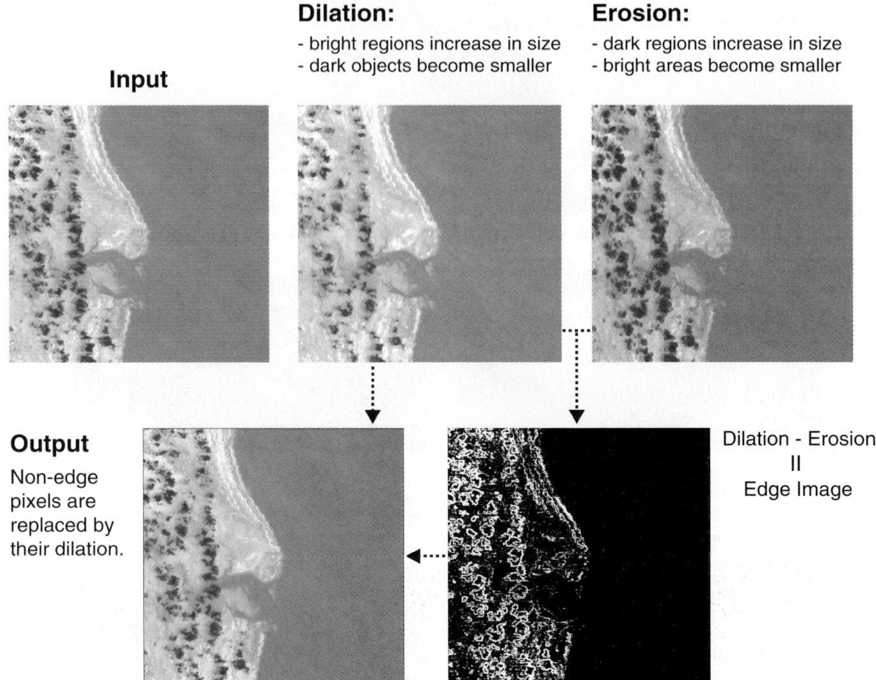

Fig. 7.10. Illustration of the proposed filtering scheme for the denoising of satellite images

by 9 detectors, so if one of these 9 does not work well we receive images typically with bands or stripes (e.g. Fig. 7.11 (a) and (c)). In this section we illustrate how MM can be used to reduce this kind of noise. As already mentioned in Sect. 7.1 and 7.3 the basic morphological operators (dilation and erosion) transfer an image into another image, using a structuring element that can be chosen by the user. The goal is to retrieve additional information from the images. One tool to extract the edge information of an image is the morphological gradient.

Definition 7.4 *Let A be a greyscale image and B a binary structuring element. The t-morphological gradient $G_t(A, B)$ is a greyscale image given by*

$$G_t(A, B) = D_t(A, B) - E_t(A, B).$$

Our destripping method uses the following greyscale morphology operators: gradient $G_t(A, B)$, dilation $D_t(A, B)$ and erosion $E_t(A, B)$. As mentioned before the effect of the dilation is that bright regions increase in size while black objects become smaller. The effect of the erosion is that dark regions increase in size while bright areas become smaller. The effect of a gradient operator is that edges become visible, i.e. the brighter a pixel the more this pixel can be seen as an important edge, and the darker a pixel the less important this pixel is for the edge detection. We use MM to reduce artefacts like bands or stripes in digital images. The idea is to replace the noisy image by the dilation, but in order to preserve the important edge structures we do not want to change the pixel if it is an edge. Therefore we use the gradient image $G_t(A, B)$ in order to control the amount of filtering. We use a transferred gradient image $\tilde{G}_t(A, B)$, i.e., we transfer the image so that all pixels can only have values in the unit interval $[0, 1]$. This transferred image can be seen as a fuzzy set *edge*, where a membership degree one (zero) indicates that the corresponding pixel is a(n) (non-)edge pixel quite sure. Our method works as follows:

1. We calculate the dilated $D_t(A, B)$ and eroded image $E_t(A, B)$ (with a typical structuring element of size three by three) .
2. Using these two images we can calculate the transferred gradient image $\tilde{G}_t(A, B)$ as shown in Fig. 7.10.

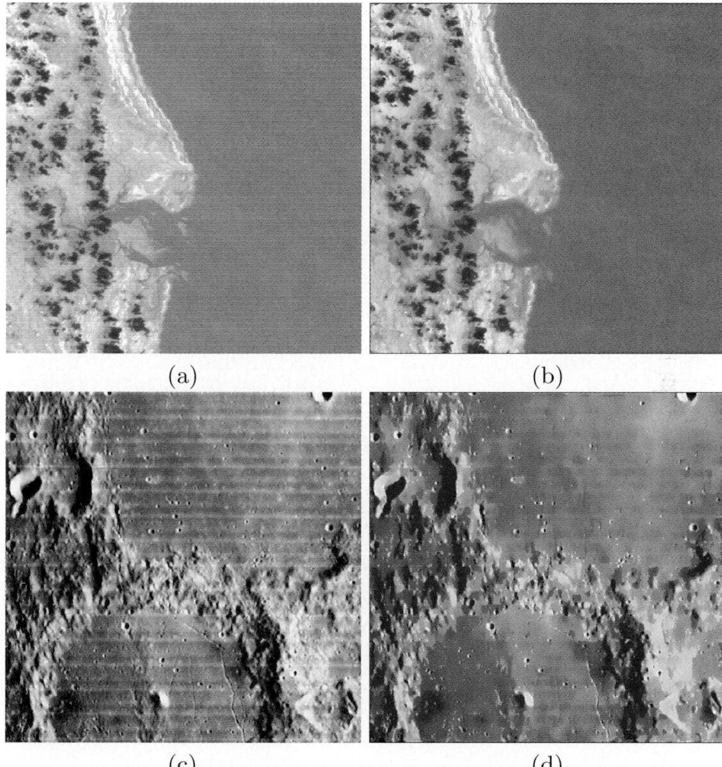

(a)   (b)

(c)   (d)

Fig. 7.11. Example of the destripping performance of the proposed method for two SAR images

3. Non-edge pixels (i.e., pixels that have a low membership degree in the fuzzy set *edge*) are finally replaced by their dilation (because the noisy stripes are black lines and will be reduced by the dilation) so that the stripes or bands disappear.

In Fig. 7.11 (b) and (d) we illustrate the performance of the proposed method for two SAR images that are corrupted with stripes. We can observe that MM can be used to eliminate such kinds of unwanted effects.

References

1. Baets M, Kerre EE, Gupta MM (1995) The fundamentals of fuzzy mathematical morphology. part 2: idempotence, convexity and decomposition. International Journal of General Systems 23:307–322
2. College of Saint Benedict - Saint John's University Color Depth and Color Spaces. http://www.csbsju.edu/itservices/teaching/c_space/colors.htm
3. De Baets B (1997) Fuzzy morphology: a logical approach. In: Ayyub A, Gupta M (eds) Uncertainty Analysis in Engineering and Sciences - Fuzzy Logic, Statistics, and Neural Network Approach pp 53-67. Kluwer Academic Press, Boston
4. De Baets B, Kerre EE, Gupta MM (1995) The fundamentals of fuzzy mathematical morphology. part 1: basic concepts. International Journal of General Systems 23:155–171
5. Heijmans HJAM (1994) Morphological Image Operators - Advances in Electronics and Electron Physics. Academic Press, London
6. Heijmans HJAM, Ronse C (1990) The algebraic basis of mathematical morphology I - dilations and erosions. Computer Vision, Graphics and Image Processing 50(3):245–295
7. Kerre EE (1998) Fuzzy Sets and Approximate Reasoning. Xian Jiaotong University Press
8. Louverdis G, Vardavoulia M, Andreadis I, Tsalides P (2002) A new approach to morphological color image processing. Pattern Recognition 35(8):1733–1741
9. Maragos P (2005) Lattice image processing: a unification of morphological and fuzzy algebraic systems. Journal of Mathematical Imaging and Vision 22:333–353
10. Nachtegael M, Kerre EE (2001) Connections between binary, gray-scale and fuzzy mathematical morphologies. Fuzzy Sets and Systems 124:73–86
11. Ronse C, Heijmans HJAM (1991) The algebraic basis of mathematical morphology II - openings and closings. Computer Vision, Graphics and Image Processing 54(1):74–97
12. Sangwine SJ, Horne REN (1998) The Colour Image Processing Handbook. Chapman & Hall, London
13. Serra J (1994) The 'centre de morphologie mathématique': an overview. In: Serra J and Soille P (eds) Mathematical Morphology and Its Applications to Image Processing pp 369-374. Kluwer Academic Press, Boston
14. Sharma G (2003) Digital Color Imaging Handbook. CRC Press, Boca Raton
15. Soille P (1999) Morphological Image Analysis: Principles and Applications. Springer-Verlag, Berlin

Morphological and Certain Fuzzy Morphological Associative Memories for Classification and Prediction

Peter Sussner and Marcos Eduardo Valle

Institute of Mathematics, Statistics, and Scientific Computing, State University of Campinas `sussner@ime.unicamp.br, mevalle@ime.unicamp.br`

Summary. Morphological associative memories (MAMs) are based on a lattice algebra known as minimax algebra. In previous papers, we gained valuable insight into the storage and recall phases of gray-scale autoassociative memories. This article extends these results to the heteroassociative and to the fuzzy case in view of the fact that a gray-scale MAM model can be converted into a fuzzy MAM model that coincides with the Lukasiewicz IFAM by applying an appropriate threshold. The article includes experimental results concerning applications of MAM and fuzzy MAM models in classification and prediction.

8.1 Introduction

A number of recent approaches to neurocomputing are either explicitly or implicitly rooted in lattice theory [6]. These approaches include fuzzy lattice neural networks [31], morphological neural networks [34, 35], and fuzzy min-max neural networks [38, 39]. The morphological associative memory (MAM) model that we discuss in this article belongs to the class of morphological neural networks (MNNs).

The mathematical foundations of MNNs can be found in mathematical morphology [18] which represents a set theoretic approach to image processing. Mathematical morphology (MM) can be conducted very generally in a complete lattice setting [36]. In mathematical morphology, an operator that commutes with the lattice operation "meet" is called erosion and an operator that commutes with the lattice operation "join" is called dilation. An erosion can be formulated in terms of an inclusion measure and a dilation can be constructed from a given erosion via a relationship of duality. A fuzzification of an inclusion measure can be used to formulate a fuzzy erosion in the complete lattice $[0,1]^n$ and a fuzzy dilation arises as the dual of fuzzy erosion. Other operators of mathematical morphology include anti-dilation, anti-erosion, opening and closing. The four morphological operators dilation,

P. Sussner and M.E. Valle: *Morphological and Certain Fuzzy Morphological Associative Memories for Classification and Prediction*, Studies in Computational Intelligence (SCI) **67**, 149–171 (2007)
`www.springerlink.com` © Springer-Verlag Berlin Heidelberg 2007

erosion, anti-dilation, and anti-erosion can be considered to be the elementary operators of mathematical morphology [4, 5].

In the context of artificial neural networks, we speak of a MNN if the first step in computing the next state of a neuron is given by one of the four elementary morphological operators. A MNN whose neurons perform operations in the fuzzy domain is called a fuzzy MNN. Applications of morphological and hybrid morphological/linear neural nets include automatic target recognition, land mine detection, handwritten character recognition, and prediction of financial markets [2, 15, 21, 22, 30].

The morphological associative memory model represents one of the first MNN models that appeared in the literature [35]. B. Raducanu, M. Graña *et al.* have applied this type of MNN to the problems of face-localization, self-localization, and hyperspectral image analysis [16, 17, 32]. Although initial research efforts have focused on the binary autoassociative case, the MAM model was proposed from the outset as a heteroassociative memory model for the storage and the recall of real-valued patterns and a number of notable features of autoassociative morphological memories (AMMs) such as optimal absolute storage capacity and one-step convergence have been shown to hold in the general case for real-valued patterns [35]. More importantly, these results remain valid for integer-valued patterns since MAMs can be applied in this setting without any roundoff errors.

In a recent paper, we presented a thorough analysis of gray-scale AMMs [43]. Specifically, we described the fixed points and the basins of attraction of real- and integer-valued AMMs. We also introduced a modified gray-scale AMM model that produces as an output a fixed point which is closest to the input pattern with respect to the Chebyshev distance. This article generalizes some of these results to include the heteroassociative case. In particular, we obtain a theorem that characterizes the output of a MAM for every input pattern. Furthermore, we show that a slightly modified version of this theorem describes the output patterns of a certain fuzzy morphological associative memory (FMAM). We construct this fuzzy model from the MAM model by applying appropriate thresholds and we point out that the new FMAM represents a special case of an implicative fuzzy associative memory (IFAM) [44, 45]. Finally, we undermine our theoretical results by applying the MAM and FMAM models to problems in classification and prediction.

8.2 Some Background Information on Lattice Theory, Mathematical Morphology, and Minimax Algebra

In contrast to traditional semi-linear neural network models, morphological neural networks perform the morphological operations of erosion or dilation at every node. Alternatively, we can describe the operations that are performed in each layer of a morphological neural network in terms of matrix products in minimax algebra. Minimax algebra is a lattice algebra which originated from problems in operations research and machine scheduling [13, 14].

Lattice theory is concerned with algebraic structures that arise by imposing some type of ordering on a set [6, 18, 36]. A partially ordered X is called a *lattice* if and only if there exists an infimum and a supremum for every finite subset of X. The infimum of $Y \subseteq X$ is denoted by the symbol $\bigwedge Y$. Alternatively, we write $\bigwedge_{j \in J} y_j$ instead of $\bigwedge Y$ if $Y = \{y_j : j \in J\}$ for some index set J. Similar notations are used to denote the supremum of Y. We speak of a *complete lattice* X if every (finite or infinite) subset has an infimum and a supremum in X. From now on, we denote complete lattices by the symbols $\mathbb{L}$ and $\mathbb{M}$.

The elementary operations of mathematical morphology are erosion, dilation, anti-dilation, and anti-erosion [5]. In the general complete lattice setting, an *erosion* is an operator $\varepsilon : \mathbb{L} \to \mathbb{M}$ that commutes with the infimum operation [18, 37]. In other words, the operator ε represents an erosion if and only if the following equality holds for every subset $Y \subseteq \mathbb{L}$:

$$\varepsilon\left(\bigwedge Y\right) = \bigwedge_{y \in Y} \varepsilon(y) \,. \tag{8.1}$$

Similarly, an operator $\delta : \mathbb{L} \to \mathbb{M}$ that commutes with the supremum operation is called a *dilation*. In other words, the operator δ represents a dilation if and only if the following equality holds for every subset $Y \subseteq \mathbb{L}$:

$$\delta\left(\bigvee Y\right) = \bigvee_{y \in Y} \delta(y) \,. \tag{8.2}$$

Apart from erosions and dilations, we will also consider the elementary operators anti-erosion and anti-dilation that are defined as follows [5, 18]. An operator $\bar{\varepsilon}$ is called an anti-erosion if and only if (8.3) holds for every $Y \subseteq \mathbb{L}$ and an operator $\bar{\delta}$ is called a anti-dilation if and only if (8.4) holds for every subset $Y \subseteq \mathbb{L}$.

$$\bar{\varepsilon}\left(\bigwedge Y\right) = \bigvee_{y \in Y} \bar{\varepsilon}(y) \,, \tag{8.3}$$

$$\bar{\delta}\left(\bigvee Y\right) = \bigwedge_{y \in Y} \bar{\delta}(y) \,. \tag{8.4}$$

Erosions, dilations, anti-erosions, and anti-dilations exemplify the concept of morphological operator, i.e., an operator that arises in the context of mathematical morphology [18]. Banon and Barrera showed that every mapping f between complete lattices $\mathbb{L}$ and $\mathbb{M}$ can be represented either as the supremum of pair-wise infimums of erosions and anti-dilations or as the infimum of pair-wise supremums of dilations and anti-erosions [5].

In minimax algebra, we define certain algebraic structures called *belts* and *blogs* ("bounded lattice ordered groups"). The set of extended real numbers

$\mathbb{R}_{\pm\infty} = \mathbb{R} \cup \{+\infty\} \cup \{-\infty\}$, exemplifies a belt as well as a blog. Specifically, we have that $(\mathbb{R}_{\pm\infty}, \vee, +)$ and $(\mathbb{R}_{\pm\infty}, \wedge, +')$ represent belts and that $(\mathbb{R}_{\pm\infty}, \vee, \wedge, +, +')$ represents a blog. The operations "$+$" and "$+'$" act like the usual sum operation and are identical on $\mathbb{R}_{\pm\infty}$ with the following exceptions:

$$(-\infty) + (+\infty) = (+\infty) + (-\infty) = -\infty \,, \tag{8.5}$$

$$(-\infty) +' (+\infty) = (+\infty) +' (-\infty) = +\infty \,. \tag{8.6}$$

If $(\mathbb{E}, \oplus, \otimes)$ and $(\mathbb{F}, \oplus', \otimes')$ are belts then a *belt homomorphism* is a function $f : \mathbb{E} \to \mathbb{F}$ that is compatible with the operations. We refer to f as a *belt isomorphism* if f is bijective. In this article, we employ the belt isomorphism of *conjugation*, denoted by a "*" symbol, between the belts $(\mathbb{R}_{\pm\infty}, \vee, +)$ and $(\mathbb{R}_{\pm\infty}, \wedge, +')$. This isomorphism is given as follows.

$$x^* = \begin{cases} -x & \text{if } x \in \mathbb{R} \,, \\ -\infty & \text{if } x = +\infty \,, \\ +\infty & \text{if } x = -\infty \,. \end{cases} \tag{8.7}$$

We say that $(\mathbb{R}_{\pm\infty}, \vee, +)$ is the conjugate of $(\mathbb{R}_{\pm\infty}, \wedge, +')$ or simply that the blog $(\mathbb{R}_{\pm\infty}, \vee, \wedge, +, +')$ is self-conjugate. Note that $(\mathbb{Z}_{\pm\infty}, \vee, \wedge, +, +')$ also represents a self-conjugate blog.

A matrix $A \in \mathbb{R}_{\pm\infty}^{m \times n}$ corresponds to a conjugate matrix $A^* \in \mathbb{R}_{\pm\infty}^{n \times m}$. Each entry $a_{ij}^* = [A^*]_{ij}$ of A^* is given by

$$a_{ij}^* = (a_{ji})^* \,. \tag{8.8}$$

Obviously, $(A^*)^* = A$ for all $A \in \mathbb{R}_{\pm\infty}^{m \times n}$, and thus the isomorphism of conjugation $\mathbb{R}_{\pm\infty}^{m \times n} \to \mathbb{R}_{\pm\infty}^{n \times m}$ is involutive. We say that a matrix $A \in \mathbb{R}_{\pm\infty}^{m \times n}$ is *finite* if every row vector and every column vector has at least one finite entry. In particular, a vector $\mathbf{x} \in \mathbb{R}_{\pm\infty}^n$ is finite if and only if $\mathbf{x} \in \mathbb{R}^n$.

The maximum and the minimum of two matrices are performed elementwise. For matrices $A, B \in \mathbb{R}_{\pm\infty}^{m \times n}$, we have,

$$(A \vee B)^* = A^* \wedge B^* \quad \text{and} \quad (A \wedge B)^* = A^* \vee B^* \,. \tag{8.9}$$

There are two types of matrix products with entries in $\mathbb{R}_{\pm\infty}$. For an $m \times p$ matrix A and a $p \times n$ matrix B with entries from $\mathbb{R}_{\pm\infty}$, the matrix $C = A \boxvee B$, also called the *max product* of A and B, and the matrix $D = A \boxwedge B$, also called the *min product* of A and B, are defined by

$$c_{ij} = \bigvee_{k=1}^{p} (a_{ik} + b_{kj}) \quad \text{and} \quad d_{ij} = \bigwedge_{k=1}^{p} (a_{ik} +' b_{kj}) \,. \tag{8.10}$$

Suppose that A is an arbitrary matrix in $\mathbb{R}_{\pm\infty}^{m \times n}$. Consider operators ε_A and δ_A such that $\varepsilon_A(\mathbf{x}) = A \boxwedge \mathbf{x}$ and $\delta_A(\mathbf{x}) = A \boxvee \mathbf{x}$. Note that ε_A and δ_A associate elements of the complete lattice $\mathbb{R}_{\pm\infty}^n$ with elements of the complete lattice

$\mathbb{R}^m_{\pm\infty}$. Clearly, we have that ε_A is an erosion and δ_A is a dilation. Morphological associative memories employ erosions and dilations of this form.

For appropriately sized matrices A and B with entries in $\mathbb{R}_{\pm\infty}$, we obtain the following equalities that will also be useful for describing MAMs.

$$(A \boxvee B)^* = B^* \boxwedge A^* \quad \text{and} \quad (A \boxwedge B)^* = B^* \boxvee A^*. \tag{8.11}$$

Note that the second halves of (8.9) and (8.11) are the duals of the first halves. As another example for this duality relationship, the reader may find a true statement of minimax algebra as well as the corresponding dual statement in (8.12) and (8.13) [13]. The matrices A, B, C are assumed to be appropriately sized.

$$A \boxvee (B \wedge C) \le (A \boxvee B) \wedge (A \boxvee C) \quad \forall\, A, B, C. \tag{8.12}$$
$$A \boxwedge (B \vee C) \ge (A \boxwedge B) \vee (A \boxwedge C) \quad \forall\, A, B, C. \tag{8.13}$$

Finally, note that (8.9) and (8.11) imply that every statement in minimax algebra induces a dual statement which simply arises by replacing each "$\wedge$" symbol with a "$\vee$" symbol and vice versa, and by reversing each inequality. Taking advantage of this fact, we only need to present primal statements on MAMs and we may omit the corresponding dual statements.

8.3 A Brief Review of Morphological Associative Memories

Morphological associative memories (MAM) were originally conceived as simple matrix memories endowed with recording recipes that are similar to correlation recording. Suppose that we want to record k vector pairs $\left(\mathbf{x}^1, \mathbf{y}^1\right), \ldots, \left(\mathbf{x}^k, \mathbf{y}^k\right)$ using a *morphological associative memory* [35]. Let X denote the matrix in $\mathbb{R}^{n \times k}$ whose column vectors are the vectors $\mathbf{x}^\xi \in \mathbb{R}^n$ and let Y denote the matrix in $\mathbb{R}^{m \times k}$ whose column vectors are the vectors $\mathbf{y}^\xi \in \mathbb{R}^m$, where $\xi = 1, \ldots, k$. For simplicity, we write $X = [\mathbf{x}^1, \ldots, \mathbf{x}^k]$ and $Y = [\mathbf{y}^1, \ldots, \mathbf{y}^k]$. The first recording scheme consists in constructing an $m \times n$ matrix W_{XY} as follows:

$$W_{XY} = Y \boxwedge X^*. \tag{8.14}$$

In other words, the entry w_{ij} of the matrix W_{XY} is given by the equation

$$w_{ij} = \bigwedge_{\xi=1}^{k} \left(y_i^\xi - x_j^\xi\right). \tag{8.15}$$

The second, dual scheme consists in constructing an $m \times n$ matrix M_{XY} of the form $M_{XY} = Y \boxvee X^*$. Note that the identity $(W_{XY})^* = M_{YX}$ can be deduced from (8.14) and (8.11).

If the matrix W_{XY} receives a vector $\mathbf{x}$ as input then the product $W_{XY} \boxdot \mathbf{x}$ is formed. Dually, if the matrix M_{XY} receives a vector $\mathbf{x}$ as input then the product $M_{XY} \boxbox \mathbf{x}$ is formed.

We speak of a binary MAM if $X \in \{0,1\}^{n \times k}$ and $Y \in \{0,1\}^{m \times k}$. In the special case that $X = Y$, we obtain the *autoassociative morphological memories* (AMMs) W_{XX} and M_{XX} [43]. If $X \neq Y$, we have a *heteroassociative morphological associative memory*.

From now on, we will focus on the MAM W_{XY}. Results concerning the dual model M_{XY} can be obtained in a similar fashion by applying the duality relationship given by (8.9) and (8.11).

Example 8.1

$$X = \begin{pmatrix} 0 & 3 & -3 & 1 \\ 8 & 6 & -4 & 5 \\ 6 & 3 & -3 & -4 \end{pmatrix}, \tag{8.16}$$

$$W_{XX} \boxdot X = X = M_{XX} \boxbox X. \tag{8.17}$$

Note that although the number of stored patterns exceeds the length of the patterns in this example, we have perfect recall for undistorted patterns. We will see in the next section that the absolute storage capacity for autoassociative morphological memories is unlimited, i.e., as many patterns as desired can be stored in an AMM with perfect recall.

Example 8.2 *Let X be as in Example 8.1 and let Y be as follows.*

$$Y = \begin{pmatrix} 3 & 3 & -1 & 1 \\ -1 & -4 & -2 & -5 \\ 0 & 1 & 4 & -1 \\ -1 & -4 & -5 & -5 \end{pmatrix}, \tag{8.18}$$

$$W_{XY} \boxdot X = \begin{pmatrix} 3 & 3 & -3 & 1 \\ -1 & -4 & -10 & -5 \\ 0 & 1 & -5 & -1 \\ -1 & -4 & -10 & -5 \end{pmatrix}. \tag{8.19}$$

Note the difference between Y and $W_{XY} \boxdot X$ and the fact that $Y \geq W_{XY} \boxdot X$. Unlike AMMs, heteroassociative morphological memories (HMMs) are not capable of storing an arbitrary number of patterns.

8.4 Fundamental Results on Gray-scale Autoassociative Morphological Memories

Autoassociative models represent an important special case for associative memories [9, 19, 47]. In this section, we review some fundamental results concerning gray-scale autoassociative morphological memories that have recently appeared in the literature [43]. We only need to formulate the results for the AMM W_{XX} since similar results for the dual model M_{XX} can be obtained by applying the relationship of duality that was discussed at the end of Sect. 8.2.

We begin by providing a powerful theorem that yields a complete characterization of the fixed points and basins of attraction of gray-scale AMMs. We say that a pattern $\mathbf{x} \in \mathbb{R}^n_{\pm\infty}$ is a *fixed point* of the AMM W_{XX} if and only if $W_{XX} \boxtimes \mathbf{x} = \mathbf{x}$. Similarly, we say that a pattern $\mathbf{x} \in \mathbb{R}^n_{\pm\infty}$ is a *fixed point* of the AMM M_{XX} if and only if $M_{XX} \boxtimes \mathbf{x} = \mathbf{x}$. We denote *the set of finite fixed points* of W_{XX} using the symbol $F(W_{XX})$ and we denote the set of finite fixed points of M_{XX} using the symbol $F(M_{XX})$.

Theorem 8.1 *For $X \in \mathbb{R}^{n \times k}$, the sets $F(W_{XX})$ and $F(M_{XX})$ coincide. If $\mathcal{F}$ denotes this set then $\mathcal{F}$ consists exactly of the following expressions:*

$$\bigvee_{i=1}^{n} \bigwedge_{\xi=1}^{k} (a_i^\xi + \mathbf{x}^\xi), \quad where \quad a_i^\xi \in \mathbb{R}. \tag{8.20}$$

Alternatively, the set $\mathcal{F}$ can be characterized as the set of all expressions of the form

$$\bigwedge_{j=1}^{r} \bigvee_{\xi=1}^{k} (c_j^\xi + \mathbf{x}^\xi), \quad where \quad c_j^\xi \in \mathbb{R} \quad and \quad r \in \mathbb{N}. \tag{8.21}$$

Moreover, given an arbitrary pattern $\mathbf{x} \in \mathbb{R}^n$, we have

$$W_{XX} \boxtimes \mathbf{x} = \hat{\mathbf{x}} \quad and \quad M_{XX} \boxtimes \mathbf{x} = \check{\mathbf{x}}, \tag{8.22}$$

where $\hat{\mathbf{x}}$ is the supremum of $\mathbf{x}$ in $\mathcal{F}$ and where $\check{\mathbf{x}}$ is the infimum of $\mathbf{x}$ in $\mathcal{F}$.

Corollary 8.1 *If $\mathcal{F}$ denotes $F(W_{XX}) = F(M_{XX})$ where $X \in \mathbb{R}^{n \times k}$ then we have*

$$\mathcal{F} = \{ \bigwedge_{j=1}^{n} \bigvee_{\xi=1}^{k} (c_j^\xi + \mathbf{x}^\xi) : c_j^\xi \in \mathbb{R} \}. \tag{8.23}$$

The proof of Theorem 8.1 involves some theorems on eigenvectors and eigenvalues in minimax algebra [43]. Another, different proof of the first statement of this theorem was independently presented in [33].

Theorem 8.1 has several important consequences, most notably the unlimited absolute storage capacity and one-step convergence of AMMs. These facts are formally expressed in the following corollaries.

Corollary 8.2 *For all $X \in \mathbb{R}^{n \times k}$, the fixed points of W_{XX} include the patterns $\mathbf{x}^1, \dots, \mathbf{x}^k$.*

Corollary 8.3 *Let $X \in \mathbb{R}^{n \times k}$. The set of finite fixed points of W_{XX} consists of all $W_{XX} \boxtimes \mathbf{x}$ such that $\mathbf{x} \in \mathbb{R}^n$. Moreover, if $\mathbf{x}$ is attracted to $\mathbf{x}^\xi$ for some $\xi \in \{1, \dots k\}$ under an application of W_{XX} then $\mathbf{x}$ is an eroded version of $\mathbf{x}^\xi$, i.e., $\mathbf{x} \leq \mathbf{x}^\xi$.*

Theorem 8.1 also induces necessary and sufficient conditions for the perfect recall of an original pattern $\mathbf{x}^\gamma$ [43]. These conditions are formulated in Theorem 8.2.

Theorem 8.2 *Let $X \in \mathbb{R}^{n \times k}$ and let $\mathbf{x} \in \mathbb{R}^n$. The equality $W_{XX} \boxtimes \mathbf{x} = \mathbf{x}^\gamma$ holds if and only if $\mathbf{x} \le \mathbf{x}^\gamma$ and there is no "linear combination" $\mathbf{l} = \bigvee_{\xi=1}^{k}(c_\xi + \mathbf{x}^\xi) \ne \mathbf{x}^\gamma$ such that $\mathbf{x} \le \mathbf{l} \le \mathbf{x}^\gamma$.*

Example 8.3 *Figure 8.1 depicts four images of size 64×64 with 256 gray levels (these images represent downsized versions of images contained in the database of the Computer Vision Group, University of Granada, Spain). For each of these image, we generated a vector $\mathbf{x}^\xi$ of length 4096. We synthesized the weight matrices W_{XX} and $M_{XX} = -W_{XX}^t$ of size 4096×4096, applied them to the original patterns $\mathbf{x}^\xi$, and we confirmed that perfect recall was achieved as we had pointed out in Corollary 8.2. We also stored the vectors $\mathbf{x}^\xi$, $\xi = 1, \ldots, 4$, using the* optimal linear associative memory *(OLAM)* [23], kernel associative memory *(KAM)* [47] the generalized BSB *model of Costantini* et al. *[12], and the* complex-valued Hopfield net *of Müezzinoğlu* et al. *[28].*

Example 8.4 *In this experiment, we probed the associative memory models under consideration with incomplete patterns which arose from leaving away substantial parts of the original images. The outcome of this experiment is visualized in Fig. 8.2. Table 8.1 lists the resulting NMSEs produced by the morphological memory W_{XX}, the OLAM, and the generalized BSB model of Costantini* et al. *for each partial image. Since the KAM model performed very poorly in this experiment, we refrained from displaying the output of the KAM in Fig. 8.2.*

We would like to clarify that we did not conduct this experiment using the complex-valued Hopfield net for the following reasons. Due to computational limitations, the complex-valued Hopfield net can only store small segments of the images. In this experiment, we may have an input segment that contains no information at all, making it impossible to recover the desired image segment.

Fig. 8.1. Original images that were used in constructing the memories W_{XX} and M_{XX}. Presenting the corresponding patterns as inputs to either one of W_{XX} or M_{XX} results in perfect recall ·

Fig. 8.2. The images in the top row represent severely incomplete versions of original face images. The following rows show - from top to bottom - the corresponding recalled patterns using the morphological memory W_{XX}, the OLAM, and the generalized BSB model

Table 8.1. NMSEs produced by AM models in applications to incomplete patterns of Fig. 8.4

	AMM W_{XX}	OLAM	KAM	Generalized BSB
Tree	0.0017	0.2302	1.0000	0
Lena	0.0137	0.4588	1.0000	0.0449
Cameraman	0.0088	0.6017	1.0000	0.1721
Church	0.0030	0.7448	1.0000	0.2332

Example 8.5 *Theorem 8.1 and the dual version of Theorem 8.2 indicate that the AMM M_{XX} exhibits tolerance with respect to dilative noise. In order to exemplify this type of noise tolerance, we added the absolute value of gaussian noise with zero mean and with variance 0.1 to the original patterns $\mathbf{x}^\xi$, $\xi = 1, \dots, 4$. We compared $\mathbf{x}^\xi$ with the patterns that were retrieved by the M_{XX} memory and the other associative memory models. Table 8.2 displays the resulting NMSEs for each model in 100 experiments for each pattern $\mathbf{x}^\xi$, $\xi = 1, \dots, 4$. Figure 8.3 provides for a visual interpretation of this simulation.*

Table 8.2. NMSEs produced by AM models in applications to patterns that were corrupted by adding the absolute value of gaussian noise with zero mean and variance 0.1

	M_{XX}	OLAM	KAM	Gen. BSB	Compl. Hopf.
Dil. Gauss.	0.0247	0.4185	0.9950	0.4538	0.5460

Fig. 8.3. The images in the top row display the original Lena image, a corrupted image that was generated by adding dilative gaussian noise, and the output of the morphological memory M_{XX}. The images in the bottom row show the corresponding recalled patterns using - from left to right - the OLAM, the generalized BSB, and the complex-valued Hopfield model

Since the KAM model performed very poorly in this experiment, we refrained from displaying the output of the KAM in Fig. 8.3.

Theorems 8.1 and 8.2 imply that the AMMs W_{XX} and M_{XX} are not suited for dealing with arbitrary noise, i.e., noise that is neither (mostly) erosive nor (mostly) dilative. To overcome these limitations of the original AMM models, we introduced modified versions of W_{XX} and M_{XX} that we denoted using the symbols $W_{XX} + \nu$ and $M_{XX} + \mu$. The AMMs $W_{XX} + \nu$ and $M_{XX} + \mu$ exhibit a much better tolerance with respect to arbitrary noise compared to W_{XX} and M_{XX} while maintaining the properties of optimal absolute storage capacity and one-step convergence. For further details, we refer the reader to a recent journal article [43].

8.4.1 Applications of gray-scale autoassociative morphological memories to classification problems

Autoassociative memory models such as the morphological models W_{XX} and M_{XX} can be applied to solve multi-class classification problems. Suppose that X^j represents the matrix that consists of all training patterns belonging to class j. For a given test pattern $\mathbf{x}$, we compute the Chebyshev distance $\zeta(\mathbf{x}, \mathbf{x}^{(j)})$ where $\mathbf{x}^{(j)}$ denotes $W_{X^j X^j} \boxtimes \mathbf{x}$. The smallest error $\zeta(\mathbf{x}, \mathbf{x}^{(j)})$ indicates the class that corresponds to $\mathbf{x}$. Obviously, the same principle of classification can be applied to other AMM models. For instance, we can employ autoassociators such as the OLAM and the KAM together with the Euclidean distance.

Example 8.6 *Let us consider the image segmentation problem that is available from the UCI Repository of Machine Learning Databases [1]. In this problem, we have 19 continuous attributes concerning a 3×3 region of a hand-segmented image. The instances were drawn randomly from a database of 7 outdoors images, namely, brickface, sky, foliage, cement, window, path, and grass. The data set contains 30 instances per class for training and 300 instances per class for testing. The data was standardized before processed.*

Table 8.3 displays the errors of classification that we obtained using the autoassociative morphological memories W_{XX} and M_{XX}, the OLAM, and the KAM model. The classifiers based on W_{XX} and M_{XX} yield the same result. Table 8.3 also includes the image segmentation results obtained by a fuzzy lattice neural network (FLNN) [31] and a support vector machine (SVM) with gaussian kernel and one-against-one method. Note that the AMMs outperformed the other classifiers except for the SVM model which succeeded in providing the correct classification for 4 more patterns than the AMM models W_{XX} and M_{XX}.

The implementation of the SVM model that we used in this experiment is available on the Internet [7]. We chose to adopt the default parameters. Finally, recall that a particular FLNN model or FLR classifier depends on the choice of a positive valuation function [3, 20]. We employed the linear positive valuation function $v(x) = (x - x_{min})/(x_{max} - x_{min})$, where $x \in [x_{min}, x_{max}]$. We also conducted experiments considering the sigmoid valuation function $v(x) = 1/(1 - \exp[-\lambda(x - x_{med})])$, where $x_{med} = (x_{max} - x_{min})/2$. The error of classification obtained for $\lambda = 1$ and $\lambda = 0.1$ were 12.14% and 12.52%, respectively.

Table 8.3. Results of the image segmentation problem

Classifier	Error Rate	Classifier	Error Rate
AMM W_{XX}	11.05%	AMM M_{XX}	11.05%
SVM	10.86%	FLNN	12.00%
KAM	82.52%	OLAM	82.95%

Table 8.4. Results of the glass classification problem

Classifier	Error Rate	Classifier	Error Rate
W_{XX}	34.8 ± 4.1	M_{XX}	34.8 ± 4.1
KAM	37.9 ± 4.2	KAA-2	37.4 ± 5.4
MLP	56.9 ± 6.1	SVM	42.2 ± 5.9

Example 8.7 *Let us consider the Glass Recognition problem, another classification problem that can be found in the UCI Repository of Machine Learning Databases [1]. The data set consists of six types of glass. Each type has 70, 17, 76, 13, 9, or 27 instances. The goal is to determine the glass type from nine attributes.*

Zhang et al. have considered this problem in a recent paper [48]. Several classifiers such as multi-layer perceptrons and support vector machines were tested using two-fold cross-validation. The data were normalized to the range $[-1, 1]$ in order to remove the scale effect, and each network was fine tuned. Table 8.4 shows the results of the experiment. The acronym KAA-2 denotes an extension of the KAM that was introduced by H. Zhang et al. The error rates concerning the KAM, the KAA-2, the MLP, and the SVM model were taken from [48].

Table 8.4 also displays the results obtained via applications of the morphological autoassociative memories. Note that the AMM based models outperformed the other classifiers including the SVM model. Moreover, an application of the AMM model does not require any fine tuning of the network.

8.5 Heteroassociative and Fuzzy MAMs

Heteroassociative morphological memories (HMMs) naturally extend the autoassociative models that we reviewed in the previous section. Heteroassociative morphological memories have proved to be useful in several applications [16, 17, 32]. Additional motivation for discussing HMMs can be drawn from the fact that the following theorems on HMMs have some important consequences for the fuzzy domain.

In contrast to binary HMMs [41], gray-scale HMMs have yet to be studied extensively. In fact, only two theorems on gray-scale HMMs concerned with conditions for perfect recall and tolerance to noise are known [35]. Unfortunately, these conditions are rather complicated and hard to understand. Therefore, we consider it timely to present new results which offer considerable insight into the functionality of HMMs. To this end, we will generalize the fundamental results on gray-scale AMMs that we presented in the previous section.

First, let us introduce a few pertinent notations. The symbol $O(W_{XY})$ denotes the set of all $W_{XY} \boxtimes \mathbf{x}$ such that $\mathbf{x} \in \mathbb{R}^n$. Similarly, the symbol $O(M_{XY})$ denotes the set of all $M_{XY} \boxtimes \mathbf{x}$ such that $\mathbf{x} \in \mathbb{R}^n$. Theorem 8.1 and

Corollary 8.3 imply that the set $O(W_{XX})$ consists of the expressions given by (8.20) or by (8.21). The first two statements of Theorem 8.3 generalize this result to include the heteroassociative case.

Theorem 8.3 *For $X \in \mathbb{R}^{n \times k}$ and $Y \in \mathbb{R}^{m \times k}$, the sets $O(W_{XY})$ and $O(M_{XY})$ coincide. If $\mathcal{O}$ denotes this set then $\mathcal{O}$ consists exactly of the following expressions:*

$$\bigvee_{i=1}^{n} \bigwedge_{\xi=1}^{k} (a_i^\xi + \mathbf{y}^\xi), \quad where \quad a_i^\xi \in \mathbb{R}. \tag{8.24}$$

Alternatively, the set $\mathcal{O}$ can be characterized as the set of the following expressions:

$$\bigwedge_{j=1}^{r} \bigvee_{\xi=1}^{k} (c_j^\xi + \mathbf{y}^\xi), \quad where \quad c_j^\xi \in \mathbb{R} \quad and \quad r \in \mathbb{N}. \tag{8.25}$$

Moreover, for arbitrary $\mathbf{x} \in \mathbb{R}^n$, the pattern $W_{XY} \boxtimes \mathbf{x}$ equals the smallest expression given by (8.24) such that $\bigvee_{i=1}^{n} \bigwedge_{\xi=1}^{k} (a_i^\xi + \mathbf{x}^\xi)$ is the supremum of $\mathbf{x}$ in $\mathcal{F}$. In addition, the pattern $W_{XY} \boxtimes \mathbf{x}$ equals the smallest expression given by (8.25) such that $\bigwedge_{j=1}^{r} \bigvee_{\xi=1}^{k} (c_j^\xi + \mathbf{x}^\xi)$ is the supremum of $\mathbf{x}$ in $\mathcal{F}$.

Proof.   Let $\mathbf{x} \in \mathbb{R}^n$ be an arbitrary input pattern. Consider the following matrix $Z \in \mathbb{R}^{p \times k}$ and the following vector $\mathbf{z} \in \mathbb{R}_{\pm\infty}^p$, where $p = n + m$.

$$Z = \begin{pmatrix} X \\ Y \end{pmatrix} \quad \text{and} \quad \mathbf{z} = \begin{pmatrix} \mathbf{x} \\ -\infty \end{pmatrix} . \tag{8.26}$$

Here $-\infty$ denotes the constant vector $(-\infty, \ldots, -\infty)^t$ of length m.

Note that $W_{ZZ} \in \mathbb{R}^{p \times p}$ can be written in block matrix form as follows.

$$W_{ZZ} = \begin{pmatrix} W_{XX} & W_{YX} \\ W_{XY} & W_{YY} \end{pmatrix} . \tag{8.27}$$

Computing the max product $W_{ZZ} \boxtimes \mathbf{z}$ yields

$$W_{ZZ} \boxtimes \mathbf{z} = \begin{pmatrix} W_{XX} \boxtimes \mathbf{x} \vee W_{YX} \boxtimes (-\infty) \\ W_{XY} \boxtimes \mathbf{x} \vee W_{YY} \boxtimes (-\infty) \end{pmatrix} = \begin{pmatrix} W_{XX} \boxtimes \mathbf{x} \\ W_{XY} \boxtimes \mathbf{x} \end{pmatrix} . \tag{8.28}$$

The resulting vector is finite since W_{XX}, W_{XY}, and $\mathbf{x}$ are finite. Moreover, the vector $W_{ZZ} \boxtimes \mathbf{z}$ represents a fixed point of W_{ZZ} because $W_{ZZ} \boxtimes W_{ZZ} = W_{ZZ}$ and because the max product is associative [14]. Identity $W_{ZZ} \boxtimes W_{ZZ} = W_{ZZ}$ follows from Theorem 11 of [43] and the fact that W_{ZZ} has a zero diagonal [40].

Let $\mathbf{f}$ denote $W_{ZZ} \boxtimes \mathbf{z} \in \mathcal{F}_Z = F(W_{ZZ}) = F(M_{ZZ})$. By Theorem 8.1, the elements of $\mathcal{F}_Z$ are described by (8.20) and (8.21) (where $\mathbf{z}^\xi$ replaces $\mathbf{x}^\xi$). Comparing these equations with (8.28) reveals that $W_{XY} \boxtimes \mathbf{x}$ can be expressed in terms of (8.24) and (8.25).

Regarding the proof of the last two statements, we observe that $\mathbf{f}$ represents the supremum of $\mathbf{f}$ in $\mathcal{F}_Z$. As mentioned before, the set $\mathcal{F}_Z$ consists of the expressions given by (8.20) and (8.21) with $\mathbf{z}^\xi$ replacing $\mathbf{x}^\xi$. Recall that $\mathbf{f}$ can also be written in the form $(W_{XX} \boxtimes \mathbf{x}, W_{XY} \boxtimes \mathbf{y})^t$. Therefore, we have

$$\mathbf{f} = \bigvee_{i=1}^{n} \bigwedge_{\xi=1}^{k} (a_i^\xi + \mathbf{z}^\xi) = \bigwedge_{j=1}^{r} \bigvee_{\xi=1}^{k} (c_j^\xi + \mathbf{z}^\xi), \qquad (8.29)$$

where $\bigvee_{i=1}^{n} \bigwedge_{\xi=1}^{k} (a_i^\xi + \mathbf{x}^\xi) = \bigwedge_{j=1}^{r} \bigvee_{\xi=1}^{k} (c_j^\xi + \mathbf{x}^\xi)$ is the supremum of $\mathbf{x}$ in $\mathcal{F} = F(W_{XX}) = F(M_{XX})$. Therefore, $W_{XY} \boxtimes \mathbf{x}$ is as stated in the theorem.

Theorem 8.4 *If $\mathcal{O}$ denotes $O(W_{XY}) = O(M_{XY})$ where $X \in \mathbb{R}^{n \times k}$ and $Y \in \mathbb{R}^{m \times k}$ then we have*

$$\mathcal{O} = \{ \bigwedge_{j=1}^{n} \bigvee_{\xi=1}^{k} (c_j^\xi + \mathbf{y}^\xi) : c_j^\xi \in \mathbb{R} \}. \qquad (8.30)$$

For arbitrary $\mathbf{x} \in \mathbb{R}^n$, the pattern $W_{XY} \boxtimes \mathbf{x}$ equals the smallest expression given by (8.30) such that $\bigwedge_{j=1}^{n} \bigvee_{\xi=1}^{k} (c_j^\xi + \mathbf{x}^\xi)$ is the supremum of $\mathbf{x}$ in $\mathcal{F}$.

Proof.   The proof of this theorem employs Corollary 8.1 and resembles the proof of Theorem 8.3.

Corollary 8.4 *For $X \in \mathbb{R}^{n \times k}$, $Y \in \mathbb{R}^{m \times k}$, and let $\mathbf{x} \in \mathbb{R}^n$ be such that $W_{XX} \boxtimes \mathbf{x} = \mathbf{x}^\gamma$. We have $W_{XY} \boxtimes \mathbf{x} = \mathbf{y}^\gamma$ if and only if the following implication holds for all $c^\xi \in \mathbb{R}$ where $\xi = 1, \ldots, k$.*

$$\mathbf{x}^\gamma \leq \bigvee_{\xi=1}^{k} c^\xi + \mathbf{x}^\xi \Rightarrow \mathbf{y}^\gamma \leq \bigvee_{\xi=1}^{k} c^\xi + \mathbf{y}^\xi. \qquad (8.31)$$

Proof.   Let us again consider the matrix Z and the vector $\mathbf{z}$ represented in (8.26). Let $\mathbf{x}$ be as stated above.

Suppose that $W_{XY} \boxtimes \mathbf{x} = \mathbf{y}^\gamma$ which – in view of (8.28) – implies that $W_{ZZ} \boxtimes \mathbf{z} = \mathbf{z}^\gamma$. In other words, we have $\hat{\mathbf{z}} = \mathbf{z}^\gamma$, where $\hat{\mathbf{z}}$ denotes the supremum of $\mathbf{z}$ in $\mathcal{F}_Z$. By Corollary 8.2, $\mathcal{F}_Z$ consists of all patterns $\bigwedge_{j=1}^{n} \bigvee_{\xi=1}^{k} (c_j^\xi + \mathbf{z}^\xi)$ such that $c_j^\xi \in \mathbb{R}$.

Consider arbitrary scalars c^ξ, where $\xi = 1, \ldots, k$, such that the left hand side of (8.31) holds. If $\mathbf{u}$ denotes $\bigvee_{\xi=1}^{k} (c^\xi + \mathbf{z}^\xi)$ then $\mathbf{u} \in \mathcal{F}_Z$ represents an upper bound of $\mathbf{z} = (\mathbf{x}, -\infty)^t$. We obtain the following relationships.

$$\begin{pmatrix} \mathbf{x}^\gamma \\ \mathbf{y}^\gamma \end{pmatrix} = \mathbf{z}^\gamma = W_{ZZ} \boxtimes \mathbf{z} \leq W_{ZZ} \boxtimes \mathbf{u} = \mathbf{u} = \bigvee_{\xi=1}^{k} \left[c^\xi + \begin{pmatrix} \mathbf{x}^\xi \\ \mathbf{y}^\xi \end{pmatrix} \right]. \qquad (8.32)$$

In particular, we recognize that $\mathbf{y}^\gamma \leq \bigvee_{\xi=1}^{k} (c^\xi + \mathbf{y}^\xi)$ which shows that the (8.31) is a necessary condition for $W_{XY} \boxtimes \mathbf{x} = \mathbf{y}^\gamma$.

Let us now prove that (8.31) implies $W_{XY} \boxdot \mathbf{x} = \mathbf{y}^\gamma$. Let $\bigwedge_{j=1}^{r} \bigvee_{\xi=1}^{k} (c_j^\xi + \mathbf{y}^\xi)$ denote the smallest pattern such that $\bigwedge_{j=1}^{r} \bigvee_{\xi=1}^{k} (c_j^\xi + \mathbf{x}^\xi) = \hat{\mathbf{x}}$, the supremum of $\mathbf{x}$ in $\mathcal{F} = F(W_{XX})$.

By Theorem 8.3, we have $W_{XY} \boxdot \mathbf{x} = \bigwedge_{j=1}^{r} \bigvee_{\xi=1}^{k} (c_j^\xi + \mathbf{y}^\xi)$. Theorem 8.1 states that $\hat{\mathbf{x}} = W_{XX} \boxdot \mathbf{x}$ which equals $\mathbf{x}^\gamma$ by assumption. Thus, we obtain the following identity

$$\bigwedge_{j=1}^{r} \bigvee_{\xi=1}^{k} (c_j^\xi + \mathbf{x}^\xi) = \mathbf{x}^\gamma . \tag{8.33}$$

Equation (8.33) implies that the patterns $\bigvee_{\xi=1}^{k}(c_j^\xi + \mathbf{x}^\xi)$ are bounded from below by $\mathbf{x}^\gamma$ for all $j = 1, \dots, n$. From (8.31) we infer that

$$\bigvee_{\xi=1}^{k} (c_j^\xi + \mathbf{y}^\xi) \geq \mathbf{y}^\gamma \quad \forall j = 1, \dots, r . \tag{8.34}$$

$$\Leftrightarrow \bigwedge_{j=1}^{r} \bigvee_{\xi=1}^{k} (c_j^\xi + \mathbf{y}^\xi) \geq \mathbf{y}^\gamma \tag{8.35}$$

$$\Leftrightarrow \bigwedge_{j=1}^{r} \bigvee_{\xi=1}^{k} (c_j^\xi + \mathbf{y}^\xi) \wedge \mathbf{y}^\gamma = \mathbf{y}^\gamma . \tag{8.36}$$

Obviously, we have $\bigwedge_{j=1}^{r} \bigvee_{\xi=1}^{k}(c_j^\xi + \mathbf{y}^\xi) \wedge \mathbf{y}^\gamma \leq \bigwedge_{j=1}^{r} \bigvee_{\xi=1}^{k}(c_j^\xi + \mathbf{y}^\xi)$ and $\bigwedge_{j=1}^{r} \bigvee_{\xi=1}^{k}(c_j^\xi + \mathbf{x}^\xi) \wedge \mathbf{x}^\gamma = \hat{\mathbf{x}} \wedge \mathbf{x}^\gamma = \hat{\mathbf{x}}$ since $\hat{\mathbf{x}} = \bigwedge_{j=1}^{r} \bigvee_{\xi=1}^{k}(c_j^\xi + \mathbf{x}^\xi) = W_{XX} \boxdot \mathbf{x} = \mathbf{x}^\gamma$. Therefore, $\bigwedge_{j=1}^{r} \bigvee_{\xi=1}^{k}(c_j^\xi + \mathbf{y}^\xi) \wedge \mathbf{y}^\gamma$ and $\bigwedge_{j=1}^{r} \bigvee_{\xi=1}^{k}(c_j^\xi + \mathbf{y}^\xi)$ coincide since the former is bounded from above by the latter pattern and since the latter pattern $\bigwedge_{j=1}^{r} \bigvee_{\xi=1}^{k}(c_j^\xi + \mathbf{y}^\xi)$ represents the smallest pattern such that $\bigwedge_{j=1}^{r} \bigvee_{\xi=1}^{k}(c_j^\xi + \mathbf{x}^\xi) = \hat{\mathbf{x}}$. In view of (8.36), we are able to finish the proof of the theorem as follows:

$$W_{XY} \boxdot \mathbf{x} = \bigwedge_{j=1}^{r} \bigvee_{\xi=1}^{k} (c_j^\xi + \mathbf{y}^\xi) = \bigwedge_{j=1}^{r} \bigvee_{\xi=1}^{k} (c_j^\xi + \mathbf{y}^\xi) \wedge \mathbf{y}^\gamma = \mathbf{y}^\gamma . \tag{8.37}$$

Theorem 8.3 and Corollary 8.4 completely characterize the output of HMMs for any input pattern $\mathbf{x}$. The rest of the article is concerned with the implications of these results with respect to fuzzy morphological associative memory (FMAM) models.

Note that the fuzzy domain $[0,1]^n$ represents a complete lattice. An erosion $\varepsilon : [0,1]^n \to [0,1]^m$ is called a *fuzzy erosion*. In a similar vein, we speak of *fuzzy dilations, fuzzy anti-dilations,* and *fuzzy anti-erosions.* In *fuzzy mathematical morphology,* a fuzzy erosion can be defined in terms of a fuzzy inclusion measure and a fuzzy dilation can be defined in terms of a fuzzy intersection measure [29, 42]. We consider fuzzy morphological neural networks to be models of artificial neural networks that calculate an elementary operation of mathematical morphology such as fuzzy erosion or fuzzy dilation at each node. FMAMs belong to this class of models.

Let us consider a particular FMAM model. Let $X \in [0,1]^{n \times k}$, $Y \in [0,1]^{m \times k}$, and let W_{XY} be defined as before. The symbols $\mathbf{0}_{m \times n}$ and $\mathbf{0}_m$ stand for the zero matrix of size $m \times n$ and the zero vector of length m. If W denotes the $m \times n$ matrix $W_{XY} \wedge \mathbf{0}_{m \times n}$ then the *Lukasiewicz FMAM* is defined in terms of the following relationship between an input pattern $\mathbf{x} \in [0,1]^n$ and an output pattern $\mathbf{y} \in [0,1]^m$.

$$\mathbf{y} = (W \boxdot \mathbf{x}) \vee \mathbf{0}_m . \tag{8.38}$$

The Lukasiewicz FMAM performs a fuzzy dilation $[0,1]^m \to [0,1]$ at every node. This fuzzy dilation can also be expressed in terms of a supremum of fuzzy Lukasiewicz conjunctions, hence the name Lukasiewicz FMAM. Furthermore, it can be shown that (8.38) yields the Lukasiewicz IFAM model [44, 45].

Recall that $W_{XY} = \bigvee_{\xi=1}^{k} [\mathbf{y}^\xi \boxtimes (\mathbf{x}^\xi)^*]$. Hence, the probability that $W_{XY} \in [-1,0]^{m \times n}$ increases as more and more fundamental memories $(\mathbf{x}^\xi, \mathbf{y}^\xi)$ are stored in the network. If $W_{XY} \in [-1,0]^{m \times n}$ then W_{XY} equals W, the weight matrix of the Lukasiewicz FMAM, and we may apply Theorem 8.3 to Lukasiewicz FMAMs which leads to the following corollary.

Corollary 8.5 *Let $X \in [0,1]^{n \times k}$ and $Y \in [0,1]^{m \times k}$ such that $W_{XY} \in [-1,0]^{m \times n}$. For an arbitrary input pattern $\mathbf{x} \in [0,1]^n$, an application of the Lukasiewicz FMAM produces the maximum of $\mathbf{0}_m$ and the smallest expression given by (8.24) such that $\bigvee_{i=1}^{n} \bigwedge_{\xi=1}^{k} (a_i^\xi + \mathbf{x}^\xi)$ is the supremum of $\mathbf{x}$ in $\mathcal{F}$.*

Corollary 8.6 *Let $X \in [0,1]^{n \times k}$ and $Y \in [0,1]^{m \times k}$ be such that $W_{XY} \in [-1,0]^{m \times n}$. Suppose that $\mathbf{x} \in [0,1]^n$ satisfies $W_{XX} \boxdot \mathbf{x} = \mathbf{x}^\gamma$. We have $(W \boxdot \mathbf{x}) \vee \mathbf{0}_m = \mathbf{y}^\gamma$ if the following implication holds for all $c^\xi \in \mathbb{R}$ where $\xi = 1, \ldots, k$.*

$$\mathbf{x}^\gamma \leq \bigvee_{\xi=1}^{k} c^\xi + \mathbf{x}^\xi \Rightarrow \mathbf{y}^\gamma \leq \bigvee_{\xi=1}^{k} c^\xi + \mathbf{y}^\xi . \tag{8.39}$$

Proof.   Let W_{XY} and $\mathbf{x} \in [0,1]^n$ be as stated above. By Corollary 8.4, the assumption that (8.31) holds for all $c^\xi \in \mathbb{R}$, where $\xi = 1, \ldots, k$, implies that $W_{XY} \boxdot \mathbf{x} = \mathbf{y}^\gamma$ which belongs to $[0,1]^m$. Therefore, we have $(W_{XY} \boxdot \mathbf{x}) \vee \mathbf{0}_m = \mathbf{y}^\gamma$. Finally, note that W_{XY} can be replaced by W since $W_{XY} \in [-1,0]^{m \times n}$.

8.5.1 Applications of FMAMs in prediction

Fuzzy associative memories such as the FMAM can be used to implement mappings of fuzzy rules. In this case, a set of rules in the form of human-like

IF-THEN conditional statements are stored. In this subsection, we present two applications of the FMAM model to the problem of forecasting time-series.

Example 8.8 *Let us consider the problem presented in [10] and discussed latter in [44, 45]. This problem consists of assessing the manpower requirement in steel manufacturing industry in the state of West Bengal, India. Initially, we have five linguistic values representing concepts such as "the requirement in manpower is* large*". A set of fuzzy conditional statements such as "If the manpower requirement of year n is* large*, then that of year n+1 is very* large*" is obtained from the past values. We converted these conditional statements into the set of input-output pairs that are represented in Table 8.5 and we generated the following matrix* W_{XY}.

$$W_{XY} = \begin{bmatrix} -0.5 & -1.0 & -1.0 & -1.0 & -1.0 \\ 0 & -0.5 & -1.0 & -1.0 & -1.0 \\ -0.5 & -0.5 & -0.5 & -1.0 & -1.0 \\ -1.0 & -1.0 & -1.0 & -.5 & -.5 \\ -1.0 & -1.0 & -1.0 & -0.5 & 0 \end{bmatrix}. \tag{8.40}$$

Since $W_{XY} \in [-1,0]^{5\times5}$, *the matrices* W *and* W_{XY} *coincide. Therefore, Corollaries 8.5 and 8.6 can be applied. For example, if* $\mathbf{x} = [1, 0.6, 0, 0, 0]^T$ *then the supremum of* $\mathbf{x}$ *in* $\mathcal{F}$ *is* $W_{XX} \boxtimes \mathbf{x} = [1, 0.6, 0.1, 0, 0]^T = \mathbf{x}^1 \vee [(0.1+\mathbf{x}^1)\wedge\mathbf{x}^2]$. *Note that the latter expression represents a meet of joins that can be easily brought into the form* $\bigvee_{i=1}^{5} \bigwedge_{\xi=1}^{5}(a_i^{\xi} + \mathbf{x}^{\xi})$ *by adding some coefficients* $a_i^{\xi} \geq 1$. *By Corollary 8.5, we have* $(W_{XY} \boxtimes \mathbf{x}) \vee \mathbf{0}_5 \leq \mathbf{y}^1 \vee [(0.1 + \mathbf{y}^1) \wedge \mathbf{y}^2] \vee \mathbf{0}_5 = [0.5, 1, 0.5, 0, 0]^T$. *In fact, we calculate* $(W_{XY} \boxtimes \mathbf{x}) \vee \mathbf{0}_5 = [0.5, 1, 0.5, 0, 0]^T$.

Now, let us consider Corollary 8.6. On one hand, Corollary 8.2 implies that $W_{XX} \boxtimes \mathbf{x}^2 = \mathbf{x}^2$. *On the other hand, an application of the Lukasiewicz FMAM to* $\mathbf{x}^2$ *yields* $[0, 0.5, 0.5, 0, 0]^T \neq \mathbf{y}2$. *Therefore, there are some* c^{ξ} *such that the implication in (8.39) does not hold. We have, for example, that* $\mathbf{x}^2 \leq \mathbf{x}^3$ *whereas* $\mathbf{y}^2 \nleq \mathbf{y}^3$.

Table 8.5. Set of input and output pairs used in the forecasting application

ξ	$\mathbf{x}^{\xi}$	$\mathbf{y}^{\xi}$
1	$[1.0, 0.5, 0, 0, 0]^T$	$[0.5, 1.0, 0.5, 0, 0]^T$
2	$[0.5, 1.0, 0.5, 0, 0]^T$	$[0.5, 1.0, 0.5, 0, 0]^T$
3	$[0.5, 1.0, 0.5, 0, 0]^T$	$[0, 0.5, 1.0, 0.5, 0]^T$
4	$[0, 0.5, 1.0, 0.5, 0]^T$	$[0.5, 1.0, 0.5, 0, 0]^T$
5	$[0, 0.5, 1.0, 0.5, 0]^T$	$[0, 0.5, 1.0, 0.5, 0]^T$
6	$[0, 0.5, 1.0, 0.5, 0]^T$	$[0, 0, 0.5, 1.0, 0.5]^T$
7	$[0, 0, 0.5, 1.0, 0.5]^T$	$[0, 0, 0.5, 1.0, 0.5]^T$
8	$[0, 0, 0.5, 1.0, 0.5]^T$	$[0, 0, 0, 0.5, 1.0]^T$
9	$[0, 0, 0, 0.5, 1.0]^T$	$[0, 0, 0, 0.5, 1.0]^T$

Table 8.6. Average errors in forecasting manpower

Method	Average Error
Lukasiewicz FMAM	2.29%
Kosko's FAM	2.67%
Lukasiewicz GFAM	2.67%
Gödel IFAM	2.73%
Max-min FAM with threshold	2.73%
Goguen IFAM	2.99%
AM of Wang and Lu	2.99%
ARIMA2	5.48%
ARIMA1	9.79%

If $\mathbf{x}$ *represents a fuzzy set corresponding to the manpower of year n then (8.38) can be used to forecast the manpower of year $n + 1$ by means of the Lukasiewicz FMAM. Specifically, a defuzzification of the output pattern* $\mathbf{y}$ *according to the rule described in [10] yields the prediction for year $n + 1$. Table 8.6 displays the average errors in the predictions that were obtained by means of the Lukasiewicz FMAM and several other methods that can be found in the literature [11, 24, 25, 44, 45, 46]. Figure 8.4 plots the manpower data of the years 1984 through 1995. The actual values are compared to the predictions obtained by some of these methods. Note that the Lukasiewicz FMAM outperformed the other models.*

Example 8.9 *In this example, we applied the Lukasiewicz FMAM to the problem of forecasting the average monthly streamflow of a large hydroelectric plant called Furnas, that is located in southeastern Brazil. This problem was previously discussed in [26, 27].*

Note that the seasonality of the monthly streamflow suggests the use of 12 different models, one for each month of the year. Let s_ξ, for $\xi = 1, \ldots, q$, be samples of a seasonal streamflow time series. The goal is to estimate the value of s_γ from a subsequence of $(s_1, s_2, \ldots, s_{\gamma-1})$. Here, we employ subsequences that correspond to a vector of the form

$$\mathbf{p}^\gamma = (s_{\gamma-h}, \ldots, s_{\gamma-1})^T , \tag{8.41}$$

where $h \in \{1, 2, \ldots, \gamma - 1\}$. In this experiment, our FMAM based model only uses a fixed number of three antecedents. For example, the values of January, February, and March were considered for predicting the streamflow of April.

The uncertainty that is inherent in hydrological data suggests the use of fuzzy sets to model the streamflow samples. For $\xi < \gamma$, a fuzzification of $\mathbf{p}^\xi$ and s^ξ using Gaussian membership functions yields fuzzy sets $\mathbf{x}^\xi : \mathcal{U} \to [0, 1]$ and $\mathbf{y}^\xi : \mathcal{V} \to [0, 1]$ respectively, where $\mathcal{U}$ and $\mathcal{V}$ represent finite universes of discourse. A subset S of the resulting input-output pairs $\{(\mathbf{x}^\xi, \mathbf{y}^\xi), \xi < q\}$ is implicitly stored in the Lukasiewicz FMAM (we only construct the parts of the

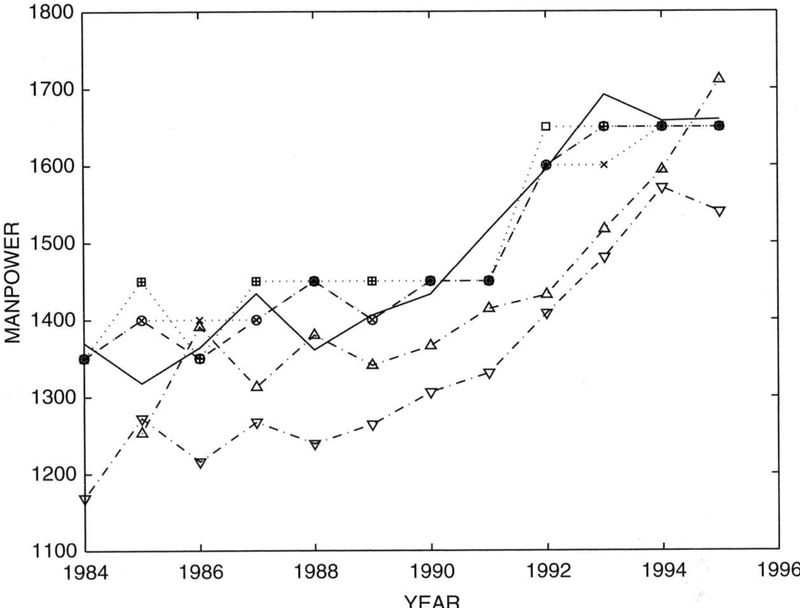

Fig. 8.4. Predictions of manpower. The continuous line represents the actual manpower. The dashed line marked by 'o' corresponds to the Lukasiewicz FMAM model, the dotted line marked by '×' corresponds to Kosko's FAM model and the Lukasiewicz Generalized FAM, the dotted line marked by '+' corresponds to Max-min FAM with threshold and Gödel IFAM, and the dotted line marked by '□' corresponds to the Associative Memory model of Wang and Lu and the Goguen IFAM. The lines marked by '△' and '▽' represent ARIMA1 and ARIMA2

weight matrix that are actually used in the recall phase). We employed the subtractive clustering method to determine the set S [8]. Feeding the pattern $\mathbf{x}^\gamma$ into the FMAM model, we retrieved the corresponding output pattern $\mathbf{y}^\gamma$. For computational reasons, $\mathbf{x}^\gamma$ is modeled as a discrete Dirac-δ (impulse) function. A defuzzification of $\mathbf{y}^\gamma$ using the mean of maximum yields s_γ.

Figure 8.5 shows the forecasted streamflows estimated by the prediction model based on the FMAM for the Furnas reservoir from 1991 to 1998. Table 8.7 compares the errors that were generated by the FMAM model and several other models [26, 27]. In contrast to the FMAM-based model, the MLP, NFN, and FPM-PRP models were initialized by optimizing the number of the parameters for each monthly prediction. For example, the MLP considers 4 antecedents to predict the streamflow of January and 3 antecedents to predict the streamflow for February. Moreover, the FPM-PRP model also takes into account slope information which requires some additional "fine tuning". We experimentally determined a variable number of parameters (including slopes) for the FMAM model such that MSE $= 0.88 \times 10^5$, MAE $= 157$, *and* MPE $= 15$.

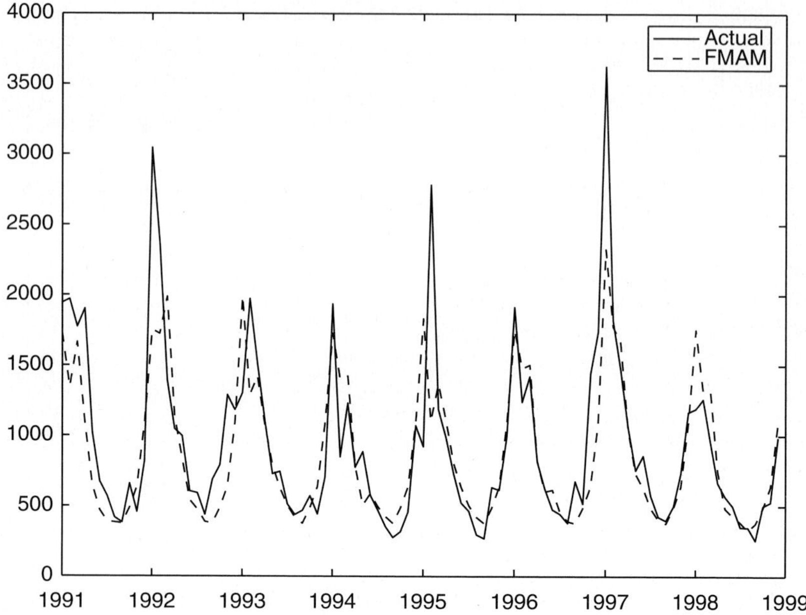

Fig. 8.5. The streamflow prediction for the Furnas reservoir from 1991 to 1998. The continuous line corresponds to the actual values and the dashed line corresponds to the predicted values

Table 8.7. Mean square, mean absolute, and mean relative percentage errors produced by the prediction models

Methods	MSE ($\times 10^5$)	MAE (m^3/s)	MPE (%)
FMAM	1.42	226	22
PARMA	1.85	280	28
MLP	1.82	271	30
NFN	1.73	234	20
FPM-PRP	1.20	200	18

References

1. UCI Repository of Machine Learning Databases. University of California, Irvine, Dept. of Information and Computer Sciences. Available at: http://www.ics.uci.edu/~mlearn/MLRepository.html
2. Araújo RA, Madeiro F, Sousa RP, Pessoa LFC, Ferreira TAE (2006) An evolutionary morphological approach for financial time series forecasting. In: Proc IEEE Congress on Evolutionary Computation pp 2467–2474
3. Athanasiadis I, Kaburlasos VG (2006) Air quality assessment using fuzzy lattice reasoning (FLR). In: Proc IEEE International Conference on Fuzzy Systems pp 231–236

4. Banon GJF, Barrera J (1991) Minimal representation for translation invariant set mappings by mathematical morphology. SIAM J Appl Math 51(6): 1782–1798
5. Banon GJF, Barrera J (1993) Decomposition of mappings between complete lattices by mathematical morphology: part I - general lattices. Signal Processing 30:299–327
6. Birkhoff G (1993) Lattice Theory. American Mathematical Society, 3 ed.
7. Canu S, Grandvalet Y, Guigue V, Rakotomamonjy A (2005) SVM and Kernel Methods Matlab Toolbox. Perception Systèmes et Information, INSA de Rouen, Rouen, France
8. Chiu S (1994) Fuzzy model identification based on cluster estimation. Journal of Intelligent and Fuzzy Systems 2(3)
9. Chiueh T, Goodman R (1991) Recurrent correlation associative memories. IEEE Trans on Neural Networks 2(2):275–284
10. Choudhury J, Sarkar B, Mukherjee S (2002) Forecasting of engineering manpower through fuzzy associative memory neural network with arima: a comparative study. Neurocomputing 47(1-2):241–257
11. Chung F, Lee T (1996) On fuzzy associative memory with multiple-rule storage capacity. IEEE Transactions on Fuzzy Systems 4(3):375–384
12. Costantini G, Casali D, Perfetti R (2003). Neural associative memory storing gray-coded gray-scale images. IEEE Transactions on Neural Networks 14(3):703–707
13. Cuninghame-Green R (1979) Minimax Algebra. Lecture Notes in Economics and Mathematical Systems 166. Springer-Verlag, New York
14. Cuninghame-Green R (1995) Minimax algebra and applications. In: Hawkes P (ed) Advances in Imaging and Electron Physics 90:1–121. Academic Press, New York
15. Gader PD, Khabou M, Koldobsky A (2000) Morphological regularization neural networks. Pattern Recognition (special issue on Mathematical Morphology and Its Applications) 33(6):935–945
16. Graña M, Gallego J, Torrealdea FJ, D'Anjou A (2003a) On the application of associative morphological memories to hyperspectral image analysis. Lecture Notes in Computer Science 2687:567–574
17. Graña M, Sussner P, Ritter G (2003b) Innovative applications of associative morphological memories. Mathware and Softcomputing 10(3):155–168
18. Heijmans H (1994) Morphological Image Operators. Academic Press, New York
19. Hopfield JJ (1982) Neural networks and physical systems with emergent collective computational abilities. In: Proc National Academy of Sciences 79:2554–2558
20. Kaburlasos V, Petridis V (2000) Fuzzy lattice neurocomputing (FLN) models. Neural Networks 13(10):1145–1170
21. Khabou M, Gader P (2000) Automatic target detection using entropy optimized shared-weight neural networks. IEEE Transactions on Neural Networks 11(1):186–193
22. Khabou M, Gader PD, Keller JM (2000) LADAR target detection using morphological shared-weight neural networks. Machine Vision and Applications 11(6):300–305
23. Kohonen T (1984) Self-Organization and Associative Memory. Springer-Verlag
24. Kosko B (1992) Neural Networks and Fuzzy Systems: A Dynamical Systems Approach to Machine Intelligence. Prentice Hall, Englewood Cliffs, New Jersey

25. Liu P (1999) The fuzzy associative memory of max-min fuzzy neural networks with threshold. Fuzzy Sets and Systems 107(2):147–157
26. Magalhães M (2004) Redes neurais, metodologias de agrupamento e combinação de previsores aplicados a previsão de vazões naturais. Master's Thesis, State University of Campinas
27. Magalhães M, Ballini R, Gonçalves R, Gomide F (2004) Predictive fuzzy clustering model for natural streamflow forecasting. In: Proc IEEE International Conference on Fuzzy Systems pp 390–394
28. Müezzinoğlu M, Güzeliş C, Zurada J (2003) A new design method for the complex-valued multistate Hopfield associative memory. IEEE Trans on Neural Networks 14(4):891–899
29. Nachtegael M, Kerre E (2001) Connections between binary, gray-scale and fuzzy mathematical morphologies. Fuzzy Sets and Systems 124(1):73–85
30. Pessoa L, Maragos P (2000) Neural networks with hybrid morphological/rank/linear nodes: a unifying framework with applications to handwritten character recognition. Pattern Recognition 33(6):945–960
31. Petridis V, Kaburlasos V (1998) Fuzzy lattice neural network (FLNN): a hybrid model for learning. IEEE Transactions on Neural Networks 9(5):877–890
32. Raducanu B, Graña M, Albizuri XF (2003) Morphological scale spaces and associative morphological memories: results on robustness and practical applications. Journal of Mathematical Imaging and Vision 19(2):113–131
33. Ritter GX, Gader PD (2006) Fixed points of lattice transforms and lattice associative memories. In: Hawkes P (ed) Advances in Imaging and Electron Physics 144:165–242. Elsevier, Amsterdam, The Netherlands
34. Ritter GX, Sussner P (1996) An introduction to morphological neural networks. In: Proc 13th International Conference on Pattern Recognition pp 709–717
35. Ritter GX, Sussner P, de Leon JLD (1998) Morphological associative memories. IEEE Transactions on Neural Networks 9(2):281–293
36. Ronse C (1990) Why mathematical morphology needs complete lattices. Signal Processing 21(2):129–154
37. Serra J (1988) Image Analysis and Mathematical Morphology, vol. 2: Theoretical Advances. Academic Press, New York
38. Simpson P (1992) Fuzzy min-max neural networks – part 1: classification. IEEE Transactions on Neural Networks 3(5):776–786
39. Simpson P (1993) Fuzzy min-max neural networks – part 2: clustering. IEEE Transactions on Neural Networks 1(1):32–35
40. Sussner P (2000) Fixed points of autoassociative morphological memories. In: Proc International Joint Conference on Neural Networks pp 611–616
41. Sussner P (2005) New results on binary auto- and heteroassociative morphological memories. In: Proc International Joint Conference on Neural Networks pp 1199–1204
42. Sussner P, Valle ME (2005) A brief account of the relations between grayscale mathematical morphologies. In: Proc Brazilian Symposium on Computer Graphics and Image Processing pp 79–86
43. Sussner P, Valle ME (2006a) Grayscale morphological associative memories. IEEE Transactions on Neural Networks 17(3):559–570
44. Sussner P, Valle ME (2006b) Implicative fuzzy associative memories. IEEE Transactions on Fuzzy Systems 14(6):793–807

45. Valle ME, Sussner P, Gomide F (2004) Introduction to implicative fuzzy associative memories. In: Proc IEEE International Joint Conference on Neural Networks pp 925–931
46. Wang S, Lu H (2004) On new fuzzy morphological associative memories. IEEE Transactions on Fuzzy Systems 12(3):316–323
47. Zhang BL, Zhang H, Ge SS (2004) Face recognition by applying wavelet subband representation and kernel associative memory. IEEE Transactions on Neural Networks 15(1):166–177
48. Zhang H, Huang W, Huang Z, Zhang B (2005) A kernel autoassociator approach to pattern classification. IEEE Transactions on Systems, Man and Cybernetics, Part B 35(3):593–606

Part III

Machine Learning Applications

The Fuzzy Lattice Reasoning (FLR) Classifier for Mining Environmental Data

Ioannis N. Athanasiadis

Istituto Dalle Molle di Studi sull'Intelligenza Artificiale,
Galleria 2, CH-6928 Manno, Lugano, Switzerland
`ioannis@athanasiadis.info`

Summary. This chapter introduces a rule-based perspective on the framework of fuzzy lattices, and the Fuzzy Lattice Reasoning (FLR) classifier. The notion of fuzzy lattice rules is introduced, and a training algorithm for inducing a fuzzy lattice rule engine from data is specified. The role of positive valuation functions for specifying fuzzy lattices is underlined and non-linear (sigmoid) positive valuation functions are proposed, that is an additional novelty of the chapter. The capacities for learning of the FLR classifier using both linear and sigmoid functions are demonstrated in a real-world application domain, that of air quality assessment. To tackle common problems related to ambient air quality, a machine learning approach is demonstrated in two applications. The first one is for the prediction of the daily vegetation index, using a dataset from Athens, Greece. The second concerns with the estimation of quartely ozone concentration levels, using a dataset from Valencia, Spain.

9.1 Introduction

The framework of fuzzy lattices has been utilized lately in machine learning applications, mainly by utilizing artificial neural network architectures, i.e. as in [9, 10, 11, 12, 14, 15]. In this chapter, the fuzzy lattice framework is approached from a rule-based, reasoning perspective. Two issues related to fuzzy lattice reasoning are discussed here. The first one is the foundation of the fuzzy lattice rule engines, as a remedy for classification problems. The notion of fuzzy lattice rule is introduced, which employes fuzzy lattice elements as the rule antecedents, while the fuzzy inclusion measure serves as a truth function for deriving to rule consequences (conclusions). On top of the fuzzy lattice rule, a fuzzy lattice rule engine is specified, with which classification tasks can be performed. A fuzzy lattice rule engine induction algorithm is presented in this chapter as well. The second issue deals with the positive valuation functions for defining fuzzy lattices from partially ordered sets. Previous works have employed only linear positive valuation functions for defining fuzzy lattices. In this work, non-linear positive valuation functions

I.N. Athanasiadis: *The Fuzzy Lattice Reasoning (FLR) Classifier for Mining Environmental Data*, Studies in Computational Intelligence (SCI) **67**, 175–193 (2007)
`www.springerlink.com`

are investigated and a sigmoid one is introduced. The sigmoid positive valuation functions are immidiately applicable to the framework of fuzzy lattices, and its variety of applications. Non-linear positive valuation functions can be considered as an extention to existing artificial neural network architectures based on fuzzy lattices, as the Fuzzy Lattice Neural Networks (FLNN) [14], and the Fuzzy Lattice Neurocomputing models (FLN) [11]. In this chapter, the employment of non-linear positive valuation functions is demonstrated in the context of the Fuzzy Lattice Reasoning classifier presented. Specifically, the Fuzzy Lattice Reasoning classifier is used for addressing classification tasks related to ambient air quality, is comparison with other rule-based classifiers. In conclusion, the FLR classifier turned out with credible models both for the prediction of the daily vegetation index in the metropolitan area of Athens, Greece, and the estimation of quartely ozone concentration levels in the region of Valencia, Spain. In both cases the use of sigmoid positive valuation functions improved the performance of the classifier, while it didn't increase the complexity of model. Actually, in one of the cases it was reduced significantly.

The rest of the chapter is organized as follows. Section 9.2 summarizes briefly the required mathematics, and is provided for the reader of the stand alone chapter. Readers who are are comfortable with the terms and notations of the book and can proceed to Sect. 9.3, where the Fuzzy Lattice Reasoning classifier is presented. The introduction of a sigmoid positive valuation function is detailed in Sect. 9.4. Section 9.5 demonstrates the two test-cases and presents comparative results with other classification methods. The main findings of this chapter are discussed in the last Sect. 9.6.

9.2 Mathematical Background

A **lattice** L is a partially ordered set (poset), so that any two of its elements $a, b \in \mathsf{L}$ have a greatest lower bound (or meet) denoted by $a \wedge b := \mathtt{inf}\{a, b\}$ and a least upper bound (or join) denoted by $a \vee b := \mathtt{sup}\{a, b\}$. A lattice L is called complete when each of its subsets has a least upper bound and a greatest lower bound in L. A non-void complete lattice has a least element and a greatest element denoted by O and I, respectively.

The Cartesian product $\mathsf{L} = \mathsf{L}_1 \times \ldots \times \mathsf{L}_N$ of N constituent lattices $\mathsf{L}_1 \ldots \mathsf{L}_N$ (**product lattice**) is a lattice [7]. In a product lattice $\mathsf{L} = \mathsf{L}_1 \times \ldots \times \mathsf{L}_N$ inclusion can be defined as:

$$(x_1, \ldots, x_N) \le (y_1, \ldots, y_N) \iff (x_1 \le y_1) \& \ldots \& (x_N \le y_N) \qquad (9.1)$$

The meet in a product lattice $\mathsf{L} = \mathsf{L}_1 \times \mathsf{L}_N$ is given by $(x_1, \ldots, x_N) \wedge (y_1, \ldots, y_N) = (x_1 \wedge y_1, \ldots, x_N \wedge y_N)$, whereas the join is given by $(x_1, \ldots, x_N) \vee (y_1, \ldots, y_N) = (x_1 \vee y_1, \ldots, x_N \vee y_N)$ [7, 8].

A product lattice could combine diverse constituent lattices thus implying the potential to deal either separately and/or jointly, in any combination, with disparate types of data such as vectors of real numbers, propositions, (fuzzy) sets, events in a probability space, symbols, graphs, etc.

A **fuzzy lattice** is a pair $\langle L, \mu \rangle$, where L is a lattice and $(L \times L, \mu)$ is a fuzzy set with membership function $\mu : L \times L \to [0, 1]$ such that $\mu(a, b) = 1 \iff a \leq b$ [11].

The set of all fuzzy lattices $\langle L, \mu \rangle$ is called framework of fuzzy lattices and has been used for decision-making in various applications [10, 11]. This paper approaches the fuzzy lattice framework from a rule-based perspective as presented in Sect. 9.3 below. Some more usefull instruments of the fuzzy lattice framework are the followings.

A **valuation function** $v : L \to R$ is defined on a lattice L as any real function that satisfies: $v(a) + v(b) = v(a \wedge b) + v(a \vee b), \forall a, b \in L$. A valuation function is called **positive** if and only if $a < b \iff v(a) < v(b)$ [7]. Linear positive valuations function have been used in previous works [10] for defining an inclusion measure (σ) in a complete lattice L.

In general, an inclusion measure on a complete lattice L is defined as a real mapping function $\sigma : L \times L \to [0, 1]$, such that for each $a, b, x \in L$ the following conditions are satisfied:

$$\sigma(a, O) = 0, \forall a \neq O \tag{9.2}$$

$$\sigma(a, a) = 1 \tag{9.3}$$

$$a < b \Rightarrow \sigma(x, a) < \sigma(x, b) \tag{9.4}$$

$$a \wedge b < a \Rightarrow \sigma(a, b) < 1 \tag{9.5}$$

Given a lattice L and an inclusion measure $\sigma : L \times L \to [0, 1]$ it turns out that $\langle L, \sigma \rangle$ is a fuzzy lattice, with σ the membership function.

Another useful tool implied by a positive valuation in a general lattice L is a metric distance function $d : L \times L \to R$ defined as $d(x, y) = v(x \vee y) - v(x \wedge y)$.

A positive valuation function $v : L \to R$ in a lattice L with $v(O) = 0$ is a sufficient condition for two inclusion measures [10]:

$$k(a, b) = \frac{v(b)}{v(a \vee b)} \tag{9.6}$$

$$s(a, b) = \frac{v(a \wedge b)}{v(a)} \tag{9.7}$$

Ultimately, given a lattice L, for which a positive valuation function $v : L \to R$ can be defined with $v(O) = 0$, then both $\langle L, k \rangle$ and $\langle L, s \rangle$ are fuzzy lattices. It becomes apparent that the only requirement for specifing a fuzzy lattice on a lattice L is the selection of an appropriate positive valuation function $v(\cdot)$. Based on this remark, any kind of partially ordered data, as numbers, sets, graphs, etc that can define a lattice, to which if a positive valuation function is ascribed, then it becomes available in the framework of fuzzy lattices.

The framework of fuzzy lattices has been extended to lattices of closed intervals [11], which are of particular interest for the introduction of fuzzy lattice rules in Sect. 9.3. In a complete lattice L, a closed interval of lattice elements is defined as

$$[a, b] = \{x | a \leq x \leq y\} \tag{9.8}$$

A singleton interval is defined as $[a, a] = a, \forall a \in \mathsf{L}$. The set $\tau(\mathsf{L})$ of all closed interval of lattice elements in L (including singletons) is also a complete lattice with upper bound $[O, I]$ and lower bound $[I, O]$. In $\tau(\mathsf{L})$ an ordering relation is defined as:

$$[a, b] \leq [c, d] \equiv \{c \leq a \quad \& \quad b \leq d\} \tag{9.9}$$

Join of two $\tau(\mathsf{L})$ elements $[a, b]$ and $[c, d]$ is defined as:

$$[a, b] \vee [c, d] = [a \wedge c, b \vee d] \quad (join) \tag{9.10}$$

And meet of two $\tau(\mathsf{L})$ elements $[a, b]$ and $[c, d]$ is defined as:

$$[a, b] \wedge [c, d] = \begin{cases} [a \vee c, b \wedge d], & \text{if} \quad a \vee c \leq b \wedge d \\ 0, & \text{otherwise} \end{cases} \quad (meet) \tag{9.11}$$

The definition of a valuation function v_τ for the lattice of closed intervals $\tau(\mathsf{L})$ has been discussed in [11, 15]. Based on the positive valuation function $v : \mathsf{L} \to \mathsf{R}$ of lattice L and an isomorphic function $\theta : \mathsf{L}^\partial \to \mathsf{L}$, a valuation function in $\tau(\mathsf{L})$ is defined as:

$$v_\tau([a, b]) = v(\theta(a)) + v(b) \tag{9.12}$$

Both inclusion measures defined in (9.6), (9.7) using v_τ can be applied on $\tau(\mathsf{L})$:

$$k_\tau([a, b], [c, d]) = \frac{v_\tau([c, d])}{v_\tau([a, b] \vee [c, d])} \tag{9.13}$$

$$s_\tau([a, b], [c, d]) = \frac{v_\tau([a, b] \wedge [c, d])}{v_\tau([a, b])} \tag{9.14}$$

As a result it turns out that $\langle \tau(\mathsf{L}), k_\tau \rangle$, $\langle \tau(\mathsf{L}), s_\tau \rangle$ are fuzzy lattices. Note that the inclusion measure k_τ is more usable compared to s_τ, due to the conditional definition of the meet in $\tau(\mathsf{L})$, which appears in the nominator of (9.14). In the contrary, k_τ is unconditionally defined as:

$$\begin{aligned} k_\tau([a, b], [c, d]) &= \frac{v_\tau([c, d])}{v_\tau([a, b] \vee [c, d])} \\ &= \frac{v_\tau([c, d])}{v_\tau([a \wedge c, b \vee d])} \\ &= \frac{v(\theta(c)) + v(d)}{v(\theta(a \wedge c)) + v(b \vee d)} \end{aligned} \tag{9.15}$$

9.3 Fuzzy Lattice Reasoning (FLR) Classifier

Many data structures of practical interest are lattice ordered. The objective here is to present a classifier for inducing a rule-based inference engine from data, based on the instruments of the fuzzy lattice framework presented in the previous section.

9.3.1 Fuzzy lattice rule engine

A fuzzy lattice rule engine is based on fuzzy lattice rules. A fuzzy lattice rule employs a fuzzy lattice element as the rule antecedent, while the fuzzy inclusion measure serves as a truth function for deriving to rule consequences (conclusions).

A **fuzzy lattice rule** is a pair $\langle a, c \rangle$ where a is an element in a fuzzy lattice $\langle L, \mu \rangle$ and $c \in C$ is a categorical label. Note that this definition applies to the whole framework of fuzzy lattices, including product lattices and lattices of closed intervals. A fuzzy lattice rule can be considered as the mapping $a \to c$ of a fuzzy lattice $\langle L, \mu \rangle$ element a to a categorical label c, where a is the rule antecedent and c is the consequence of the rule.

Let a and b be two lattice L elements, c a categorical label in C and function k, as defined in (9.6) be a fuzzy membership function in L. We define the **degree of truth** of the fuzzy lattice rule $a \to c$ against the perception b to be defined by the fuzzy membership function of the fuzzy lattice $\langle L, \mu \rangle$, as:

$$\mu(b, a) = k(b, a) = \frac{v(a)}{v(b \vee a)} \tag{9.16}$$

Similarly holds for $\tau(L)$ using k_τ as defined in (9.13).

A **fuzzy lattice rule engine** $\mathcal{E}_{\langle L, \mu \rangle, C}$ can be considered as a set of N fuzzy lattice rules that are commonly activated:

$$\mathcal{E}_{\langle L, \mu \rangle, C} \equiv \{a_i \to c_i\}, a_i \in \langle L, \mu \rangle, c_i \in C, i = 1 \ldots N \tag{9.17}$$

Reasoning with a fuzzy lattice rule engine implies the calculation of the degree of truth for each one of engines rules. For example consider the following engine that consists of three rules:
$\mathcal{E}_{\langle L, \mu \rangle, C} = \{a_1 \to c_1, a_2 \to c_2, a_3 \to c_3\}$, where a_1, a_2, a_3, are elements of a fuzzy lattice $\langle L, \mu \rangle$ and c_1, c_2, c_3 a set of predifined labels. Against an input element a_0, the engine will result with the following table of degree of truth for each consequence: $c_1 = \sigma(a_0, a_1)$, $c_2 = \sigma(a_0, a_2)$, and $c_3 = \sigma(a_0, a_3)$. The fuzzy lattice reasoning engine will respond with the class $c = \underset{i}{argmax}(\sigma(a_0, a_i))$. In another mode of generalization, $\mathcal{E}$ may respond with the label that is additively included the most. In this way, a fuzzy lattice reasoning engine can be used for generalization.

9.3.2 Fuzzy lattice rule induction (training)

The task of inducing a fuzzy lattice rule engine can be described as follows: Let a training set of M partially ordered objects $\{u_1, u_2, \ldots, u_M\} \in \mathcal{U}$, each one of which is associated with a class label $c \in \mathsf{C}$, where $\mathsf{C} = \{c_1, c_2, \ldots, c_K\}$ is a set of K predefined labels (classes). The objective is to induce a set of fuzzy lattice rules that implement a function $h : \mathsf{U} \to \mathsf{C}$, associating any object $u \in \mathsf{U}$ with a classification label $c \in \mathsf{C}$.

In general, the universe U of the training objects can include any type of complex data structures, as vectors of real numbers, graphs or sets. Obviously, U is a complete lattice. Given a positive valuation function $v : \mathsf{U} \to \mathsf{R}$, an inclusion measure $\sigma : \mathsf{U} \times \mathsf{U} \to [0, 1]$ can be defined in U, as denoted above in (9.6), (9.7), (9.13), (9.14), which implies a fuzzy membership function $\mu : \mathsf{U} \times \mathsf{U} \to [0, 1]$. In this respect, it turns out that $\langle \mathsf{U}, \mu \rangle$ is a fuzzy lattice. The classifier to be built is equivalent to a map $h' : \langle \mathsf{U}, \mu \rangle \to \mathsf{C}$, which is a set of fuzzy lattice rules, i.e. a fuzzy lattice rule engine: $h \equiv h' \equiv \mathcal{E}_{\langle \mathsf{U}, \mu \rangle, \mathsf{C}}$.

Each object u of the training set is an element of U and each training pair $\langle u, c \rangle$ can be expressed as a fuzzy lattice rule $u \to c$, where u is an element of the fuzzy lattice $\langle \mathsf{U}, \mu \rangle$ and c the corresponding class. This means that the instances of a training set could be treated as fuzzy lattice rules. For example consider the simple case where the universe of the training instances is a closed interval of real numbers $[O, I]$. Then any training pair $\langle x, c \rangle$ where $x \in [O, I]$ and $c \in \mathsf{C}$ can be expressed as a fuzzy lattice rule consisted from a lattice interval singleton mapped to class c, as: $\langle x, c \rangle \equiv \langle [x, x], c \rangle$. Likewise for alternative universes of discourse.

A **naive fuzzy lattice reasoning classifier** that can be induced directly from a set of M training pairs $(u_1, c_1), \ldots, (u_M, c_M)$, $u_i \in \mathsf{U}$, and $c_i \in \mathsf{C}$, is the one that memorizes all training instances as fuzzy lattice rules. Given a positive valuation function v, each training element u_i is an element of the fuzzy lattice $\langle \mathsf{U}, \sigma \rangle$, where σ is an inclusion measure defined in (9.6), (9.7), (9.13). In this way, the most simple fuzzy lattice rule engine will consist at most out of M (trivial) rules and will be: $\mathcal{E} = \{u_1 \to c_1, \ldots, u_i \to c_i, \ldots, u_M \to c_M\}$, where $u \in \langle \mathsf{U}, \sigma \rangle$ and $c \in \mathsf{C}$.

A **training process** for inducing a fuzzy lattice rule engine is based on joining lattice rules pointing to the same class for formulating lattice rules of higher size, and potentially higher ability for generalization. The training procedure for inducing the classifier h, through single pass iteration over all training instances is presented below. Note that a simplified version of the FLR algorithm was presented previously implemented as neural network σ-FLN architecture [10, 11].

FLR training algorithm

Step-0: Let a fuzzy lattice rule engine $\mathcal{E}_{\langle \mathsf{L}, \sigma \rangle, \mathsf{C}} = \{a_1 \to c_1, \ldots, a_R \to c_R\}$ of size R.

Note that $\mathcal{E}_{\langle L,\sigma\rangle,C}$ could be initially empty, i.e. $R = 0$, and a user-defined threshold size D_{crit}.

Step-1: Present the next training pair $\langle u, c\rangle$, in the form of a fuzzy lattice rule $u \rightarrow c$ to the initially set rules in $\mathcal{E}_{\langle L,\sigma\rangle,C}$.

Step-2: If no more rules in $\mathcal{E}$ are set then append input rule $u \rightarrow c$ in $\mathcal{E}$ and go to Step-1.

Else, compute the fuzzy degree of inclusion $\sigma(u \leq a_r)$, $\forall l = 1 \ldots R$ of the antecedent u to the antecedents of all the set rules in $\mathcal{E}$.

Step-3: Competition among the set rules in $\mathcal{E}$. Winner is the rule $a_J \rightarrow c_J$, where

$$J = arg \max_{r \in \{1,\ldots R\}} \sigma(u \leq a_r) \tag{9.18}$$

Step-4: If both $c = c_J$ and $diag(u \vee a_J) < D_{crit}$ (assimilation condition), then replace the antecedent a_J of the winner rule $a_J \rightarrow c_J$ by the join-lattice $u \vee a_J$, i.e. with the rule: $u \vee a_J \rightarrow c_J$. Go to Step-1.

Else, reset the winner rule $a_J \rightarrow c_J$, and go to Step-2.

Previous works has employed for the algorithm tuning, instead of D_{crit}, the dimensionless vigilance parameter:

$$\rho_{crit} = \frac{N}{N + D_{crit}} \Leftrightarrow D_{crit} = \frac{N(1 - \rho_{crit})}{\rho_{crit}} \tag{9.19}$$

Note that ρ_{crit} varies in the interval $[0.5, 1]$ for any number of dimensions N as shown in [11]. In the following experiments ρ_{crit} has been employed, as its range is not related to the dimension of the lattice.

9.3.3 Decision making with fuzzy lattice rules (testing)

The decision making process (**testing phase**) of an (induced) fuzzy rule engine $\mathcal{E}_{\langle L,\sigma\rangle,C}$ of size R, involves the competition of its rules over a perception $x \in U$, of unknown label. The element x is presented to each rule of the engine: $a_r \rightarrow c_r$, and the inclusion measure $\sigma(x \leq a_r) \equiv \sigma(x, a_r)$ is calculated. Finally, x is assigned to the category c_J, where

$$J = arg \max_{r \in \{1,\ldots,R\}} \sigma(x \leq a_r) \tag{9.20}$$

A second mode of reasoning for the an (induced) fuzzy rule engine $\mathcal{E}_{\langle L,\sigma\rangle,C}$ of size R may involve a contributing competition, where all rules pointing to the same class c_i will add-up their inclusion measures and the perception $x \in U$ will be assigned to the class that additively includes it the most:

$$J = arg \max_{c_i \in C} \sum_{r=\{1,\ldots,R/c_r=c_i\}} \sigma(x \leq a_r) \tag{9.21}$$

In principal, in any universe of partially ordered data, that can be formalized as lattices, product lattices or lattices of intervals a fuzzy lattice reasoning classifier can be induced. A similar lattice algorithm, namely Find-S algorithm, has been presented in a machine learning context [13], but without an employment of positive valuation functions. In the following section the capacity of the algorithm is further broaden, by introducing non-linear positive valuation functions.

9.4 Non-linear Positive Valuation Functions

The whole procedure for inducing an FLR classifier is related to the selection of an appropriate valuation function in U, that implies the formation of the fuzzy lattice $\langle \mathsf{U}, \sigma \rangle$, where σ is an inclusion measure as those defined in (9.6), (9.7), (9.13), (9.14). This remark holds also for other decision making schemes built upon the framework of fuzzy lattices that map data to lattices. Typically, prior works within the Framework of Fuzzy Lattices [2, 10, 12, 14, 15] have focused in forming fuzzy lattices from numerical datasets by employing linear valuation functions. In cases that data reside in the N-dimensional unit hypercube $\mathsf{I}^N = [0,1] \times [0,1] \times \ldots \times [0,1]$, the positive valuation function selected for each consistuent lattice is the simple function $v_i(x) = x$. In other cases, where data reside in R^N the training dataset can be formulated as $\mathsf{T} = [O_1, I_1] \times [O_2, I_2] \times \ldots \times [O_N, I_N]$, and the positive valuation function for each constituent lattice is given by the following equation that which linearly scales T to the N-dimensional unit hypercube I^N:

$$v_i(x) = \frac{(x - O)}{(I - O)} \tag{9.22}$$

In this paper, both linear and non-linear positive valuation functions are considered for inducing an FLR classifier from a numerical dataset. The sigmoid function is an example non-linear increasing function with range $[0, 1]$ that could be used as a positive valuation function for mapping an interval of real numbers to a fuzzy lattice. In the particular case of lattice I a non-linear positive valuation function is defined by:

$$v_\lambda(x) = \frac{1}{1 + e^{-\lambda(x - 1/2)}}, \lambda > 0 \tag{9.23}$$

In generic the case of data residing within the interval $[O, I]$, a positive valuation function can be defined by the sigmoid function:

$$v_\varsigma(x) = \frac{1}{1 + e^{-\lambda(x - x_{med})}}, \tag{9.24}$$

where

$$x_{med} = \frac{I + O}{2}, \quad \lambda = \frac{\varsigma}{I - O}, \quad \varsigma > 0$$

The single parameter ς can be used for tuning the slope of $v_\varsigma(x)$. Figure 9.1 plots function $v_\varsigma(x)$ for various values of ς, in contrast with a linear valuation function $v(x) = x$.

The capacity of non-linear positive valuation functions to improve performance has been demonstrated lately in classification and regression applications [1, 6]. In the following section, the capacity for learning of the FLR classifier is evaluated in two air quality data sets, by employing both linear and sigmoid positive valuation functions.

9.5 Application on Environmental Datasets

In this work, the problem of operational decision support related to air pollution is tackled by utilizing a machine learning approach. Specifically, the FLR Classifier is demonstrated in comparison with other state-of the-art algorithms in two application cases related to urban air quality assessment. The first one concerns with the prediction of the daily vegetation index in the metropolitan area of Athens, Greece, while the second one is for the estimation of ambient ozone concentration levels, in a rural area in Valencia, Spain.

9.5.1 Air quality assessment

Ambient air quality assessment and management is characterized by complexity and uncertainty mainly due to the difficulties of atmospheric chemistry and physics and the stochastic processes involved in air pollutant generation. These boundaries raise the major obstacles in building simple models for credible prediction. In most cases, decision making relies on human expertise, as analytical models are too complex and slow for operational decision support. Legislation in Europe, the US, and elsewhere, define environmental quality indicators, which could be communicated to the public on-time (or even in advance) for informing population about air quality, especially in urban areas.

In both application cases, focus is given on ambient ozone, which is a secondary pollutant formed as a result of catalytic reactions between pollutants emitted from industrial sources and automobiles. In the presence of sunlight (ultra-violet radiation) and under suitable meteorological conditions, the precursors react photo-chemically to *produce* ozone. Due to the chemical reaction dynamics, the analytical models for describing ozone formation in ambient air are very complex. As a consequence, simple, yet credible prediction models are required for achieving both the requirements of accurate air quality assessment and capabilities for fast decision making (in contrast with the analytical complex models). These properties can be realized by learning from data, using knowledge discovery techniques as discussed in previous works [3, 4, 5], and presented below.

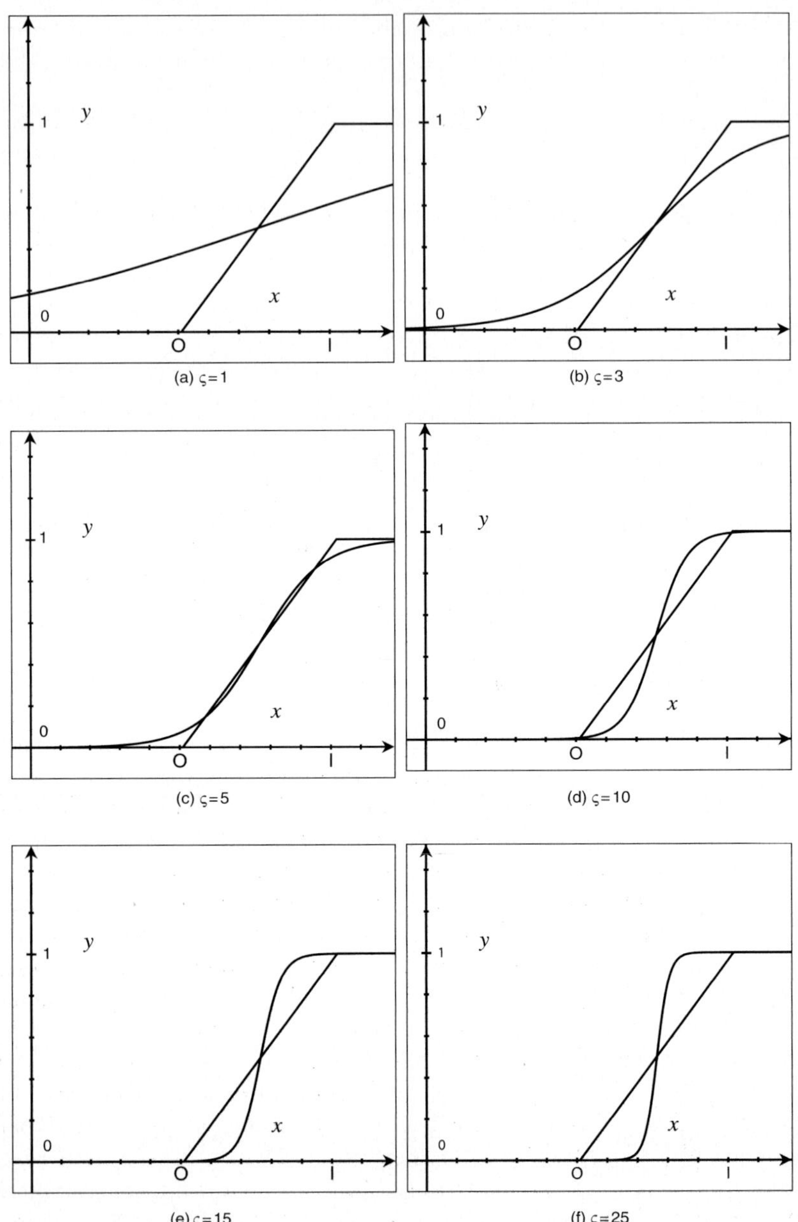

Fig. 9.1. Sigmoid positive valuation functions $v_\varsigma(x)$ illustrated in the interval $[O, I]$ in contrast with the linear positive valuation function $v_i(x)$ for various values of the normalized parameter ς

9.5.2 Daily vegetation index prediction in Athens, Greece

The first demonstration case concentrates in the metropolitan area of Athens, Greece, that suffers from air-pollution problems, mainly due to the traffic and industrial emission, but also because of the urban landscaping. A measure of the impact of air pollution to human quality of life is the daily vegetation index introduced by the European Commission with the Directive 92/72/EEC. It is an early warning indicator of the overall air quality and specifies a threshold on the ozone's mean concentration (O_3) over a 24 hours period. The same directive sets the daily vegetation threshold at the limit of 65 $\mu g/m^3$.

The prediction of the daily vegetation index threshold (exceeded or not) is the actual goal of the first demonstration case, where FLR has been employed. Specifically, a dataset was available that contained daily observations from the Maroussi station of the Ministry of Environment lined up with meteorological data from the Athens National Observatory. The selection of the monitoring station was based on the frequency of high exceedances of the selected index. The dataset covers a 3.5 year period (January 1999 - June 2001) and has been split in two parts: one for training (that corresponds to the period January 1999 - December 2001) and one for testing (January - June 2002). The purpose of this selection was the ability to make comparisons with the statistical methods used for the same test case as previously reported [3]. Tables 9.1, 9.2 present the dataset attributes and statistics.

Table 9.1. Athens dataset attributes

	Attribute	Symbol	Datatype	Units
1	Carbon monoxide	CO	real number	mg/m^3
2	Nitrogen oxide	NO	real number	$\mu g/m^3$
3	Nitrogen dioxide	NO_2	real number	$\mu g/m^3$
4	Nitrogen oxides	NO_x	real number	$\mu g/m^3$
5	Sulfur dioxide	$SO2$	real number	$\mu g/m^3$
6	Ozone	$O3$	real number	$\mu g/m^3$
7	Air temperature	Ta	real number	$\deg C$
8	Soil temperature	Ts	real number	$\deg C$
8	Relative humidity	RH	real number	
9	Wind speed	WS	real number	m/s
10	Wind direction	WD	real number	rad
11	Mean temperature	$meantemp$	real number	$\deg C$
12	Max temperature	$maxtemp$	real number	$\deg C$
13	Min temperature	$mintemp$	real number	$\deg C$
14	Solar radiation	ws	real number	Wm^{-2}
16	Vegetation index	O_3 alert	class label	Yes/No

Table 9.2. Athens dataset statistics

	Records in class	
	Yes	No
Training set	571	525
Testing set	62	120

Table 9.3. Results for the Athens dataset

Model	Accuracy (%)	False positive rate (%)
LCA	53.37	0.37
PCA	21.47	0.55
ARIMA	31.58	2.94
FLR (sigmoid)	84.53	5.88
FLR (linear)	80.11	8.4
Nnge	59.68	8.4
Conjunctive Rule	64.52	10.92
OneR	70.97	11.76
Decision Table	69.35	12.61
IBk	59.68	15.13
Voted Perceptron	80.65	15.97
ADTree	85.48	17.65
NaiveBayes	80.65	17.65
C4.5 (J48)	75.81	19.33
KStar	61.67	43.7

9.5.3 Comparative results for the Athens dataset

A set of cross-evaluation experiments were conducted for the Athens test-case. FLR with both linear and sigmoid valuation functions has been applied for predicting the daily vegetation index, for a range of values for the parameters ρ and ς. For comparison purposes, results are presented in Table 9.3 along with previous results obtained for the same test case with statistical methods (LRA, ARIMA and PCA) [16] and other ten classification algorithms and neural networks (ADTree, C4.5, Conjunctive Rule, Decision Table, IBk, KStar, NaiveBayes, Nnge, OneR, Voted Perceptron) [3]. Note that WEKA platform [17] implementations of the algorithms have been used. The statistical methods resulted low overall accuracy rates (less than 60%), with high false positive rates, i.e. there are several alarms missed, but the confidence on those identified is very high. On the contrary, classification techniques manage higher classification accuracies, with the cost of a lower credibility on their decisions.

The application of FLR with the linear valuation function resulted up to an overall classification accuracy of 80.11% with a fuzzy lattice rule engine of 212 rules. The accuracy compared to the rest classification algorithms is relatively good (ADTree and NaiveBayes performed better). Also, the false positive rates of FLR with linear positive valuation function is the best achieved among classification algorithms. In this terms FLR performance is competitive.

Next, the FLR with a sigmoid positive valuation function was employed, resulting a Fuzzy Lattice Rule Engine of 123 rules. In this case, the overall accuracy improved to 84.53%, while the false positive rate decreased significantly to 5.88%. Overall, the introduction of a sigmoid positive valuation function achieved to decrease the number of extracted rules almost to half, result an overall accuracy similar to that of ADTree, while arriving the optimal false positive rate. Based on these remarks, the FLR with a sigmoid positive valuation function can be considered as the most credible model for predicting the daily vegetation index in this particular application.

9.5.4 Ozone level estimation in Valencia, Spain

The second test-case concerns with the peri-urban area of Valencia, Spain where ambient air quality is diminished due to industrial activities. The goal here is to identify the level of ozone concentration, a critical photochemical pollutant, which is commonly used as an indicator of the overall ambient air quality. In this case, the objective is to estimate the ozone concentration levels from the concurrent observations of other pollutants and meteorological attributes, a task related with both quality assurance and control activities and operational decision making.

In this case, data were available from a single metrological stations, that monitors eight parameters, including both meteorological attributes and air-pollutant concentrations, as shown in Table 9.4. Data are sampled on a quarter-hourly basis during the year 2001. In total there are available 35,040 data vectors, out of which 565 records have the ozone label missing, and thus where excluded in the analysis below. Values were missing in other attributes, and in total there are 6,020 records (that is around 17% of the total) with at least one missing value. In the following experiments, both the original dataset with missing values and a preprocessed one that excluded all records

Table 9.4. Valencia dataset attributes

	Attribute	Symbol	Datatype	Units
1	Sulfur dioxide	SO_2	real number	$\mu g/m^3$
2	Nitrogen oxide	NO	real number	$\mu g/m^3$
3	Nitrogen dioxide	NO_2	real number	$\mu g/m^3$
4	Nitrogen oxides	NO_x	real number	$\mu g/m^3$
5	Wind velocity	VEL	real number	m/s
6	Temperature	TEM	real number	deg C
7	Relative humidity	HR	real number	%
8	Ozone level	O_3	class label	low, med

Note: Class low corresponds to concentration levels in range $0 - 60$ $\mu g/m^3$, and class med(ium) to $60 - 100$ $\mu g/m^3$.

Table 9.5. Valencia dataset statistics for (a) the dataset without missing values, and (b) the dataset with missing values (original)

	Records in class				Records in class	
	low	medium			low	medium
Training set	6,865	4,761		Training set	9,472	6,074
Testing set	12,256	5,138		Testing set	13,483	5,446
(a)				(b)		

with missing values. Data collected from January 1, 2001 until mid June have been used for training, whereas the remaining data until year end have been used for testing. The corresponding numbers of data vectors available in classes `low` and `med(ium)`, respectively, are shown in Table 9.5.

9.5.5 Comparative results for the Valencia dataset

For estimating the ozone concentration level three classifiers were employed: (a) The C4.5 classifier, (b) The FLR classifier, with a linear positive valuation function, and (c) The FLR classifier, with a sigmoid positive valuation function. Two series of experiments have been carried out: first, using the dataset without missing values and, second, the original set including the ones with missing values.

First, the C4.5 classifier has been employed on a standard software platform (WEKA platform [17]), for generating decision trees, in which the internal nodes specify inequalities for the values of environmental attributes, moreover the tree leaves specify an output class. Initially, the C4.5 classifier has been applied on the data without missing values, without pruning, resulting in a decision tree with 1393 leaves (rules). The corresponding classification accuracy on the training set reached 94.8%, whereas on the testing set it was only 64.85%. Similar results have been obtained for the dataset with no missing values. Obviously, C4.5 over-fits the training data, therefore two pruning methods have been employed: (1) Confidence Factor Pruning (CFP), and (2) Reduced Error Pruning (REP). Results are shown in Tables 9.6 and 9.9 for selected pruning parameter values. The highest accuracy achieved on the testing split was 73.74% and 77.56% respectively for each dataset.

The FLR classifier has been implemented on the same software platform (WEKA) using both linear and sigmoid valuation functions. Initially, the FLR Classifier has been employed using a linear valuation function. In this case, the valuation function used was $v_i(x) = \frac{(x-O)}{(I-O)}$, where [O, I] are the minimum and maximum values of the training data in each dimension. Results are presented in Tables 9.7 and 9.10 for selected values of the vigilance parameter ρ. The FLR Classifier achieved a classification accuracy of 83.23% with only three rules for the dataset without missing values and 84.60% with 19 rules for the dataset with missing values. Note that the FLR classifier outperforms C4.5.

Table 9.6. Results with C4.5 for the Valencia dataset without missing values

Parameter value	Classification accuracy (%)		No. of Rules (Tree leaves)
	Training set	Test set	
Unpruned			
-	94.80	64.85	1393
Confidence factor pruning (parameter: CF)			
0.1	91.33	67.31	575
0.2	92.87	66.71	823
0.3	93.92	67.40	1055
0.4	94.10	67.39	1101
0.5	94.31	67.19	1169
Reduced error pruning (parameter: no. of Folds)			
2	89.31	63.71	507
10	89.01	71.85	465
50	85.05	60.62	251
100	83.33	73.74	131
300	81.55	69.98	75
500	77.73	72.48	31

Table 9.7. Results with FLR with linear valuation function for the Valencia dataset without missing values

Parameter ρ	Classification accuracy (%)		No. of Rules (Tree leaves)
	Training set	Test set	
0.5	59.16	70.46	2
0.6	64.73	83.23	3
0.7	73.68	74.85	20
0.8	67.43	72.59	139

Then, experiments have been conducted for the FLR Classifier using the sigmoid function of (9.24) on both datasets. In this case the FLR Classifier has been tuned using two parameters: The vigilance parameter ρ and the slope parameter ς of the sigmoid valuation function. Results obtained by FLR with sigmoid valuation function are presented in Tables 9.8 and 9.11. For the dataset without missing values the FLR with sigmoid positive valuation function achieved a classification accuracy of 85.22% with three rules. Note that using the sigmoid positive valuation function the best performance has improved by 2% without increasing the number of induced rules. For the dataset with missing values, the best accuracy improved by nearly 1%, again without increasing the number of rules, as shown in Tables 9.8 and 9.11.

Table 9.8. Results with FLR with sigmoid valuation function for the Valencia dataset without missing values

Parameter		Classification accuracy (%)		No. of Rules
ς	ρ	Training set	Test set	(Tree leaves)
1	0.5	59.16	70.46	2
	0.6	59.16	70.46	2
	0.7	59.16	70.46	2
	0.8	62.73	85.22	3
5	0.5	59.16	70.46	2
	0.6	65.40	82.70	3
	0.7	70.48	79.64	19
	0.8	67.53	78.72	40
10	0.5	59.16	70.46	2
	0.6	64.27	83.43	3
	0.7	65.77	74.89	34
	0.8	69.56	82.87	115
15	0.5	59.16	70.46	2
	0.6	64.73	83.24	3
	0.7	68.85	78.88	23
	0.8	70.39	81.54	112

Table 9.9. Results with C4.5 for the Valencia dataset with missing values

Parameter value	Classification accuracy (%)		No. of Rules
	Training set	Test set	(Tree leaves)
Unpruned			
-	94.80	64.85	1393
Confidence factor pruning (parameter: CF)			
0.1	89.14	60.26	279
0.2	89.98	59.19	368
0.3	90.81	59.44	463
0.4	91.37	59.30	542
0.5	91.59	59.32	598
Reduced error pruning (parameter: no. of Folds)			
2	88.14	64.91	318
10	88.28	59.19	288
50	85.44	60.17	144
100	84.01	61.36	84
300	82.48	77.56	44
500	81.33	70.19	32

9.6 Discussion

This chapter introduced the Fuzzy Lattice Reasoning (FLR) classifier, by considering Fuzzy Lattices as the foundation for specifying rules, both for Fuzzy Lattices and Fuzzy Lattices of intervals. Modes of generalization in a Fuzzy Lattice Rule Engine have been identified and a training procedure was

Table 9.10. Results with FLR with linear valuation function for the Valencia dataset with missing values

| Parameter | Classification accuracy (%) | | No. of Rules |
ρ	Training set	Test set	(Tree leaves)
0.5	60.99	71.22	5
0.6	60.99	71.22	8
0.7	63.48	84.60	19
0.8	69.00	66.54	43

Table 9.11. Results with FLR with sigmoid valuation function for the Valencia dataset with missing values

| Parameter | | Classification accuracy (%) | | No. of Rules |
ς	ρ	Training set	Test set	(Tree leaves)
1	0.5	60.99	73.37	2
	0.6	60.99	73.37	2
	0.7	60.99	73.37	3
	0.8	60.99	73.37	4
5	0.5	60.99	71.22	4
	0.6	60.99	71.22	6
	0.7	60.99	71.23	9
	0.8	65.34	85.53	19
10	0.5	60.99	71.22	6
	0.6	60.99	71.23	9
	0.7	60.99	71.22	14
	0.8	63.55	82.55	26
15	0.5	60.99	71.22	6
	0.6	60.99	71.23	10
	0.7	60.99	71.23	17
	0.8	64.00	82.59	31

detailed. Also, here non-linear positive valuation functions are introduced as an instrument for further improving the capacity for decision-making within the framework of Fuzzy Lattices. The FLR classifier was demonstrated for assessing ambient air quality on two test-cases. Results obtained with FLR Classifier for the case of the prediction of the daily vegetation index in Athens have compared favorably with the results obtained by other state-of-the-art classifiers and statistical approaches used in previous works. The FLR Classifier achieved the best performance in terms of false positive rates (5.88%), while keeping the overall accuracy at very high levels (84%). The introduction of the non-linear positive valuation function in this case resulted to an improvement of performance by 5%, while reducing the number of the model complexity (induced rules) to half. In the case of the estimation of ozone level concentrations in Valencia, the FLR Classifier resulted positively, with respect to the performance achieved with C4.5 decision trees. The FLR classifier with linear positive valuation function, compared to C4.5, improved classification accuracy by 9.5% for the dataset without missing values and by 7% for the dataset with missing values. Furthermore, the employment of a sigmoid positive valuation function by the FLR classifier achieved further improvement without increasing the complexity (number of induced rules) of the model. Finally, the approach presented here for tackling with the complexity and the uncertainties of the air quality assessment by using machine-learning techniques and in particular the FLR classifier rendered with trustworthy and credible results with a great potential for the application domain.

References

1. Athanasiadis IN, Kaburlasos VG (2006) Air quality assessment using Fuzzy Lattice Reasoning (FLR). In: Proc World Congress Computational Intelligence (WCCI) FUZZ-IEEE Program pp 231–236
2. Athanasiadis IN, Kaburlasos VG, Mitkas PA, Petridis V (2003) Applying machine learning techniques on air quality data for real-time decision support. In: First Intl Symposium Information Technologies in Environmental Engineering (ITEE-2003) ICSC-NAISO Academic Press p 51
3. Athanasiadis IN, Karatzas KD, Mitkas P (2006) Classification techniques for air quality forecasting. In: Fifth ECAI Workshop on Binding Environmental Sciences and Artificial Intelligence, 17th European Conf on Artificial Intelligence
4. Athanasiadis IN, Mitkas PA (2004) Supporting the decision-making process in environmental monitoring systems with knowledge discovery techniques. Vol III Knowledge-based Services for the Public Sector Symposium: KDnet pp 1–12
5. Athanasiadis IN, Mitkas PA (accepted) Knowledge discovery for operational decision support in air quality management. J of Environmental Informatics
6. Cripps A, Nguyen N (2007) Fuzzy lattice reasoning (FLR) classification using similarity measures. This volume, chapter 13
7. Birkhoff G (1967) Lattice Theory. Colloquium Publications, Providence, RI
8. Davey B, Priestley H (1990) Introduction to Lattices and Order. Cambridge University Press, Cambridge, UK
9. Georgiopoulos M, Fernlund H, Bebis G, Heileman G (1996) Order of search in fuzzy ART and fuzzy ARTMAP: effect of the choice parameter. Neural Networks 9(9):1541–1559
10. Kaburlasos V (2006) Towards a Unified Modeling and Knowledge-Representation Based on Lattice Theory: Computational Intelligence and Soft Computing Applications, ser Studies in Computational Intelligence 27. Springer, Heidelberg, Germany
11. Kaburlasos VG, Petridis V (2000) Fuzzy lattice neurocomputing (FLN) models. Neural Networks 13(10):1145–1170
12. Kaburlasos VG, Petridis V (2002) Learning and decision-making in the framework of fuzzy lattices. In: Jain LC, Kacprzyk J (eds) New Learning Paradigms in Soft Computing, ser Studies in Fuzziness and Soft Computing 84:55–96. Physica-Verlag, Heidelberg, Germany
13. Mitchell T (1997) Machine Learning. McGraw-Hill, New York, NY

14. Petridis V, Kaburlasos VG (1998) Fuzzy lattice neural network (FLNN): a hybrid model for learning. IEEE Trans Neural Networks 9(5):877–890
15. Petridis V, Kaburlasos V (1999) Learning in the framework of fuzzy lattices. IEEE Trans Fuzzy Systems 7(4):422–440
16. Slini T, Karatzas K, Moussiopoulos N (2005) Ozone forecasting supported by data mining statistical methods. In: R Sokhi, J Brexhler (eds) 5th International Conference on Urban Air Quality Measurement, Modelling and Management
17. Witten IH, Frank E (1999) Data Mining: Practical Machine Learning Tools and Techniques with Java Implementations. Morgan Kaufmann, USA

Machine Learning Techniques for Environmental Data Estimation

Vassilios Petridis[1] and Vassilis Syrris[2]

[1] Department of Electrical and Computer Engineering, Aristotle University of Thessaloniki, Greece `petridis@eng.auth.gr`
[2] Department of Electrical and Computer Engineering, Aristotle University of Thessaloniki, Greece `vsyrris@auth.gr`

Summary. In this paper we investigate the issue of wind speed prediction at a particular location in the urban area of Thessaloniki, Greece, based on the historical data containing wind parameter values at two other different locations. We evaluate the performance of two significant machine learning methodologies, the Fuzzy Lattice Neurocomputing (FLN) and the Support Vector Regression (SVR). The results of the specific applications are compared with past work on the same data set, and a discussion upon the exhibited features is carried out.

10.1 Introduction

Wind prediction (short/long-term) is of great significance for the wind energy production which is the fastest growing type of renewable energy in Europe. The energy produced strongly depends on the actual wind speed at a certain location, so the power output cannot be guaranteed at all times. The objective for the energy market is the secure and economical management of a power system [13], thus there is an urgent need for the development of flexible computational techniques capable of controlling and governing efficiently the energy resources.

Several researchers studied wind parameters at a certain location [9, 17]. However the performance of such an approach was not particularly remarkable, at short-time prediction mainly, especially when compared with simple techniques (e.g. Persistence). Wind forecasting is a stochastic process with a high level of non-stationarity. This fact prompted researchers to take into account *spatial correlation* of wind parameters at different locations [3, 19, 21].

This article describes two computational representations the first originating from the Fuzzy Lattice Neurocomputing (FLN) framework introduced in [11, 12, 20] and the second from Support Vector Machines and the Kernel-based framework. FLN is a scheme which combines elements from the Adaptive Resonance Theory (ART) [6], the mathematical theory of lattices [4] and

V. Petridis and V. Syrris: *Machine Learning Techniques for Environmental Data Estimation*, Studies in Computational Intelligence (SCI) **67**, 195–214 (2007)
`www.springerlink.com`

the theory of fuzzy sets [25]. The FL-framework can handle families of intervals. Also it can cope with disparate types of data including vectors, Boolean data, symbols, images, text, graphs, etc. On the other hand, we have a promising regression technique, the Support Vector Regression (SVR) which has been introduced by V. Vapnik and his collaborators [16, 23, 24]. It is deemed as a universal learning machine founded on the principle of Structural Risk Minimization which derives from statistical theory.

The goal herein is to test and validate the prediction hypothesis resolved by both a FLN clustering/regression and a SVR approach. We examine the case of real-time learning in a significantly non-stationary problem and we demonstrate how the spatial correlation improves the final learning performance. The application domain is the forecasting of wind speed at a meteorological station based on values also measured by two other stations presenting spatial correlation with the one in discussion.

10.2 Case Description

The study area is around the Thermaikos gulf, in the city of Thessaloniki, Greece (Fig. 10.1); it is a region rather flat and almost at sea level. The type of wind we are interested in is medium to high-speed. The prevailing winds blow from the north-northwest. Therefore, two meteorological stations have been situated at locations S_1 and S_2 that together with our location of interest, the local station S_0, are along the axis N-NW where strong winds appear. The wind speed distribution is similar in all three stations. Their distances are: $(S_1 S_2) = 27$ km, $(S_2 S_0) = 12$ km and obviously $(S_1 S_0) = 39$ km.

One minute measurements of wind speed and direction were collected at the aforementioned three locations for a year [1] and processed in order to remove non valid values and errors. The remaining set was averaged for every 15 minutes eliminating thus the measurement noise and the sudden wind variations. The resulting measurements set consists of 6 measurements at each instant $(v_{S_1}(t), \varphi_{S_1}(t), v_{S_2}(t), \varphi_{S_2}(t), v_{S_0}(t), \varphi_{S_0}(t))$ where v and φ denote speed and direction respectively. There are 3260 time instants.

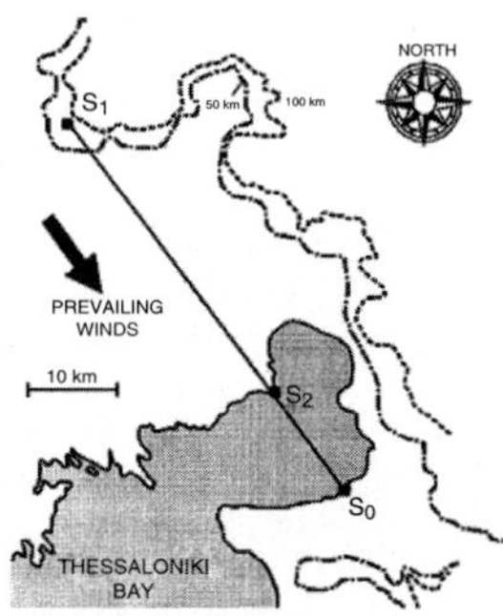

Fig. 10.1. The map shows the location of the three meteorological stations

We point out that the location of the three stations allows the exploitation of any *spatial correlations* among the locations involved. To this end we use neighboring areas measurements concerning wind events in order to achieve higher precision in the predictions regarding the study area. Our aim is to predict future values of site S_0 based on past values of all three stations.

In time-series prediction, the prediction origin, denoted t_0, is the time from which the prediction is generated. The time between the prediction origin and the predicted data point is the prediction horizon h or the time lead of the time series, while the stack of data used to conduct the prediction is defined as batch history data, b_{hd}, or look back length.

10.3 The FLN Forecasting Method

The reasoning behind the wind prediction problem as we formulate it is to exploit possible similarities and correlations among the three stations by means of FL-clustering application onto their historical data concerning wind speed and wind direction. Thus, the subject matter is simplified and cast as a problem of clustering; that is detecting sets of values the average of which, with the suitable adjustment, can be employed as predictor. In case of new evidence, the algorithm classifies it to a cluster. In fact this method is a regression technique involving two phases: a clustering process in the first phase and a classification process in the second phase of the algorithm. The method is applicable in a fuzzy lattice data domain and in this paper, in space R^4.

10.3.1 The FL-framework

A *lattice* L is a partially ordered set of which any two elements have a greatest lower bound (*meet*) denoted by $x \wedge y$ and a least upper bound (*join*) denoted by $x \vee y$. A lattice L is called *complete* when each of its subsets has a least upper bound and a greatest lower bound in L. A non-void complete lattice has a least element (O) and a greatest element (I).

At this point we mention a considerable asset of lattice theory in knowledge representation. A lattice L can be the Cartesian product $L = L_1 \times \cdots \times L_N$ of N constituent lattices $L_1, \cdots , L_N$. A product lattice L involving disparate constituent lattices has the potential of dealing with disparate types of data such as vectors of real numbers, propositions, fuzzy sets, events in a probability space, symbols, graphs, etc.

A useful function in a lattice L is a valuation function $v : L \to R$ defined by

$$v(x) + v(y) = v(x \wedge y) + v(x \vee y), x, y \in L.$$

A valuation is called positive if and only if $x < y \Rightarrow v(x) < x(y)$. An *inclusion measure* function σ is defined on a complete lattice L as a map $\sigma : L \times L \to [0, 1]$ such that for $u, w, x \in L$ the following three axioms are satisfied:

(Axm1) $\sigma(x, O) = 0, x \neq O$
(Axm2) $\sigma(x, x) = 1, \forall x \in L$
(Axm3) $u \leq w \Rightarrow \sigma(x, u) \leq \sigma(x, w)$ *Consistency Property*

Given a positive valuation function $v(.)$ in a lattice L an inclusion measure can be defined by the ratio $\sigma(x, u) = \frac{v(u)}{v(x \vee u)}$ [11]. The complete lattice $\tau(L)$ of intervals of lattice elements has been analyzed in [11]. In that work it was shown that an inclusion measure in a lattice L implies a fuzzy lattice, which can be defined as follows: *A fuzzy lattice is a pair $< L, \mu >$, where L is a crisp lattice and $(L \times L, \mu)$ is a fuzzy set with membership function $\mu : L \times L \to [0, 1]$ such that $\mu(x, y) = 1$ if and only if $x \leq y$. It turns out that $< L, \sigma >$ is a fuzzy lattice.*

Finally the interval size is defined given a positive valuation function $v(.)$ in a lattice L. The size of an interval $x = [a, b] \in \tau(L)$ is a function $Z : L \to R$ as expressed in $Z([a, b]) = v(b) - v(a)$.

10.3.2 Application of the method

In FL-data formulation we represent both parameters (speed and direction) of each station as a 2-dimensional vector: $x_{S_0}(t) = [v_{S_0}(t), \varphi_{S_0}(t)], x_{S_1}(t) = [v_{S_1}(t), \varphi_{S_1}(t)], x_{S_2}(t) = [v_{S_2}(t), \varphi_{S_2}(t)]$. The data set D_{FL} consists of $x_{S_0}(t)$, $x_{S_1}(t), x_{S_2}(t)$ for 3260 time instants. So the reference domain is considered to be the complete product lattice $L = R^2 \times R^2$ (called a rectangle), where a constituent complete lattice is the interval of real numbers. The prediction scheme takes the form:

$$\tilde{x}_{S_0}(t_0 + h) = f(x_{S_0}(t_0), ..., x_{S_0}(t_0 - (b_{hd} - 1)), x_{S_1}(t_0), ...,$$
$$x_{S_1}(t_0 - (b_{hd} - 1)), x_{S_2}(t_0), ..., x_{S_2}(t_0 - (b_{hd} - 1))) \tag{10.1}$$

During the experiments we set up the variables h and b_{hd} to be equal.

A positive valuation function in R is given by $v(x) = x$. The *inclusion measure* $\sigma : R \times R \to [0, 1]$ is defined as:

$$\sigma(x, y) = \frac{v(y)}{v(x \vee y)} = \frac{y}{\texttt{max}(x, y)}, x, y \in R.$$

In order to determine the inclusion measure of a point $u = (x, y)$ into the rectangle $w = (x_1, y_1) \times (x_2, y_2)$, we represent the former by its respective degenerate rectangle $u = (x, y) \times (x, y)$ and then we introduce the following formula that calculates the inclusion measure of u in w:

$$\sigma(u, w) = \frac{1}{2} \left(\frac{mean(x_1, x_2)}{\texttt{max}(x, mean(x_1, x_2))} + \frac{mean(y_1, y_2)}{\texttt{max}(y, mean(y_1, y_2))} \right) \tag{10.2}$$

where $mean(a, b) = \frac{a+b}{2}, a, b \in R.$

It is worth mentioning that all the points of the constituent lattices are comparable as we can see in the following example: consider the points $u = (0, 4)$ and $w = (3, 2)$. Their degenerate rectangles are: $u = (0, 4) \times (0, 4)$ and $w = (3, 2) \times (3, 2)$. According to the relationship (10.2) their respective inclusion measures are: $\sigma(u, w) = 1.5/2 = 0.75$ and $\sigma(w, u) = 1/2 = 0.5$ and consequently, we infer that $u \leq w$.

10.3.3 Algorithm description

The prediction scheme of (10.1) is implemented by means of the algorithm described in this subsection. Before we proceed to the essence of the FL learning mechanism, we should mention two crucial points for the fine tuning of the model. A major parameter is the *vigilance parameter* $\rho \in R$ which is necessary for the clustering procedure in order to adjust the size of the clusters. The vigilance parameter specifies the algorithm sensitivity. As ρ increases, the calculated rectangles size increases too. Another basic factor is the *correction* of the function outcome (10.1). After many experiments we obtain the best results when we adjust the vigilance parameter as $\rho = s$ (standard deviation of the sample/time window), while the prediction improves if we take its average with the last measured value, i.e. if $x_{S_0}(t)$ is the last known value at station S_0 measured at time t and $\tilde{x}_{S_0}(t+h)$ is the estimation at time $t+h$ given by the (10.1), then the final prediction is given by:

$$p_{S_0}(t+h) = \frac{x_{S_0}(t) + \tilde{x}_{S_0}(t+h)}{2} \tag{10.3}$$

Moreover, there is no need for any normalization of the experimental values.

The algorithm consists of two phases, the clustering procedure and the prediction function:

The clustering phase as an iterative procedure:

C1. A data set of n vectors is given.

C2. At the initial pass of the algorithm, the first vector is presented and its degenerate rectangle constitutes the first cluster. As the algorithm proceeds a set of clusters CR is created $CR = \{CR_1, ..., CR_L\}$ as described in the sequel. A cluster is represented by a rectangle in $R^2 \times R^2$.

C3. At each iteration a new vector $u = [x, y]$ is fed to the algorithm.

C4. The inclusion measure of $u = [x, y]$ is calculated with respect to each rectangle $CR_1, ..., CR_L$.

C5. Rectangles $CR_1, ..., CR_L$ compete over input $u = [x, y]$. Winner is the rectangle CR_W which includes u the most, i.e. it has the largest inclusion measure.

C6. Winner rectangle CR_W is augmented tentatively so as to include the degenerate rectangle $u = [x, y] \times [x, y]$. For instance, if $CR_W = [x_1, y_1] \times [x_2, y_2]$ then the union rectangle CR_U is calculated as:

$$CR_U = [\min(x, x_1), \min(y, y_1)] \times [\max(x, x_2), \max(y, y_2)]$$

C7. If size of CR_U is smaller than or equals the vigilance parameter then the rectangle CR_W is replaced by CR_U. Otherwise, reset occurs (i.e. the size of CR_W is reverted to its previous state and it is excluded from the list of the candidate rectangles to be compared with u), and the next winner is selected among the remaining rectangles.

C8. If all the rectangles have been reset, then input $u = [x, y]$ is learned as a new rectangle, $CR_{new} = [x, y] \times [x, y]$.

The prediction phase as a categorization procedure:

1. We set both the time horizon to be h steps from the last known measurement at S_0 (at the moment $t = 0$) and the look back length to be b_{hd}. This means that we want to estimate $x_{S_0}(t_0 + h)$ based on b_{hd} vectors.
2. The three vectors $x_{S_0}(t = 0) = [v_{S_0}(t = 0), \varphi_{S_0}(t = 0)], x_{S_1}(t = 0) = [v_{S_1}(t = 0), \varphi_{S_1}(t = 0)]$ and $x_{S_2}(t = 0) = [v_{S_2}(t = 0), \varphi_{S_2}(t = 0)]$ of the data set D_{FL} are selected.
3. The standard variation s of the above vectors is calculated.
4. For the clustering procedure [C1–C8] we use as data set the vectors $x_{S_0}(t = 0), x_{S_1}(t = 0), x_{S_2}(t = 0)$. The result is the generation of the set of cluster/s referred to as CR.
5. An input vector W is presented to the algorithm. When $t = 0$ then $W = x_{S_0}(t = 0)$. The inclusion measure of W is calculated for each rectangle of step 4 resulting thus in the inclusion measure vector.
6. The values of the inclusion measure vector are normalized to give the sum of 1 producing the column vector $[NIM(CR_i)]$, where $NIM(CR_i)$ is the normalized inclusion measure of rectangle CR_i, $i = 1, 2, ..., L$.
7. The mean of a rectangle $CR_i = [x_{1i}, y_{1i}] \times [x_{2i}, y_{2i}]$ is a point $M_i = \left(\frac{x_{1i}+x_{2i}}{2}, \frac{y_{1i}+y_{2i}}{2}\right) i = 1, 2, ..., L$. The estimation of $x_{S_0}(t_0 + h)$ is calculated by multiplying the mean of each rectangle with its respective normalized inclusion measure and then by summing each partial result, i.e. $\tilde{x}_{S_0}(t_0 + h) = [M_i] \times [NIM(CR_i)], i = 1, 2, ..., L$.
8. The estimate is corrected by averaging the produced result from step 7 with the last measurement (at the station S_0) as shown in (10.3). The outcome $p_{S_0}(t = 1)$ is the first estimation for one step ahead $h = 1$.
9. We check if we have reached the desired time horizon. If so, then the algorithm stops otherwise it continues to the next step.
10. A new data set is created using the vectors $x_{S_0}(t = 0), x_{S_1}(t = 0), x_{S_2}(t = 0)$, the estimate $p_{S_0}(t = 1)$ and the vectors for one step back (except the case where we have reached the look back length b_{hd}) $x_{S_0}(t = -1), x_{S_1}(t = -1), x_{S_2}(t = -1)$, i.e. $NewData = \{x_{S_0}(t_{-1}), x_{S_1}(t_{-1}), x_{S_2}(t_{-1}), x_{S_0}(t_0), x_{S_1}(t_0), x_{S_2}(t_0), \tilde{x}_{S_0}(t_1)\}$.
11. The vigilance parameter is taken equal to the standard variation of the $NewData$ data set.
12. We apply the clustering procedure [C1-C8] to the data set defined in step 10. New rectangles are formed representing the knowledge generated by the contribution of new evidence.
13. The flow of the algorithm goes to stage 5 by using as new input the estimated value $p_{S_0}(t = 1)$ (in general, by using the last estimated value), i.e. $W = p_{S_0}(t = 1)$, and CR is the set of clusters generated in step 12.

10.4 The SVR Method

The theory of Support Vector Machines has been extended in order to cope with regression problems. Suppose that we have a data set $\{(x_i, y_i), i = 1, 2, ..., n\}$ of measurements where $x_i \in X$ and $y_i \in R$ (X denotes the space of the input patterns and R is the set of real numbers). We consider that the tuples (x_i, y_i) are taken from an unknown distribution $P(x, y)$.

10.4.1 Linear regression

The formulation of the linear regression task stated by Vapnik [23] is the following *convex optimization problem*:

$$\text{minimize } \tfrac{1}{2} \|w\|^2 + C \sum_{i=1}^{n} (\xi_i + \xi_i^*)$$

$$\text{subject to } \begin{cases} y_i - \langle w, x_i \rangle - b \le \varepsilon + \xi_i \\ \langle w, x_i \rangle + b - y_i \le \varepsilon + \xi_i^* \\ \xi_i, \xi_i^* \ge 0 \end{cases} \tag{10.4}$$

where $\langle \cdot, \cdot \rangle$ refers to the inner product in X. $C > 0$ is a pre-specified value that determines the trade-off between the flatness of regressor and the amount up to which deviations larger than ε are tolerated and ξ_i, ξ_i^* are slack variables interpreted as: ξ for exceeding the target value by more than ε and ξ^* for being more than ε below the target value (soft margin regression).

In addition to (10.4) a loss function is minimized. For our SV-models we choose the linear ε-insensitive loss function :

$$L_\varepsilon = \begin{cases} 0 \text{ for } |y_i - (\langle w, x_i \rangle - b)| \le \varepsilon \\ |y_i - (\langle w, x_i \rangle - b)| - \varepsilon \text{ otherwise} \end{cases}$$

This defines an ε tube in such a way that if the predicted value is within the tube then the loss is zero, whereas if the predicted point is outside the tube, the loss is the magnitude of the difference between the predicted value and the radius ε of the tube.

To assure that the training data appear as inner products among the vectors and to better handle the constraints, the problem is transformed into a Lagrangian formulation named the primal form:

$$L_P = \tfrac{1}{2} \|w\|^2 + C \sum_{i=1}^{n} (\xi_i + \xi_i^*) - \sum_{i=1}^{n} (\eta_i \xi_i + \eta_i^* \xi_i^*)$$

$$- \sum_{i=1}^{n} a_i \left(\varepsilon + \xi_i - (y_i - \langle w, x_i \rangle - b) \right) \tag{10.5}$$

$$- \sum_{i=1}^{n} a_i^* \left(\varepsilon + \xi_i^* - (\langle w, x_i \rangle + b - y_i) \right)$$

where $a_i, a_i^*, \eta_i, \eta_i^* \ge 0$ are the Lagrange multipliers.

Differentiating with respect to w, b, ξ_i, ξ_i^* gives:

$$\frac{\partial L}{\partial w} = w - \sum_{i=1}^{n} (a_i - a_i^*)\, x_i = 0$$

$$\frac{\partial L}{\partial b} = \sum_{i=1}^{n} (a_i^* - a_i) = 0$$

$$\frac{\partial L}{\partial \xi_i} = C - a_i - \eta_i = 0$$

$$\frac{\partial L}{\partial \xi_i^*} = C - a_i^* - \eta_i^* = 0 \tag{10.6}$$

Substituting for w into (10.5) and using the relations (10.6) the dual optimization problem is obtained:

$$L_D = -\tfrac{1}{2} \sum_{i=1}^{n} \sum_{j=1}^{n} (a_i - a_i^*)\,(a_j - a_j^*)\,\langle x_i, x_j \rangle$$

$$- \varepsilon \sum_{i=1}^{n} (a_i + a_i^*) + \sum_{i=1}^{n} (a_i - a_i^*)\, y_i \tag{10.7}$$

This function is maximized subject to:

$$\sum_{i=1}^{n} (a_i - a_i^*) = 0 \text{ and } 0 \le a_i, a_i^* \le C \tag{10.8}$$

Solving the first equation of (10.6) for w gives: $w = \sum_{i=1}^{n} (a_i - a_i^*)\, x_i$ so the candidate linear function takes the form:

$$f(x) = wx + b = \sum_{i=1}^{n} (a_i - a_i^*)\, \langle x_i, x \rangle + b \tag{10.9}$$

The Karush-Kuhn-Tucker complementary conditions [14, 15] are:

$$a_i \left(\varepsilon + \xi_i - (y_i - \langle w, x_i \rangle - b) \right) = 0$$
$$a_i^* \left(\varepsilon + \xi_i^* - (\langle w, x_i \rangle + b - y_i) \right) = 0$$
$$\eta_i \xi_i = (C - a_i)\, \xi_i = 0$$
$$\eta_i^* \xi_i^* = (C - a_i^*)\, \xi_i^* = 0 \tag{10.10}$$

The (10.10) imply that $a_i a_i^* = 0$ which means that the set of *dual variables* can never be nonzero at the same time. Those patterns x_i with $a_i > 0$ or $a_i^* > 0$ are support vectors. If $a_i \in (0, C)$ or $a_i^* \in (0, C)$ then (x_i, y_i) lies on the boundary of the tube surrounding the regression function at distance ε. Moreover, if $a_i = C$ or $a_i^* = C$ then the point lies outside the tube. Thus, the parameter b is computed as follows:

$$b = y_i - \langle w, x_i \rangle - \varepsilon \text{ for } a_i \in (0, C)$$
$$b = y_i - \langle w, x_i \rangle + \varepsilon \text{ for } a_i^* \in (0, C)$$

10.4.2 Nonlinear regression

In this case the nonlinear function has the form: $f(x) = \langle w, \phi(x) \rangle + b$ where $\phi(x)$ is the image of input vector x in a high dimensional space. Using the trick of kernel functions [7] the dual form (10.5) is replaced by the quadratic problem:

$$L_D = -\frac{1}{2} \sum_{i=1}^{n} \sum_{j=1}^{n} \left(a_i - a_i^*\right) \left(a_j - a_j^*\right) K\left(x_i, x_j\right)$$
$$-\varepsilon \sum_{i=1}^{n} \left(a_i + a_i^*\right) + \sum_{i=1}^{n} \left(a_i - a_i^*\right) y_i \tag{10.11}$$

where $K\left(x_i, x_j\right) = \langle \phi(x_i), \phi(x_j) \rangle$ is a *kernel* function satisfying Mercer's conditions [8]. The relationship (10.11) is maximized subject to the constraints (10.8). Finally, the regressor takes the form:

$$f(x) = \sum_{i=1}^{n} \left(a_i - a_i^*\right) K\left(x_i, x\right) + b \tag{10.12}$$

In that way, we manage to apply linear regression not in the low dimensional input space but in a high dimensional (feature) space via the kernel function which makes the mapping implicitly, i.e. without knowing $\phi(x)$.

10.4.3 The prediction of SV-models

As analyzed in previous section, the objective of SVR is to construct a hyperplane that lies "close" to as many of the data points as possible [22]. The solution is obtained as a set of support vectors that can be sparse. These lie on the boundary and as such summarize the information required to separate the data.

SVR has an advantage over other function estimation methods: it is capable of controlling the capacity of the hypothesis; the algorithm selects the subspace of the hypothesis space that is optimal in terms of some bound on the generalization of the hypothesis. This fact leads to the choice of the kernel function, the width of the ε-insensitive zone and the capacity control C for controlling the regression model. Since there is not a rigorous way to select the finest set of hyperparameters, i.e. kernel, ε and C, we use the leave-one-out cross-validation method by constructing successive time windows of size $b_{hd} + 1$, i.e. we utilize b_{hd} vectors in order to predict one value h steps ahead.

After a sufficient number of experiments, we obtained the best results when we used the RBF kernel $K\left(x_i, x_j\right) = \langle \phi(x_i), \phi(x_j) \rangle = e^{-\frac{(x_i - x_j)^2}{2\sigma^2}}$, where the width of radial basis functions equals to $\sigma = 0.08$ and the other parameters were tuned to the values: $\varepsilon = 0.05$, $C = 1$.

We developed two different models :

$M1$. Solve the quadratic optimization problem (10.11), calculate the Lagrange multipliers and the bias and finally use the regressor (10.12) to get the prediction.

$M2$. Solve the quadratic optimization problem (10.11) and express the final prediction as a linear dependence of the arisen support vectors.

Each of the above models has two alternative versions concerning the input data type and the output:

a. We define the sets $V_1 = \{x_i(t) = [v_{S_0}(t) \; v_{S_1}(t) \; v_{S_2}(t)]$ for $t = 1, 2, ..., 3260\}$ (it consists of 3260 3-dimensional vectors) and $V_2 = \{v_{S_0}(t), v_{S_1}(t), v_{S_2}(t)$ for $t = 1, 2, ..., 3260\}$ (it consists of 3×3260 scalars). Thus, the inputs of the models are either i) 3-dimensional vectors that is $x_i \in V_1$ or ii) 1-dimensional vectors that is $x_i \in V_2$.

b. The outcome of the models $\tilde{x}_{S_0}(t_0 + h)$ can either i) remain intact or ii) be averaged with the last measured value $x_{S_0}(t_0)$ at S_0, i.e.

$$pred_{S_0}(t_0 + h) = \frac{x_{S_0}(t_0) + \tilde{x}_{S_0}(t_0 + h)}{2} \tag{10.13}$$

For simplicity reasons, when we refer, for instance, to model $M1.a_{(i)}.b_{(ii)}$ we mean the option which uses the regressor (10.12), it takes 3-dimensional vectors as input and its output results from (10.13) (i.e. model $M1$, versions $a_{(i)}$ and $b_{(ii)}$). The same convention holds for all the other options.

10.4.4 Algorithms description

A sliding window is used containing $x(t), x(t-1), ..., x(t-b_{hd}+1)$. For example, we consider as look back length 3 steps and as future time horizon 10 steps. If we start at the moment $t_0 = 6$, we use three vectors at time instances $t_{-2} = 4$, $t_{-1} = 5$ and $t_0 = 6$ as historical data which lead to the prediction at $t_{10} = 16$. At next moment $t_0 = 7$ we use the vectors at time instances $t_{-2} = 5$, $t_{-1} = 6$ and $t_0 = 7$ to make a prediction at time $t_{10} = 17$ etc. The algorithms of the previous section are described more analytically below:

Algorithm $M1.a_{(i)}.b_{(i)}$

S1) Set future time horizon h, the look back length b_{hd} and the prediction origin t_0.

S2) Normalize data in $[-1,1]$.

S3) For $i = t_0$ to 3260:

    S3.1) The vectors x_i in regressor (10.12) are the vectors in the interval $(t_0 - b_{hd} + 1, t_0)$: $[x_{S_1}(t_0 - k), x_{S_2}(t_0 - k), x_{S_0}(t_0 - k)]$ and $y_i = x_{S_0}(t_0 - k + h)$, where $k = m - 1$ and $m = (b_{hd} - 1), ..., 2, 1$.

    S3.2) Solve the quadratic problem.

S3.3) Use the regressor function (10.12) where
$$x = [x_{S_1}(t_0 + h), x_{S_2}(t_0 + h), x_{S_0}(t_0 + h)]$$
S3.4) The regressor's output $\tilde{x}_{S_0}(t_0 + 2h)$ is the prediction for h steps ahead.

Algorithm $M2.a_{(i)}.b_{(i)}$

Until $S3.2$ the steps are the same with the previous algorithm.

— S3.3) We take into account the $x_{S_0}(t_0 + h)$ component of the input vector x. Then we consider the $x_{S_0}(t + h)$ component of the support vector SV_i, $i = 1, ..., M$ (where M is the number of support vectors) denoted by $x_{S_0}^{SV_i}$. Finally, the column vector $D_{ist} = [|x_{S_0}^{SV_1} - x_{S_0}(t_0 + h)|, ..., |x_{S_0}^{SV_M} - x_{S_0}(t_0 + h)|]^T$ is computed. The D_{ist} is normalized as $ND_{ist} = D_{ist} / \sum\limits_{i=1}^{M} |x_{S_0}^{SV_i} - x_{S_0}(t_0 + h)|$.

— S3.4) The final prediction is: $x_{S_0}(t_0 + 2h) = [x_{S_0}^{SV_1}, ..., x_{S_0}^{SV_M}] \times ND_{ist}$.

In the cases $1.a_{(ii)}$ and $2.a_{(ii)}$ we modify step $S3.1$ as:

Select all vectors in the interval $(t_0 - b_{hd} + 1, t_0)$: $[x_{S_1}(t_0 - k), x_{S_0}(t_0 - k + h)], [x_{S_2}(t_0 - k), x_{S_0}(t_0 - k + h)], [x_{S_0}(t_0 - k), x_{S_0}(t_0 - k + h)]$, where $k = m - 1$ and $m = (b_{hd} - 1), ..., 2, 1$.

In the cases $1.b_{(ii)}$ and $2.b_{(ii)}$ we change the outcome of step $S3.4$ using (10.13).

10.5 Model Evaluation

In order to test the performance of the models we compare the degree of their prediction success with other approaches such as:

Persistence: The most commonly used reference model for short term forecasting of wind is the Persistence method which assumes that: "*the conditions that existed at the beginning of the forecast period will continue or persist through to the end of the period*"[5]:

$$\tilde{y}_{t+h} = y_t \text{ where } y_t \text{ is the last measured value}$$

This method is not only the simplest modeling approach but is also the most economical to implement, and surprisingly accurate for short term forecasting (1 to 5 hours).

Moving Average: It is a widely used forecasting method constituting a generalization of the Persistence model. This simple approach is based on the average value of the variable over a specific number of preceding periods. In this paper we define the Moving Average method as

$$\tilde{y}_{t+h} = \tfrac{1}{h} \sum\limits_{j=0}^{h-1} y_{t-j}, h = 1, 2, ..., n, \text{ where } h = b_{hd}$$

As h goes to infinity the Moving Average tends to the global average: $\tilde{y}_{t+h} = \bar{y}_t$ where $\bar{y}_t$ is the average of all the available measurements until time t.

Recurrent Neural Networks: A Recurrent Neural Network employs feedback connections and has the potential to represent "certain computational structures in a more parsimonious fashion"[10]. RNNs address the temporal relationship of their inputs by maintaining an internal state. In the latest bibliography regarding wind forecasting we find the RNN algorithm in [2] and because of its having been applied on the same data set displaying good performance we compare it to the FL-model.

To measure the precision of the models (i.e. how the model output is close to the real value) we make use of three statistical types of errors:

- The Mean Absolute Error: $MAE = \frac{1}{N} \sum_{i=1}^{N} |y_i - \tilde{y}_i|$

- The Normalized Mean Square Error: $NMSE = \dfrac{\frac{1}{N} \sum_{i=1}^{N} \left\{ \left(y_i - \tilde{y}_i\right)^2 \right\}}{\frac{1}{N} \sum_{i=1}^{N} \left\{ \left(y_i - \frac{1}{N} \sum_{i=1}^{N} y_i \right)^2 \right\}}$

- The Root Mean Square Error: $RMSE = \sqrt{\frac{1}{N} \sum_{i=1}^{N} \left\{ (y_i - \tilde{y}_i)^2 \right\}}$

where y_i is the real value, $\tilde{y}_i$ is the model output and N is the number of tested values. Smaller values of the aforesaid uncertainty statistics denote better model performance. The reason for using these criteria is that they operate independently of application and target value specification while they are not biased towards models that over or under predict.

10.6 Experimental Results

The FL-Approach: Tables 1.2, 1.3 regarding both parameters of wind (speed and direction) display the comparison analysis among the reference models and the proposed approach. We use symbol "-" when the value is not provided. Explanation of the table columns is given in Table 1.1.

For the validation of the result, we adopt the *sliding window* where $2h \times 3$ is the window size for all the 3260 vectors and h is the time horizon within which the variable to be predicted lies. When we try to make a prediction h steps ahead from the moment t we take as input values the time frame $h - 1$ steps back from t, i.e. $t, t - 1, ..., t - (h - 1)$ (each step refers to a period of 15 minutes).

The SV-Approach: The accuracy of each forecasting method is determined by its specific architecture [8]. The SVR algorithms offer a sufficient number of degrees of freedom in customizing models to a particular forecasting task. As we mentioned before, in order to evaluate the candidate models and determine

TABLE 1.1: EXPLANATION OF TABLES 1.2 AND 1.3

Column	Comments
1	Time horizon ahead in minutes & hours.
2	Time horizon ahead in steps.
3	Time steps backwards. This column does not refer to the RNN approach since this model needs to be trained at a sufficient number of steps back in order to minimize the difference between real and estimated value.
4-5-6	NMSE, RMSE and MAE of the persistence method.
7-8-9	NMSE, RMSE and MAE of the moving average method.
10-11-12-13	NMSE of the moving average method upon the sample data used in [2], NMSE, percentage improvement compared to the NMSE moving average model [2] and MAE of the RNN algorithm [2].
14-15-16-17	NMSE, percentage improvement compared to the NMSE moving average model, RMSE and MAE of the FL-prediction approach based on information provided by the station S_0.
18-19-20-21	NMSE, percentage improvement compared to the NMSE moving average model, RMSE and MAE of the FL-prediction approach based on information provided by all stations S_0, S_1, S_2.

TABLE 1.2
COMPARISON OF THE FL-APPROACH WITH THE REFERENCE MODELS IN PREDICTING WIND SPEED

Minutes ahead	Steps ahead	Look back length	Persistence			Moving Average			RNN				FL-Clustering Approach							
													S0				S0, S1, S2			
			NMSE	RMSE	MAE	NMSE	RMSE	MAE	MA NMSE	NMSE	%	MAE	NMSE	%	RMSE	MAE	NMSE	%	RMSE	MAE
15	1	1	0.0327	0.7415	0.5226	0.0327	0.7415	0.5226	0.0498	0.0366	26.55	-	0.0327	0.00	0.7415	0.5226	0.0327	0.00	0.7415	0.5226
30	2	2	0.0569	1.0848	0.7680	0.0629	1.1405	0.7972	0.1008	0.0746	25.98	-	0.0569	9.54	1.0848	0.7680	0.0569	9.54	1.0848	0.7680
60-1h	4	4	0.0985	1.4601	1.0268	0.1099	1.5425	1.0905	0.1778	0.1337	24.79	-	0.0976	11.19	1.4537	1.0200	0.0874	20.47	1.3754	0.9818
90	6	6	0.1162	1.7118	1.2201	0.1352	1.8466	1.3106	0.2448	0.1860	24.01	-	0.1158	14.35	1.7091	1.2108	0.1026	24.11	1.6087	1.1454
120-2h	8	8	0.1868	1.9428	1.3856	0.2180	2.0988	1.4985	0.3239	0.2474	23.61	-	0.1852	15.05	1.9345	1.3724	0.1650	24.31	1.8259	1.2903
150	10	10	0.1938	2.1339	1.5332	0.2250	2.2993	1.6317	0.3830	0.2983	22.11	-	0.1910	15.11	2.1185	1.5066	0.1708	24.09	2.0036	1.4120
180-3h	12	12	0.2253	2.3085	1.6519	0.2588	2.4742	1.7503	0.4292	0.3392	20.98	-	0.2208	14.68	2.2851	1.6154	0.1993	22.99	2.1711	1.5157
360-6h	24	24	0.5807	3.0368	2.1845	0.7182	3.3771	2.4654	-	-	-	-	0.5855	18.48	3.0494	2.1909	0.5443	24.21	2.9399	2.0886
720-12h	48	48	0.9843	4.1798	3.1322	1.0909	4.4003	3.2207	-	-	-	-	-	-	-	-	0.8570	21.44	3.9002	2.8951

TABLE 1.3
COMPARISON OF THE FL-APPROACH WITH THE REFERENCE MODELS IN PREDICTING WIND DIRECTION

Minutes ahead	Steps ahead	Look back length	Persistence			Moving Average			RNN				FL-Clustering Approach							
													S0				S0, S1, S2			
			NMSE	RMSE	MAE	NMSE	RMSE	MAE	MA NMSE	NMSE	%	MAE	NMSE	%	RMSE	MAE	NMSE	%	RMSE	MAE
15	1	1	0.0317	0.2601	0.1288	0.0317	0.2601	0.1288	-	-	-	-	0.0317	0.00	0.2601	0.1288	0.0317	0.00	0.2601	0.1288
30	2	2	0.058	0.3687	0.1904	0.0618	0.3806	0.2018	-	-	-	-	0.0580	6.15	0.3687	0.1904	0.0580	6.15	0.3687	0.1904
60-1h	4	4	0.0964	0.4788	0.2632	0.0968	0.4800	0.2738	-	-	-	-	0.0901	6.92	0.4631	0.2609	0.0821	15.19	0.4421	0.2869
90	6	6	0.1721	0.5335	0.3014	0.1801	0.5458	0.3268	-	-	-	-	0.1602	11.05	0.5148	0.3020	0.1389	22.88	0.4793	0.3131
120-2h	8	8	0.1958	0.5874	0.3408	0.2099	0.6081	0.3792	-	-	-	-	0.1821	13.24	0.5664	0.3441	0.1518	27.68	0.5172	0.3422
150	10	10	0.2233	0.6338	0.3749	0.2444	0.6631	0.4290	-	-	-	-	0.2074	15.14	0.6108	0.3818	0.1692	30.77	0.5517	0.3693
180-3h	12	12	0.1997	0.6769	0.4117	0.2210	0.7120	0.4751	-	-	-	-	0.1850	16.29	0.6516	0.4206	0.1483	32.90	0.5833	0.3975
360-6h	24	24	0.8155	0.8783	0.5975	0.8650	0.9046	0.6678	-	-	-	-	0.7335	15.20	0.8330	0.5935	0.6003	30.60	0.7536	0.5492
720-12h	48	48	0.5219	1.0702	0.7925	0.3507	0.8772	0.6753	-	-	-	-	-	-	-	-	0.3249	7.36	0.8653	0.6770

the suitable architecture we adopt the leave-one-out cross-validation method; we select these ones with the lowest error on the validation data set.

The following Tables (2.1, 3.1, 4.1) display the performance of the proposed SV-models. We show these results which correspond to look back length $b_{hd} = 3$ and 6 and to time horizon $h = 1, 6, 20, 40, 80$ and 160 steps ahead. At each instant the first line of the tables refers to $\tilde{x}_{S_0}(t_0 + h)$ and the second line refers to $pred_{S_0}(t_0 + h)$.

TABLE 2.1

COMPARISON OF SV MODELS (1-DIMENSIONAL INPUT VECTORS) WITH THE REFERENCE MODELS

| Minutes ahead | Steps ahead | Look back length | Persistence | | | Moving Average | | | SV Models | | | | | | | | | | | |
| | | | | | | | | | S0 svs | | | S0 with kernel | | | S0, S1, S2 svs | | | S0, S1, S2 with kernel | | |
			NMSE	RMSE	MAE	NMSE	RMSE	MAE	NMSE	RMSE	MAE	NMSE	RMSE	MAE	NMSE	RMSE	MAE	NMSE	RMSE	MAE
15	1	3	0.03	0.09	0.06	0.06	0.13	0.09	0.10	0.17	0.12	2.07	0.76	0.18	0.21	0.24	0.19	0.11	0.17	0.12
15	1	3	0.03	0.09	0.06	0.06	0.13	0.09	0.05	0.11	0.08	0.54	0.39	0.11	0.07	0.14	0.11	0.05	0.12	0.08
90-1.5h	6	3	0.13	0.21	0.15	0.14	0.22	0.16	0.17	0.24	0.17	1.78	0.77	0.23	0.21	0.26	0.20	0.18	0.25	0.18
90-1.5h	6	3	0.13	0.21	0.15	0.14	0.22	0.16	0.14	0.21	0.15	0.53	0.42	0.18	0.13	0.21	0.15	0.14	0.22	0.16
300-5h	20	3	0.33	0.35	0.25	0.34	0.35	0.25	0.36	0.36	0.26	1.85	0.81	0.31	0.34	0.35	0.25	0.35	0.35	0.26
300-5h	20	3	0.33	0.35	0.25	0.34	0.35	0.25	0.33	0.35	0.25	0.70	0.50	0.27	0.30	0.33	0.23	0.33	0.34	0.25
600-10h	40	3	0.93	0.47	0.35	0.94	0.47	0.35	0.97	0.48	0.36	3.11	0.87	0.39	0.87	0.46	0.33	0.92	0.47	0.36
600-10h	40	3	0.93	0.47	0.35	0.94	0.47	0.35	0.93	0.47	0.35	1.43	0.59	0.36	0.84	0.45	0.33	0.90	0.47	0.35
1200-20h	80	3	1.52	0.61	0.44	1.51	0.60	0.44	1.53	0.61	0.44	3.64	0.94	0.47	1.35	0.57	0.42	1.43	0.59	0.44
1200-20h	80	3	1.52	0.61	0.44	1.51	0.60	0.44	1.51	0.60	0.44	1.98	0.69	0.45	1.37	0.58	0.42	1.45	0.59	0.44
2400-40h	160	3	1.56	0.56	0.43	1.52	0.55	0.42	1.52	0.56	0.42	4.11	0.91	0.45	1.43	0.54	0.40	1.43	0.54	0.42
2400-40h	160	3	1.56	0.56	0.43	1.52	0.55	0.42	1.52	0.55	0.42	2.11	0.65	0.43	1.42	0.54	0.41	1.46	0.54	0.42

The algorithms S0 svs and S0 with kernel use as input data only the measurements of station S_0. The S0 svs and S0,S1,S2 svs are $M2.a_{(ii)}$ algorithms while the S0 with kernel and S0,S1,S2 with kernel are $M1.a_{(ii)}$ algorithms.

TABLE 2.2

IMPROVEMENT (%) OF SV MODELS (1-DIMENSIONAL INPUT VECTORS) AND PERSISTENCE RELATED TO MOVING AVERAGE

| Minutes ahead | Steps ahead | Look back length | Persistence | | | SV Models | | | | | | | | | | | |
| | | | | | | S0 svs | | | S0 with kernel | | | S0, S1, S2 svs | | | S0, S1, S2 with kernel | | |
			NMSE	RMSE	MAE	NMSE	RMSE	MAE	NMSE	RMSE	MAE	NMSE	RMSE	MAE	NMSE	RMSE	MAE
15	1	3	46.60	26.93	26.51	-76.26	-32.86	-35.70	-3548.56	-504.49	-107.44	-268.02	-91.98	-114.13	-95.44	-39.80	-38.01
15	1	3	46.60	26.93	26.51	17.53	9.12	7.78	-847.58	-208.06	-27.18	-21.31	-10.23	-20.50	13.44	6.96	4.86
90-1.5h	6	3	8.29	4.24	3.78	-20.02	-9.55	-10.53	-1136.02	-251.57	-50.36	-44.34	-20.14	-26.89	-25.05	-11.74	-16.57
90-1.5h	6	3	8.29	4.24	3.78	4.63	2.34	2.23	-270.65	-92.52	-14.66	10.92	5.62	3.64	3.93	2.06	-0.96
300-5h	20	3	1.45	0.73	0.21	-6.55	-3.30	-3.71	-447.12	-134.07	-23.86	-0.52	-0.33	-1.04	-4.02	-1.99	-6.18
300-5h	20	3	1.45	0.73	0.21	1.51	0.69	0.72	-105.6	-43.49	-6.94	11.91	6.08	6.83	3.62	1.83	-0.25
600-10h	40	3	0.79	0.40	0.77	-3.56	-1.84	-2.21	-231.99	-82.34	-11.17	7.44	3.72	5.44	1.67	0.84	-1.41
600-10h	40	3	0.79	0.40	0.77	0.86	0.36	0.40	-52.10	-23.42	-1.87	10.36	5.25	7.25	3.88	1.96	1.35
1200-20h	80	3	-0.68	-0.34	-1.01	-1.59	-0.87	-0.94	-141.07	-55.38	-6.80	10.86	5.52	4.94	5.59	2.83	0.07
1200-20h	80	3	-0.68	-0.34	-1.01	0.31	0.08	-0.04	-31.24	-14.64	-1.73	9.02	4.55	5.34	4.09	2.06	0.78
2400-40h	160	3	-2.91	-1.44	-1.77	-0.46	-0.26	-0.15	-171.38	-64.78	-6.85	5.88	2.96	3.67	5.74	2.91	1.13
2400-40h	160	3	-2.91	-1.44	-1.77	0.03	-0.01	-0.01	-38.98	-17.92	-1.89	6.14	3.09	3.15	3.38	1.71	0.84

TABLE 2.3

IMPROVEMENT (%) OF SV MODELS (1-DIMENSIONAL INPUT VECTORS) AND MOVING AVERAGE RELATED TO PERSISTENCE

| Minutes ahead | Steps ahead | Look back length | Moving Average | | | SV Models | | | | | | | | | | | |
| | | | | | | S0 svs | | | S0 with kernel | | | S0, S1, S2 svs | | | S0, S1, S2 with kernel | | |
			NMSE	RMSE	MAE	NMSE	RMSE	MAE	NMSE	RMSE	MAE	NMSE	RMSE	MAE	NMSE	RMSE	MAE
15	1	3	-87.27	-36.85	-36.07	-230.09	-81.82	-84.64	-6732.72	-727.23	-182.26	-589.20	-162.72	-191.36	-266.01	-91.31	-87.79
15	1	3	-87.27	-36.85	-36.07	-54.43	-24.37	-25.48	-1674.55	-321.57	-73.05	-127.19	-50.84	-63.96	-62.10	-27.32	-29.45
90-1.5h	6	3	-9.04	-4.42	-3.93	-30.87	-14.40	-14.88	-1247.78	-267.12	-56.27	-57.39	-25.46	-31.87	-36.35	-16.69	-21.15
90-1.5h	6	3	-9.04	-4.42	-3.93	-4.00	-1.98	-1.61	-304.16	-101.04	-19.17	2.87	1.45	-0.14	-4.75	-2.28	-4.92
300-5h	20	3	-1.47	-0.73	-0.21	-8.12	-4.06	-3.92	-455.18	-135.79	-24.12	-2.00	-1.07	-1.25	-5.55	-2.74	-6.40
300-5h	20	3	-1.47	-0.73	-0.21	0.06	-0.04	0.51	-108.62	-44.54	-7.17	10.61	5.39	6.63	2.21	1.11	-0.46
600-10h	40	3	-0.80	-0.40	-0.78	-4.39	-2.25	-3.01	-234.64	-83.07	-12.04	6.71	3.34	4.70	0.89	0.45	-2.20
600-10h	40	3	-0.80	-0.40	-0.78	0.07	-0.04	-0.38	-53.31	-23.91	-2.66	9.65	4.87	6.53	3.11	1.57	0.58
1200-20h	80	3	0.68	0.34	1.00	-0.90	-0.52	0.07	-139.43	-54.85	-5.73	11.47	5.84	5.88	6.23	3.17	1.07
1200-20h	80	3	0.68	0.34	1.00	0.98	0.42	0.95	-30.34	-14.25	-0.71	9.64	4.87	6.28	4.74	2.40	1.77
2400-40h	160	3	2.83	1.42	1.74	2.38	1.17	1.58	-163.71	-62.43	-4.99	8.54	4.34	5.34	8.40	4.29	2.85
2400-40h	160	3	2.83	1.42	1.74	2.86	1.41	1.73	-35.05	-16.24	-0.12	8.79	4.47	4.83	6.11	3.11	2.56

Tables 2.2, 2.3, 3.2, 3.3, 4.2 and 4.3 provide the improvement percentage with respect to both reference models, i.e. Persistence and Moving Average. For example, the improvement percentage of SV model $S_0, S_1, S_2 svs$ over the

TABLE 3.1

COMPARISON OF SV MODELS (3-DIMENSIONAL INPUT VECTORS) WITH THE REFERENCE MODELS

Minutes ahead	Steps ahead	Look back length	Persistence			Moving Average			SV Models					
									S0,S1,S2-3d svs			S0,S1,S2-3d kernel		
			NMSE	RMSE	MAE	NMSE	RMSE	MAE	NMSE	RMSE	MAE	NMSE	RMSE	MAE
15	1	3	0.03	0.09	0.06	0.06	0.13	0.09	0.30	0.29	0.22	0.48	0.36	0.30
15	1	3	0.03	0.09	0.06	0.06	0.13	0.09	0.09	0.16	0.12	0.14	0.19	0.16
90-1.5h	6	3	0.13	0.21	0.15	0.14	0.22	0.16	0.27	0.30	0.23	0.43	0.38	0.31
90-1.5h	6	3	0.13	0.21	0.15	0.14	0.22	0.16	0.14	0.21	0.16	0.18	0.25	0.20
300-5h	20	3	0.33	0.35	0.25	0.34	0.35	0.25	0.37	0.36	0.27	0.50	0.42	0.35
300-5h	20	3	0.33	0.35	0.25	0.34	0.35	0.25	0.29	0.32	0.23	0.33	0.34	0.27
600-10h	40	3	0.93	0.47	0.35	0.94	0.47	0.35	0.84	0.45	0.33	0.96	0.48	0.40
600-10h	40	3	0.93	0.47	0.35	0.94	0.47	0.35	0.80	0.44	0.32	0.81	0.44	0.35
1200-20h	80	3	1.52	0.61	0.44	1.51	0.60	0.44	1.23	0.55	0.40	1.23	0.54	0.45
1200-20h	80	3	1.52	0.61	0.44	1.51	0.60	0.44	1.29	0.56	0.40	1.24	0.55	0.43
2400-40h	160	3	1.56	0.56	0.43	1.52	0.55	0.42	1.39	0.53	0.39	1.34	0.52	0.44
2400-40h	160	3	1.56	0.56	0.43	1.52	0.55	0.42	1.37	0.53	0.40	1.28	0.51	0.40

The S0,S1,S2-3d svs is a $M2.a_{(i)}$ algorithm and the S0,S1,S2-3d kernel is a $M1.a_{(i)}$ algorithm

TABLE 3.2

IMPROVEMENT (%) OF SV MODELS (3-DIMENSIONAL INPUT VECTORS) AND PERSISTENCE RELATED TO MOVING AVERAGE

Minutes ahead	Steps ahead	Look back length	Persistence			SV Models					
						S0,S1,S2-3d svs			S0,S1,S2-3d kernel		
			NMSE	RMSE	MAE	NMSE	RMSE	MAE	NMSE	RMSE	MAE
15	1	3	46.60	26.93	26.51	-428.68	-130.10	-149.67	-743.65	-190.68	-238.73
15	1	3	46.60	26.93	26.51	-57.94	-25.77	-35.41	-140.94	-55.34	-80.86
90-1.5h	6	3	8.29	4.24	3.78	-84.84	-35.95	-44.66	-198.45	-72.76	-101.10
90-1.5h	6	3	8.29	4.24	3.78	4.80	2.43	-1.03	-27.70	-13.01	-28.47
300-5h	20	3	1.45	0.73	0.21	-8.84	-4.40	-7.21	-47.67	-21.61	-41.31
300-5h	20	3	1.45	0.73	0.21	13.38	6.86	6.49	3.61	1.75	-8.58
600-10h	40	3	0.79	0.40	0.77	10.28	5.21	6.20	-2.25	-1.19	-14.23
600-10h	40	3	0.79	0.40	0.77	14.81	7.63	9.32	13.70	7.03	1.29
1200-20h	80	3	-0.68	-0.34	-1.01	18.27	9.53	8.71	18.39	9.82	-3.17
1200-20h	80	3	-0.68	-0.34	-1.01	14.61	7.53	8.17	17.63	9.17	3.07
2400-40h	160	3	-2.91	-1.44	-1.77	8.41	4.28	6.44	11.80	6.06	-3.98
2400-40h	160	3	-2.91	-1.44	-1.77	9.59	4.89	4.83	15.26	7.92	3.74

TABLE 3.3

IMPROVEMENT (%) OF SV MODELS (3-DIMENSIONAL INPUT VECTORS) AND MOVING AVERAGE RELATED TO PERSISTENCE

Minutes ahead	Steps ahead	Look back length	Moving Average			SV Models					
						S0,S1,S2-3d svs			S0,S1,S2-3d kernel		
			NMSE	RMSE	MAE	NMSE	RMSE	MAE	NMSE	RMSE	MAE
15	1	3	-87.27	-36.85	-36.07	-890.08	-214.89	-239.71	-1479.93	-297.78	-360.89
15	1	3	-87.27	-36.85	-36.07	-195.78	-72.11	-84.24	-351.21	-112.58	-146.08
90-1.5h	6	3	-9.04	-4.42	-3.93	-101.55	-41.97	-50.34	-225.44	-80.40	-109.00
90-1.5h	6	3	-9.04	-4.42	-3.93	-3.81	-1.89	-5.00	-39.25	-18.00	-33.51
300-5h	20	3	-1.47	-0.73	-0.21	-10.45	-5.17	-7.43	-49.84	-22.50	-41.61
300-5h	20	3	-1.47	-0.73	-0.21	12.10	6.18	6.29	2.19	1.03	-8.80
600-10h	40	3	-0.80	-0.40	-0.78	9.56	4.83	5.47	-3.06	-1.59	-15.12
600-10h	40	3	-0.80	-0.40	-0.78	14.13	7.26	8.62	13.01	6.66	0.52
1200-20h	80	3	0.68	0.34	1.00	18.83	9.84	9.62	18.94	10.13	-2.15
1200-20h	80	3	0.68	0.34	1.00	15.19	7.84	9.08	18.19	9.48	4.04
2400-40h	160	3	2.83	1.42	1.74	11.00	5.64	8.07	14.30	7.40	-2.18
2400-40h	160	3	2.83	1.42	1.74	12.15	6.25	6.48	17.65	9.23	5.42

Persistence using the criterion NMSE at 1200 minutes ahead and $\tilde{x}_{S_0}(t_0 + 80)$ (Table 2.1) is $\frac{1.52-1.35}{1.52} \times 100\% = 11.47\%$ (Table 2.3).

However, it is common that the performance varies in the data set and, indeed, there is not an a priori reason to believe that accurate predictions

TABLE 4.1

COMPARISON OF SV MODELS (1-DIMENSIONAL & 3-DIMENSIONAL INPUT VECTORS) WITH THE REFERENCE MODELS

| Minutes ahead | Steps ahead | Look back length | Persistence | | | Moving Average | | | SV Models | | | | | | | | | | | |
| | | | | | | | | | S0,S1,S2-1d svs | | | S0,S1,S2-1d kernel | | | S0,S1,S2-3d svs | | | S0,S1,S2-3d kernel | | |
			NMSE	RMSE	MAE	NMSE	RMSE	MAE	NMSE	RMSE	MAE	NMSE	RMSE	MAE	NMSE	RMSE	MAE	NMSE	RMSE	MAE
15	1	6	0.03	0.09	0.06	0.07	0.15	0.10	0.19	0.25	0.2	0.07	0.16	0.11	0.26	0.29	0.22	0.35	0.34	0.27
15	1	6	0.03	0.09	0.06	0.07	0.15	0.10	0.06	0.14	0.11	0.04	0.11	0.08	0.08	0.16	0.12	0.10	0.18	0.15
90-1.5h	6	6	0.12	0.21	0.15	0.14	0.23	0.16	0.20	0.28	0.21	0.15	0.24	0.17	0.25	0.31	0.23	0.34	0.36	0.29
90-1.5h	6	6	0.12	0.21	0.15	0.14	0.23	0.16	0.12	0.21	0.15	0.12	0.21	0.15	0.13	0.22	0.16	0.16	0.24	0.19
300-5h	20	6	0.32	0.35	0.25	0.33	0.35	0.25	0.34	0.36	0.26	0.33	0.36	0.25	0.36	0.37	0.27	0.45	0.41	0.34
300-5h	20	6	0.32	0.35	0.25	0.33	0.35	0.25	0.29	0.33	0.23	0.31	0.34	0.25	0.28	0.33	0.23	0.31	0.34	0.27
600-10h	40	6	1.00	0.47	0.35	1.02	0.48	0.35	0.95	0.46	0.33	1.02	0.48	0.36	0.91	0.45	0.33	1.01	0.48	0.39
600-10h	40	6	1.00	0.47	0.35	1.02	0.48	0.35	0.90	0.45	0.33	0.99	0.47	0.35	0.86	0.44	0.32	0.88	0.44	0.35
1200-20h	80	6	1.43	0.61	0.44	1.41	0.60	0.44	1.27	0.57	0.42	1.41	0.60	0.44	1.16	0.55	0.40	1.19	0.55	0.45
1200-20h	80	6	1.43	0.61	0.44	1.41	0.60	0.44	1.29	0.58	0.41	1.40	0.60	0.44	1.21	0.56	0.40	1.20	0.56	0.43
2400-40h	160	6	1.22	0.56	0.43	1.16	0.55	0.41	1.10	0.53	0.4	1.17	0.55	0.42	1.08	0.53	0.39	1.06	0.52	0.43
2400-40h	160	6	1.22	0.56	0.43	1.16	0.55	0.41	1.10	0.53	0.4	1.18	0.55	0.42	1.07	0.53	0.40	1.03	0.52	0.41

S0,S1,S2-1d svs: M2.a$_{(ii)}$, S0,S1,S2-1d kernel: M1.a$_{(ii)}$, S0,S1,S2-3d svs: M2.a$_{(i)}$, S0,S1,S2-3d kernel: M1.a$_{(i)}$

TABLE 4.2

IMPROVEMENT (%) OF SV MODELS (1-DIMENSIONAL & 3-DIMENSIONAL INPUT VECTORS) AND PERSISTENCE RELATED TO MOVING AVERAGE

| Minutes ahead | Steps ahead | Look back length | Persistence | | | SV Models | | | | | | | | | | | |
| | | | | | | S0,S1,S2-1d svs | | | S0,S1,S2-1d kernel | | | S0,S1,S2-3d svs | | | S0,S1,S2-3d kernel | | |
			NMSE	RMSE	MAE	NMSE	RMSE	MAE	NMSE	RMSE	MAE	NMSE	RMSE	MAE	NMSE	RMSE	MAE
15	1	6	61.43	37.90	37.93	-191.18	-70.64	-89.53	-12.82	-6.22	-5.01	-294.65	-98.66	-116.41	-432.77	-130.82	-161.50
15	1	6	61.43	37.90	37.93	6.15	3.12	-5.20	43.76	25.01	24.85	-18.08	-8.66	-17.19	-55.56	-24.72	-41.75
90-1.5h	6	6	16.86	8.82	8.52	-44.12	-20.05	-26.41	-4.17	-2.06	-2.71	-77.03	-33.05	-41.36	-143.01	-55.89	-78.34
90-1.5h	6	6	16.86	8.82	8.52	15.22	7.92	6.37	14.26	7.40	7.26	9.92	5.09	2.11	-10.05	-4.91	-17.61
300-5h	20	6	4.34	2.19	1.38	-1.61	-0.80	-1.84	-0.76	-0.38	-0.99	-7.68	-3.77	-6.94	-34.92	-16.15	-33.51
300-5h	20	6	4.34	2.19	1.38	13.67	7.09	7.72	5.16	2.61	2.27	15.42	8.03	7.63	7.28	3.71	-5.85
600-10h	40	6	2.09	1.05	1.81	7.23	3.68	5.72	-0.16	-0.08	-0.27	11.34	5.84	6.99	0.62	0.31	-11.03
600-10h	40	6	2.09	1.05	1.81	11.42	5.88	8.15	2.81	1.41	1.70	16.10	8.40	10.42	13.68	7.09	2.33
1200-20h	80	6	-1.15	-0.58	-1.73	10.30	5.29	4.50	0.11	0.05	-0.53	18.13	9.52	8.25	16.13	8.42	-3.74
1200-20h	80	6	-1.15	-0.58	-1.73	8.86	4.53	5.08	0.64	0.32	-0.36	14.44	7.50	7.77	15.22	7.92	2.22
2400-40h	160	6	-5.16	-2.55	-3.23	5.33	2.70	3.28	-0.70	-0.35	-0.86	7.24	3.69	5.68	8.84	4.52	-4.99
2400-40h	160	6	-5.16	-2.55	-3.23	5.26	2.67	2.46	-1.52	-0.75	-1.23	8.19	4.18	3.92	11.37	5.85	1.80

TABLE 4.3

IMPROVEMENT (%) OF SV MODELS (1-DIMENSIONAL & 3-DIMENSIONAL INPUT VECTORS) AND MOVING AVERAGE RELATED TO PERSISTENCE

| Minutes ahead | Steps ahead | Look back length | Moving Average | | | SV Models | | | | | | | | | | | |
| | | | | | | S0,S1,S2-1d svs | | | S0,S1,S2-1d kernel | | | S0,S1,S2-3d svs | | | S0,S1,S2-3d kernel | | |
			NMSE	RMSE	MAE	NMSE	RMSE	MAE	NMSE	RMSE	MAE	NMSE	RMSE	MAE	NMSE	RMSE	MAE
15	1	6	-159.30	-61.03	-61.10	-655.04	-174.78	-205.34	-192.55	-71.04	-69.17	-923.33	-219.90	-248.64	-1281.46	-271.68	-321.28
15	1	6	-159.30	-61.03	-61.10	-143.36	-56.00	-69.49	-45.82	-20.76	-21.08	-206.17	-74.98	-88.81	-303.36	-100.84	-128.37
90-1.5h	6	6	-20.28	-9.67	-9.31	-73.35	-31.66	-38.17	-25.29	-11.94	-12.27	-112.94	-45.92	-54.52	-192.29	-70.97	-94.94
90-1.5h	6	6	-20.28	-9.67	-9.31	-1.98	-0.98	-2.34	-3.13	-1.55	-1.37	-8.35	-4.09	-7.00	-32.37	-15.05	-28.56
300-5h	20	6	-4.53	-2.24	-1.40	-6.21	-3.06	-3.26	-5.33	-2.63	-2.40	-12.56	-6.10	-8.44	-41.03	-18.76	-35.39
300-5h	20	6	-4.53	-2.24	-1.40	9.76	5.01	6.43	0.86	0.43	0.90	11.59	5.97	6.33	3.08	1.55	-7.34
600-10h	40	6	-2.13	-1.06	-1.84	5.25	2.66	3.99	-2.29	-1.14	-2.12	9.44	4.84	5.27	-1.51	-0.75	-13.07
600-10h	40	6	-2.13	-1.06	-1.84	9.53	4.88	6.45	0.73	0.37	-0.12	14.31	7.43	8.76	11.84	6.11	0.52
1200-20h	80	6	1.14	0.57	1.70	11.33	5.83	6.12	1.25	0.63	1.18	19.06	10.03	9.81	17.09	8.94	-1.98
1200-20h	80	6	1.14	0.57	1.70	9.90	5.08	6.69	1.77	0.89	1.35	15.41	8.03	9.34	16.18	8.45	3.88
2400-40h	160	6	4.91	2.48	3.13	9.98	5.12	6.31	4.24	2.14	2.30	11.79	6.08	8.63	13.31	6.89	-1.71
2400-40h	160	6	4.91	2.48	3.13	9.91	5.08	5.51	3.46	1.75	1.94	12.70	6.56	6.92	15.71	8.19	4.87

can be made at every single time step; in a mainly non-linear system, there may be islands of predictability implanted in a sea of unpredictable or chaotic behavior [18].

10.7 Discussion

Some interesting points emerging from the comparison analysis are given in the following.

The FL-Approach:

1. The Moving Average method exhibits the poorest performance whereas the RNN and the FL_{S_0,S_1,S_2}-approach demonstrate significant improvement in comparison with it and the Persistence method. Up to a certain degree, this result is expectable since the RNN and FL_{S_0,S_1,S_2} exploit the correlation among the three meteorological stations.

2. Additionally, the FL_{S_0}-approach utilizing only the historical data of station S_0 is not better than the Persistence in predicting wind speed, while it behaves better in wind direction predictions. This can be explained by the fact that the wind direction displays noticeably less variation than wind speed.

3. In the first two steps (15 and 30 minutes) the FL-model cannot beat the Persistence method because a crucial factor in FL-approach is the formulation of rectangles and in this time horizon the created rectangles are few and contain no significant information in order to result in more precise predictions. The RNN model seems to perform better in this time window but we have to consider the fact that it exploits more information (steps back) in the training phase.

4. The percentage improvement of both FL_{S_0,S_1,S_2} and RNN algorithms compared with MA shows a slightly better performance of the former especially after 4 steps ahead. For instance, at $h = 8$ the FL_{S_0,S_1,S_2}-model presents a 24.31% improvement while the respective percentage of RNN is 23.61%. Similarly, at $h = 12$ the FL_{S_0,S_1,S_2}-model displays a 22.99% improvement while the respective percentage of RNN is 20.98%. However, the RNN model and the FL-approach are not totally comparable since they refer to slightly different samples. However, we have to stress the fact that the FL-approach does not need any training while it is easily tuned (the vigilance parameter is set equal to the standard variation of the data sample and the curve correction is eventuated by means of the last measured value) and presents both low computational and time cost.

5. Another advantage of the FL-model for the user is data compression and the effortless extraction of simple rules (symbolic knowledge which has meaning to humans) where the prediction of a new fact is calculated from the weighted average of the created rectangles and its average with the last measured value.

In Fig. 10.2, we indicate how the FL-model approximates the real curve.

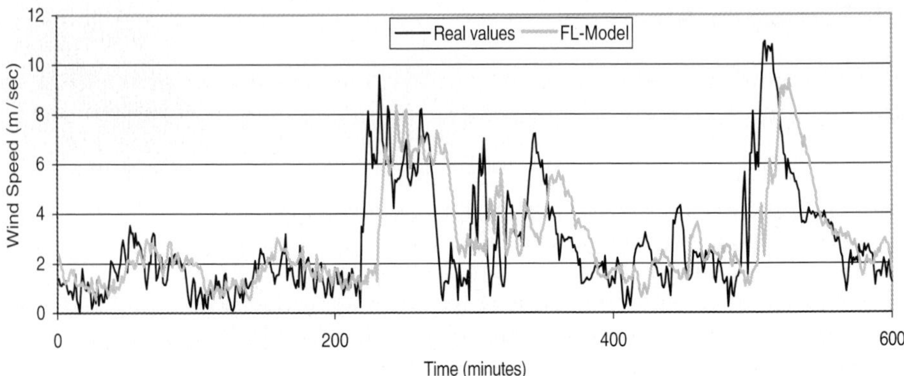

Fig. 10.2. Approximation of real curve by the FL-model

The SV-Approach: We test the SV-models only for wind parameter prediction.

1. The usage of small time frames (b_{hd}=1 or b_{hd}=6) limits the memory requirements for storage of the kernel matrix whilst the impact of past outliers is diminished. Moreover, after many tests we ascertained that greater values for b_{hd} do not give better results.
2. The two algorithms $S0$ *svs* and $S0$ *with kernel* cannot outperform any other model. This indicates the need for additional information.
3. In most cases the $pred_{S_0}(t_0 + h)$ gives better results than $\tilde{x}_{S_0}(t_0 + h)$. The latter seems to have improved after 80 steps.
4. The approach with the 3-dimensional input vectors exhibits better performance related to models with 1-dimensional input vectors.
5. The greater generalization ability is displayed by $S0, S1, S2$ *svs*, $S0, S1, S2 - 3d$ *svs* and $S0, S1, S2 - 3d$ *kernel* especially after 20 steps ahead. We note that these models take advantage of the spatial correlation among the three stations.

We point out that the performance of SV algorithms depends crucially on an appropriate choice of parameters. Although several different approaches exist for their selection, the issue of how to practically select a good set is still far from being resolved.

10.8 Conclusions

This paper applies comparatively two prediction methodologies (the Fuzzy Lattice Neurocomputing and the Support Vector Regression) to a real-world problem that of wind prediction which is a very demanding task. Short (0–6 hours) and long (>6 hours) time forecasting of wind variations is still an open problem.

FLN and SVR have sound mathematical foundations. The results produced depend on the selection of the valuation function in the FLN case and the kernel function in the SVR case. There are few parameters to adjust. In FLN we just tune the vigilance parameter while in SVR we adjust the capacity control and the kernel parameters (the width of the Gaussian kernel in RBF case). Additionally, they can be applied to different types of data such as real-valued vectors, symbols, images, text etc. Yet, only the FL technique manages to cope with different types of data simultaneously by considering the Cartesian product of multiple lattices. Both methodologies seek for a sparse representation of the final solution, i.e. FLN utilizes intervals whereas SVR uses the support vectors representing the bounds of the candidate solution regions standing for diverse classes. The problem of *missing* and *don't care* attribute values in the data has not been explicitly addressed within the SVR methodology whereas in the FL-framework *missing* values have been replaced by the least element and *don't care* values can be replaced by the greatest element of the corresponding lattice. Finally, the model selection problem is still an open research topic since both methodologies employ a quite expensive way (cross validation) to select the model parameters.

We have assessed the performance of the proposed models by comparing them with two well-known reference techniques employed widely as benchmarks to time-series analysis, the Persistence and the Moving Average method, and additionally to a recurrent neural network. In general the FL-approach demonstrates a satisfactory and constant performance. Future work involves the testing of the FL-approach in dealing with data typically containing high noise and significant non-stationarity. Finally, a synthesis of these two effective methodologies (FL and SVR) might result in superior algorithms.

References

1. Alexiadis MC (2002) Wind Speed Prediction at Aeolic Parks. PhD Thesis, Aristotle Univ of Thessaloniki, Greece
2. Barbounis A (2005) Optimized Real-Time Learning Algorithms for the Training of RNNs and Fuzzy RNNs: Application on Wind Speed and Wind Power Prediction at Aeolic Parks. PhD Thesis, Aristotle Univ of Thessaloniki, Greece
3. Beyer HG, Luther J, Steinberger-Willms R (1993) Power fluctuations in spatially dispersed wind turbine systems. Solar Energy 50:297–306
4. Birkhoff G (1967) Lattice Theory. American Math Society, Col Pub 25
5. Box J, Jenkins G (1976) Time Series Analysis, Forecasting and Control. Hol-Day
6. Carpenter G, Grossberg S (1987) A massively parallel architecture for self-organizing neural pattern recognition machine. Computer Vision, Graphics and Image Understanding 37:54–115
7. Cortes C, Vapnik V (1995) Support Vector Networks. Mach Learn 20:273–297
8. Cristianini N, Shawe-Taylor J (2000) An Introduction to Support Vector Machines (and other kernel-based learning methods). Cambridge Univ Press

9. Daniel AR, Chen AA (1991) Stochastic simulation and forecasting of hourly average wind speed sequences in Jamaica. Sol Energy 46:1–11
10. Elman JL (1991) Distributed representations, simple recurrent networks, and grammatical structure. Machine Learning 7(2/3):195–226
11. Kaburlasos VG, Petridis V (2000) Fuzzy lattice neurocomputing (FLN) models. Neural Networks 13(10):1145–1170
12. Kaburlasos VG, Petridis V (2002) Learning and decision-making in the framework of fuzzy lattices. In: Jain LC, Kacprzyk J (eds) New Learning Paradigms in Soft Computing, ser Studies in Fuzziness and Soft Computing 84:55–96. Physica-Verlag, Heidelberg, Germany
13. Kariniotakis G, Pinson P, Siebert N, Giebel G, Barthelmie R (2004) The state of the art in short-term prediction of wind power from an offshore perspective. In: Proc SeaTechWeek
14. Karush W (1939) Minima of Functions of Several Variables with Inequalities as Side Constraints. MS Thesis, Dept of Mathematics, Univ of Chicago
15. Kuhn HW, Tucker AW (1951) Nonlinear programming. In: Proc 2nd Berkeley Symp on Mathematical Statistics and Probabilistics pp 481–492. Univ of California Press
16. Müller KR, Smola AJ, Rätsch G, Schölkopf B, Kohlmorgen J, Vapnik V (1997) Prediction time series with support vector machines. In: Proc Int Conf on Artificial Neural Networks
17. Nielsen TS, Madsen H (1997) Statistical methods for predicting wind power. In: Proc European Wind Energy Association Conference (EWEC) pp 755–758
18. Packard NH (1990) A genetic learning algorithm for the analysis of complex data. Complex Systems 4(5):543–572
19. Palomino I, Martin F (1995) A simple method for spatial interpolation of the wind in complex terrain. J Appl Meteorology 34(7):1678–1693
20. Petridis V, Kaburlasos VG (1999) Learning in the framework of fuzzy lattices. IEEE Transactions on Fuzzy Systems 7(4):422–440
21. Schlueter RA, Park GL, Bouwmeester R, Shu L, Lotfalian M, Rastgoufard P, Shayanfar A (1984) Simulation and assessment of wind array power variations based on simultaneous wind speed measurements. IEEE Trans Power App Syst 103:1008–1016
22. Smola AJ and Schölkopf B (2004) A tutorial on support vector regression. Statistics and Computing 14:199–222
23. Vapnik V (1995) The Nature of Statistical Learning Theory. Springer, NY
24. Vapnik V, Golowich SE, Smola A (1997) Support vector method for function approximation, regression estimation, and signal processing. In: Adv in Neural Information Processing Systems 9:281–287. The MIT Press, Cambridge
25. Zadeh LA (1965) Fuzzy sets. Information and Control 8:338–353

Application of Fuzzy Lattice Neurocomputing (FLN) in Ocean Satellite Images for Pattern Recognition

J.A. Piedra-Fernández[1], M. Cantón-Garbín[2] and F. Guindos-Rojas[3]

[1] Department of Languages and Computation, University of Almería, 04120 Almería, Spain jpiedra@ual.es
[2] mcanton@ual.es
[3] fguindos@ual.es

Summary. The main objective of this work is to improve the automated interpretation of ocean satellite images using a fuzzy lattice system that recognizes the most important *ocean structures* in satellite AVHRR (Advanced Very High Resolution Radiometer) images. This chapter presents a hybrid model based on an expert system segmentation method, a method of correlation-based feature selection, and a few classifiers including Bayesian nets (BN) and fuzzy lattice neural networks. The results obtained by the fuzzy lattice system are clearly better than the results obtained by ANNs (Artificial Neural Nets), knowledge based reasoning systems, and graphic expert system (GES).

11.1 Introduction

During the last thirty years, a large amount of environmental data has been obtained from Earth observation satellites. At present, high resolution imaging sensors are very important in modern *remote sensing* technology. These sensors produce both multispectral and hyperspectral data. With the growth of dimensionality and higher spectral resolution, many different classes can be identified. When pattern recognition methods are applied to remote sensing problems, the limited training data, which is used for designing the classifier, is an inherent problem. Complex statistical distribution of a large number of classes is yet another important problem.

Today, the European ENVISAT satellite acquires more than 1Tb of data every day. This large amount of information requires the development of fully automatic interpretation systems to manage it efficiently.

In this work we have tried to enhance previous results in the field of automatic recognition of *meso and macroscalar ocean structures* in AVHRR satellite images. One of our main goals was to design a system which performs

J.A. Piedra-Fernández et al.: *Application of Fuzzy Lattice Neurocomputing (FLN) in Ocean Satellite Images for Pattern Recognition*, Studies in Computational Intelligence (SCI) **67**, 215–232 (2007)
www.springerlink.com

better than similar classifiers such as existing ANNs symbolic processing elements developed by our work group [1], specifically in dealing with complex remote sensing classification problems.

In this work we present an approach to solve this complex problem (the borders of the ocean structures are not precisely defined in satellite images and their morphology is highly variable) employing an automatic fuzzy lattice interpretation system developed to detect and label the most important oceanic structures for channel 4 AVHRR infrared images [2].

Our automatic interpretation system is made up of an expert system for segmentation, a filter method for feature selection and a new hybrid hierarchical classifier. This hybrid classifier consists of several BNs and a fuzzy lattice neural network (FLNN), where the BNs simplify the class selection problem inherent in FLNN design. As a result, small BNs are generated to reduce the scope of the classification problem.

Section 11.2 introduces some ocean structures with thermal expressions that are of interest for oceanographers, section 11.3 presents the overall automatic interpretation system structure.

Subsection 11.3.1 describes the method used for cloud masking. This method allows to automatically create the cloud mask and to improve the recognition of ocean structures.

Subsection 11.3.2 describes the segmentation procedure. The segmentation is the partitioning of a digital image into multiple regions and in our case the goal of segmentation is to locate ocean structures. But unfortunately, many important segmentation algorithms are too simple to solve this problem accurately. We have developed an automatic segmentation system for this problem.

Subsection 11.3.3 describes the feature selection method. In our case, we have limited data samples with a large number of features. We have used popular metrics for classification problems: correlation and mutual information, but the best result has been obtained by correlation based feature selection (CFS).

Subsection 11.3.4 shows the different oceanic structure to classify.

Section 11.4 shows the classification methods used. This section is divided into two subsection. Subsection 11.4.1 explains the first classifiers designed like a graphic expert system, Bayesian networks, neural networks and fuzzy systems. These classifiers will be compared with the fuzzy lattice classifier.

In the following subsection 11.4.2 we present the fuzzy lattice classifier and shows the advantages of the fuzzy lattice for the recognition of ocean structures. Moreover, we propose a hybrid hierarchical model based on Bayesian networks and fuzzy lattice system.

Section 11.5 and section 11.6 present, respectively, the results and conclusions of our work.

11.2 Ocean Structures

The AVHRR sensor has been a powerful tool in environmental, climatic and geophysical research tasks for over twenty years. This sensor, on board the Tiros and NOAA satellite series, have three infrared (IR) and two visible (VIS) channels. AVHRR channels 2 and 4 provide VIS and IR information in the ranges of 0.725-1.10 μm and 10.50-11.50 μm, respectively. Infrared information in particular has been used in oceanic feature identification [3, 4, 5, 6].

Our study was carried out in the region of NW Africa, Atlantic Iberian coasts and Mediterranean sea (see Fig. 11.1).

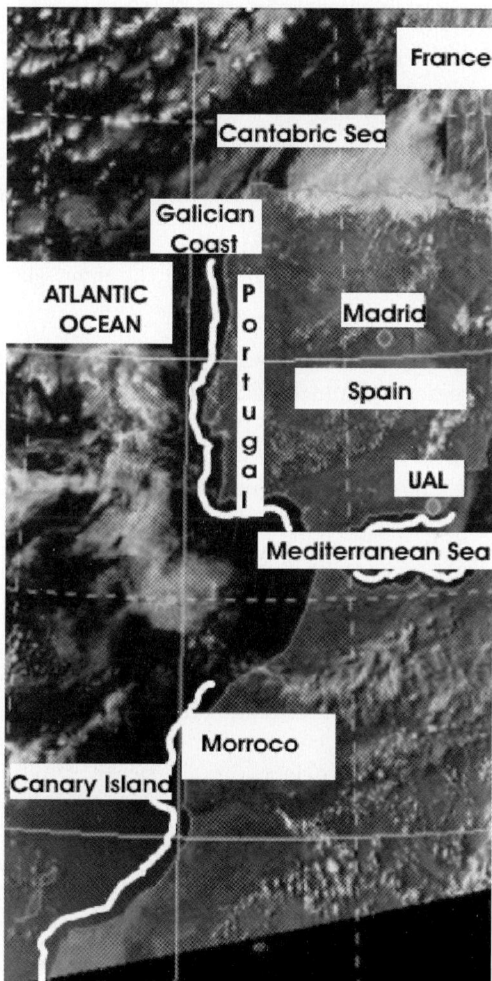

Fig. 11.1. Study area. White lines represent upwellings and white circles are zones with mesoscalar structures (From [1])

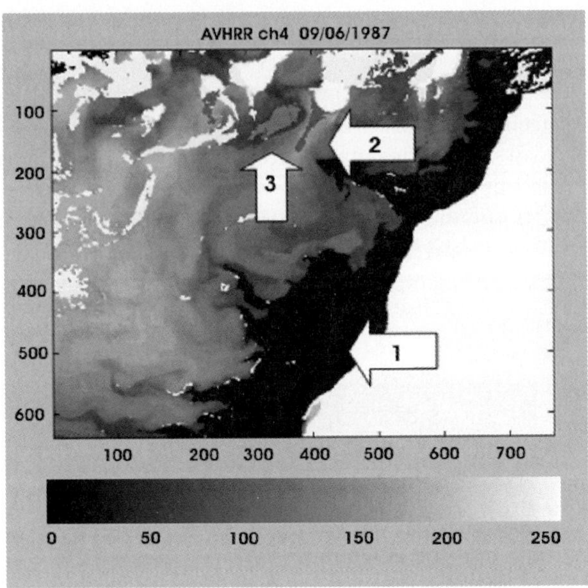

Fig. 11.2. Ch 4 AVHRR image showing: 1) the upwelling, 2) a warm wake south of Gran Canaria island and 3) a cold gyre SW of the same island.(From [1])

A detailed oceanographic description of this area can be found in [7, 8, 9, 10, 11]. In these regions the most important oceanic mesoscalar structures observable in satellite images are [8, 10, 12, 13, 14]: upwellings, cold eddies, warm eddies and warm wakes in the lee of some islands (see Fig. 11.2). All these structures have also been observed and studied also in the area by means of VIS sensors (CZCS-Coastal Zone color Scanner and SeaWiFS), ERS-1 and TOPEX altimeters and in situ and AXBTs (Aircraft Expandable Bathyther-mographs) [12, 14]. Upwelling is the term used by oceanographers to describe a structure in which cool, nutrient-rich waters from the lower layers of the ocean rise to the surface. When they reach the light, such waters becomes very fertile and a rich feeding ground for fish. Upwellings are generated when there are persistent winds parallel to the shore, and Ekman transport diverts the surface flow away from the coast permitting cold and nutrient-rich water from the bottom to reach the surface.

Prevailing NE trade winds between 30 and 60 North produce one of the biggest upwellings in the world, the NW African Upwelling (see Fig. 11.2).

Upwellings are regular phenomena off the NW African coast and other coasts like the Peruvian and Benguela (South Africa) coasts, where wind conditions are suitable. Nonetheless, upwellings may well be intermittent, depending both on the weather [8, 9] and, to a certain extent, on ocean-wide thermohaline circulation. Because of their importance to commercial fisheries, much oceanographic research has been directed at their comprehension and prediction for which remote satellite sensing is a powerful tool.

Eddies are highly morphologically and contextually variable gyres produced mainly by the hydrodynamic interaction between the cold Canary current flowing from the NE and the Canary islands [12, 13]. Similar eddies due to the same mechanism have been found in other archipelagos around the world. The temperature and salinity of the water in eddies is different from the surrounding waters and can travel great distances for a long period of time without mixing with the surrounding water. Moreover, the movement and shape of cold (cyclonic in the northern hemisphere-NH) and warm (anticyclonic) eddies in this region are controlled by the Trade Winds. Cool eddies are more intense under calm conditions, whereas with strong winds, the warm eddies are more intense and symmetrical [14]. In cool eddies vertical movement of the water is ascending: cold water, rich in nutrients, rises to the surface. However, in warm eddies water accumulates and sinks, generating transport of organic matter downward toward the ocean interior [12].

Wakes [8, 12] are warm oceanic structures leeward of islands. A wake is generated when islands form an obstacle to the prevailing wind field in this region (NE trade winds). This lowers the intensity of winds SW of the islands and warms the sea surface (see Fig. 11.2). Wakes are very thin and warmer than the surrounding waters.

11.3 The Automatic Interpretation System

Figure 11.3 shows the overall structure of the system developed in this work for improving previous results obtained with the automatic recognition systems for ocean features observed at IR AVHRR images.

In a first step, the raw image is processed by standard models for radiometric and geometrical correction and land masking. These are well known techniques also used for image analysis by human experts. On the other hand, the automatic digital image interpretation processing system works directly with the digital numbers, so it makes no sense to use image enhancement methods like histogram equalization or others.

The second step attempts to automatically detect cloud pixels that are opaque to the ocean radiance data measured in the AVHRR infrared and visible scenes and create a mask of 0s for these areas [15].

The next task, segmentation, divides the image into meaningful oceanic regions. The nature of ocean dynamics complicates this basic process, so we have designed a new iterative knowledge-driven method for it [16]. After segmentation the type of features or descriptors to describe or represent the segmented regions are selected.

The feature selection step chooses an optimal or sub-optimal feature subset from the set of features chosen previously.

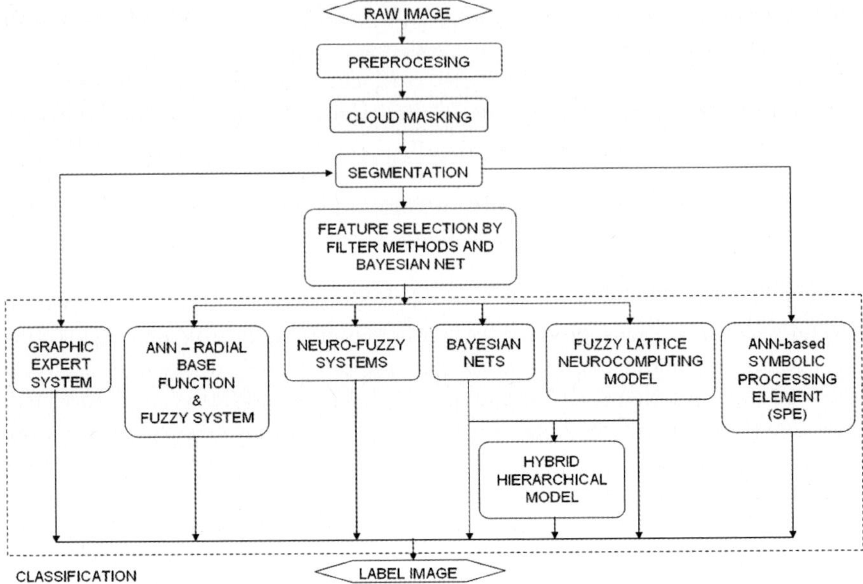

Fig. 11.3. Automatic ocean recognition system structure

In the last step of ocean region recognition, each region produced during the segmentation and represented by a selected group of descriptors is analyzed and, if the recognition is positive, it is labeled with the identifier of the matching structure (upwelling, cold or warm eddy or wake). We have implemented a redundant recognition subsystem, with different classifiers:

- Neurofuzzy system: neurofuzzy function approximator (NEFPROX) and NEFCLASS.
- Fuzzy lattice neural networks (FLNN).
- Bayesian networks (BNs) (Nave Bayes and Tree Augmented Nave Bayes).
- ANN-based symbolic processing elements (SPE)[1].
- Graphic expert system (GES).
- Hybrid hierarchical model (HHM).

One of our goals was to compare the efficiency of the different mesoscalar ocean structure recognition classifiers in which the fuzzy lattice plays an important role.

11.3.1 Cloud masking

The technique used to automatically create the cloud mask is described in detail in [15]. Cloud pixels have:

1. A very low value on channel 4 (sea surface temperature brightness) compared to their neighbors and/or

2. The digital values of the neighbor pixels have high variability in the visible
   channel 2.

The cloud mask system is composed of two back-propagation ANNs. The
first of them detect those pixels (candidate pixels) either affected by both
characteristics or very strongly by one of them in a first estimation. After
that, candidate pixels are classified by the second ANN which finds the cloud
pixels.

11.3.2 Segmentation

Segmentation is a key task for any automatic image analysis system. Final
results are good only if high-quality segmentation is achieved. But, as explained
in [16], AVHRR ocean images are very difficult to segment due to the highly
variable pixel values, yielding poor segmentation results with conventional
techniques that rely merely on gradients or textures [17, 18]. The method
that we proposed in [16] uses isothermal lines that have been proven to produce good segmentation in this kind of images. This, like any other threshold
segmentation, has the inherent drawback of threshold selection, but we have
solved this by applying the knowledge acquired in the recognition phase, thus
creating an iterative procedure. The initial threshold for segmentation of each
structure can be set at a fixed empirical value or at the water-pixel mean. This
is then used for segmentation and results are passed on to the classifier, which
labels (upwelling, cold gyre, warm gyre, wake) all the region in the image.
Then the threshold can be raised or lowered and segmentation repeated with
a new threshold.

The new region is then compared with the previous by the GES, which
recognizes the regions every time the threshold changes, and the system determines whether the threshold change was favourable, must be reversed, or the
task ended at the last value. This method yields good segmentation of AVHRR
images, producing compact regions and reducing oversegmentation resulting
from other methods tested like watersheds [17] and the Canny edge detec-
tor [18].

11.3.3 Feature selection

The most common way to procced with feature selection is to define criteria
for measuring the goodness of a set of features [19], and then use a search algorithm to find an optimal or sub-optimal set of features [20, 21]. We applied
filtering methods with good results. Filtering methods use heuristics based
on general data characteristics rather than a learning algorithm to evaluate the merit of feature subsets. We used: mutual information, Matusita distance, Kullback-Leibler, Shannon entropy, correlation based features selection
(CFS)and other. The best filtering method used was CFS [22]. CFS measures
correlation between features. The features used for each segmented region
were:

- Simple features: area, perimeter, density, volume, equivalent diameter.
- Bounding ellipse: centroid, major and minor axis, orientation, eccentricity, irradiance.
- Bounding box: height, width, area.
- Level of gray: min, max and mean level of gray, standard deviation, barycenter of grey level.
- Inertia moments: Hu's moments, Maitra's moments, Zernike's moments, tensorial moments.

CFS reduces the number of features from 80 to only 16. In order to evaluate feature selection efficiency, we cross-validated with a Naive Bayes (simple Bayesian classifier).

11.3.4 Classification

The last step in image recognition is classification. All classifiers used in this work and presented in Section 11.4 were designed and tested with the same image dataset. Figure 11.4 shows the different oceanic structure identifiers:

- Upwellings - (ID: 2, 3 and 4)
    - 2 identifies the upwellings between Cape Jubi and Cape Bojador.

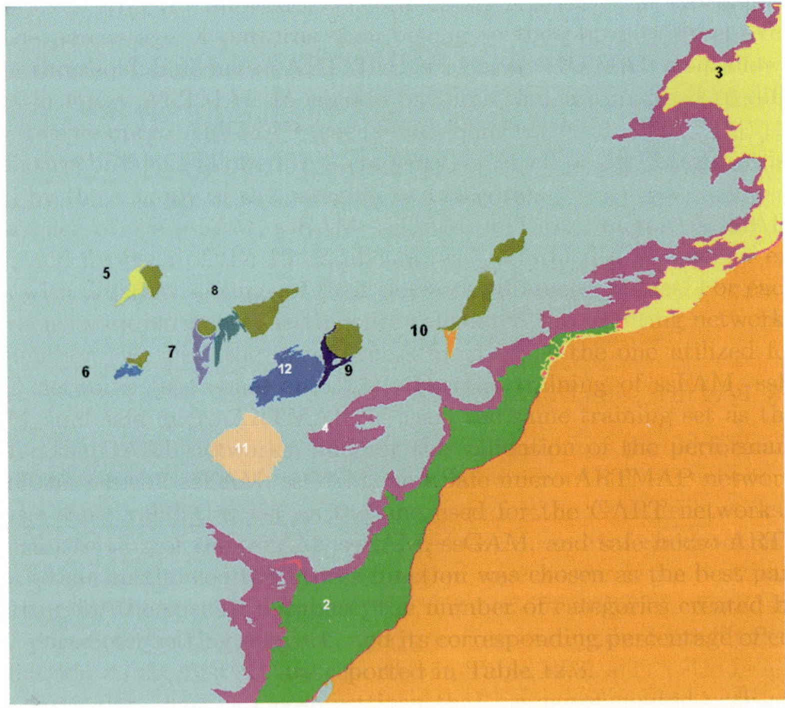

Fig. 11.4. The oceanic structure identifiers

  - 3 identifies upwellings south of Cape Bojador.
  - 4 identifies the upwellings in both coast regions.
- Wakes - (ID: 5, 6, 7, 8, 9 and 10).
  - 5 identifies the El Hierro Wake.
  - 6 identifies the Gran Canaria Wake.
  - 7 identifies the La Palma Wake.
  - 8 identifies the La Gomera Wake.
  - 9 identifies the Tenerife Wake.
  - 10 identifies Fuerteventura Wake.
- Anticyclonic eddy - (ID 11).
- Cyclonic eddy - (ID 12).
- No structure - (ID 0).

A series of factors combine to make oceanic structures in satellite images difficult to identify with classical computational techniques:

- Thermal and morphological variability in oceanic structures makes it practically impossible to construct a mathematical model incorporating the number of variables necessary for their description.
- Identification is influenced by a very significant contextual factor: the temperatures measured or features derived in a given region produces different structures depending on their geographical location and the date on which they are found.

In the next section we justify the use of fuzzy models and lattice theory for the design of different classifiers.

11.4 Classifiers

We believe that instead of selecting classifiers based only on their accuracy, it may often be more effective to attempt to select the classifiers based on the knowledge they represent. Combining classifiers is an effective way of improving classification performance. It is often possible to construct several classifiers with different characteristics.

We have implemented classifiers representing different knowledge. In the following sections, we show two classifications: the initial classifier 11.4.1 and the fuzzy lattice classifier 11.4.2. The fuzzy lattice classifier section shows the new model based on lattice theory, explains the advantage of the system used and the necessity of the combining classifiers (hybrid hierarchical model).

11.4.1 Initial classifier

We divide the classifier into two groups. The initial classifiers were GES, BNs and the neurofuzzy classifiers.

Graphic Expert System

This system is like an oceanographer, collecting knowledge about the ocean and representing it as production rules [16]. The GES filters incoming information and keeps only information relevant to the recognition of the ocean structures present in the image. In order to train this component, every ocean feature of interest has to be defined by a human expert using numerical or symbolic descriptors. The hard job here is to figure out the imprecise, sometimes intuitive, deduction scheme that human experts use to perform the task. The lack of a precise model for each feature of interest in our problem can lead to gaps or inconsistencies in the knowledge that the expert provides the system with. This is a common problem in expert systems, so some techniques have been designed to manage such imprecise or incomplete knowledge (Bayesian networks or fuzzy methods).

The main problem of this system is that it was built by human experts, that is, the learning process is manual and not automatic.

Bayesian Networks

Bayesian Networks model uncertainties. We have tested two simple Bayesian classifiers [23]: The Naive Bayes and The Tree Augmented Naive Bayes Classifier (TAN). The Naive Bayes is oriented to classification, and is based on the assumption that all the features are conditionally independent when the value of the class is known. This assumption implies that the structure of the network is rather simple, since only the arcs link the classes with features, and there are no arcs between feature variables.

The Naive Bayes is easy to construct and performs surprisingly well in classification, even though the conditional independence assumption is rarely true in real world applications. A more effective straightforward way to improve The Naive Bayes is to extend its structure to represent dependencies among attributes [24]. The TAN model is a restricted family of BNs in which the class has no parent nodes and the parents of each feature are the values of the class, or at most, another feature.

One problem of the Naive Bayes is that decoupling the class conditional feature distribution means that each distribution can be independently estimated as a one dimensional distribution. This dimensionality problem implies a need for datasets that scale exponentially with the number of features. However, the classifier is robust enough to ignore serious deficiencies in its underlying naive probability model. Other reasons for the observed success of the naive Bayes classifier are discussed in [24].

Neurofuzzy Systems

Neural networks are good for learning and classification, but do not explain how they reach their decisions (black box). Fuzzy logic systems can be well understood by humans, but cannot be built automatically. Neurofuzzy Systems combine the fuzzy set theory with ANNs with the advantages of both. In our work, we used the NEFCLASS and NEFPROX models to implement the Neurofuzzy systems.

NEFCLASS [25] is a neurofuzzy classification system derived from the generic model of a 3-layer fuzzy perceptron (may be thought of as a special 3-layer feed-forward neural network). The nodes in this network use T-norms or T-conorms instead of the activation functions common to neural networks. The first layer is associated with the input features, the hidden layer represents the fuzzy rules (each node is associated with one rule) and the third layer represents output features (in our case the classes). Fuzzy sets are encoded as connection weights. The inputs have real values. The nodes of the output layer use a T-conorm for rule aggregation, where the T-conorm is usually the maximum. The winner class is determined by the output node with the highest value.

NEFCLASS model can only be used for discrete sets (crisp classification tasks). We have therefore used NEFPROX [26] (NEuroFuzzyfunction apPROXimator). NEFPROX extends NEFCLASS using a general approach for function approximation by means of a neurofuzzy model based on supervised learning with a Mamdani fuzzy system. The main difference between them is that NEFCLASS models do not use membership functions on the consequent.

11.4.2 Fuzzy lattice classifiers

A fuzzy lattice is a lattice L such that the ordering relation has been extended to all elements in L in a fuzzy degree of truth. The inclusion relation in a lattice L can be extended to all the elements of the Cartesian product $L \times L$ using an axiomatically defined inclusion measure function $\sigma : L \times L \to [0,1]$. Learning is achieved by computing intervals of lattice elements (in the Ndimensional Euclidean space a lattice interval is an hyperbox).

The benefit of using a fuzzy lattice is that it directly enables representation and calculation with the imprecision found in the oceanic structures, which arises from variability in morphological measurement due to errors, structural defects, and contextual characteristics. It is important to emphasize that the creation of intervals by means of the fuzzy lattice neural networks in the numeric descriptors improve the understandability of the rules set by the human experts.

Fuzzy Lattice Neural Networks

The fuzzy lattice neural network (FLNN) [27] emerges as a connectionist paradigm in the framework of fuzzy lattices, which have the advantages of being able to deal rigorously with different types of data, such as numerical and linguistic data, value intervals, missing and don't care data.

The σ-FLNMAP (σ Fuzzy Lattice Neural MAP) classifier was introduced in [27, 28, 30]. The σ-FLNMAP classifier is a relationship between two FLNN models: FLNNa and FLNNb. The two models together produce a σ-FLNMAP classifier. FLNNa clusters the input data, while FLNNb has the classes (category labels), and an intermediate MAP field is used to associate clusters in FLNNa with clusters in FLNNb, as described in [28]. That is, the MAP field assigns a class (in FLNNb) to a data cluster (in FLNNa). The rules extracted with σ-FLNMAP are represented as hyperboxes. The vigilance parameter specifies the σ-FLNMAP's sensitivity. As the vigilance parameter increases, hyperbox size decreases. At the end of learning, there are M hyperboxes partitioned into K classes, where K is the cardinality of the finite set of category labels. As soon as learning is complete, the set of calculated hyperboxes can be implemented as either a decision tree or a neural network [27, 28].

Lately, Fuzzy Lattice Reasoning (FLR) algorithm has been introduced as a successor to the FLN [31]. In fact, the FLN is a (limited) neural network implementation of the more flexible FLR.

Hybrid Hierarchical Model

In our work, we propose a hybrid hierarchical model, which has three classifiers: two Bayesian Nets (Naive Bayes) and one Fuzzy Lattice Neural Network (σ-FLNMAP) as the best solution for automatic ocean structure recognition. This model has a hierarchical structure of cluster classes which improves or maintains the accuracy rate of other classifiers used and reduces the number of rules managed by the system. This classifier works on two levels (see Fig. 11.5) .

In the first level the cluster classes used were: upwellings, eddies, wakes and no oceanic structure. In the second level the cluster classes used were:

- Upwellings between Cape Jubi and Cape Bojador
- Upwellings south of Cape Bojador
- Upwellings in both coast regions
- The El Hierro Wake
- The Gran Canaria Wake
- The La Palma Wake
- The La Gomera Wake
- The Tenerife Wake
- The Fuerteventura Wake
- Anticyclonic eddy
- Cyclonic eddy

On the first level, the Naive Bayes is the classifier. The Naive Bayes chosen has 3 classes to classify. On the second level there are two classifiers: Naive Bayes and σ-FLNMAP. The classes used for each classifier are (see Fig. 11.5):

- Nave Bayes: 3 classes of upwellings (between Cape Jubi and Cape Bojador, south of Cape Bojador and both coastal regions)

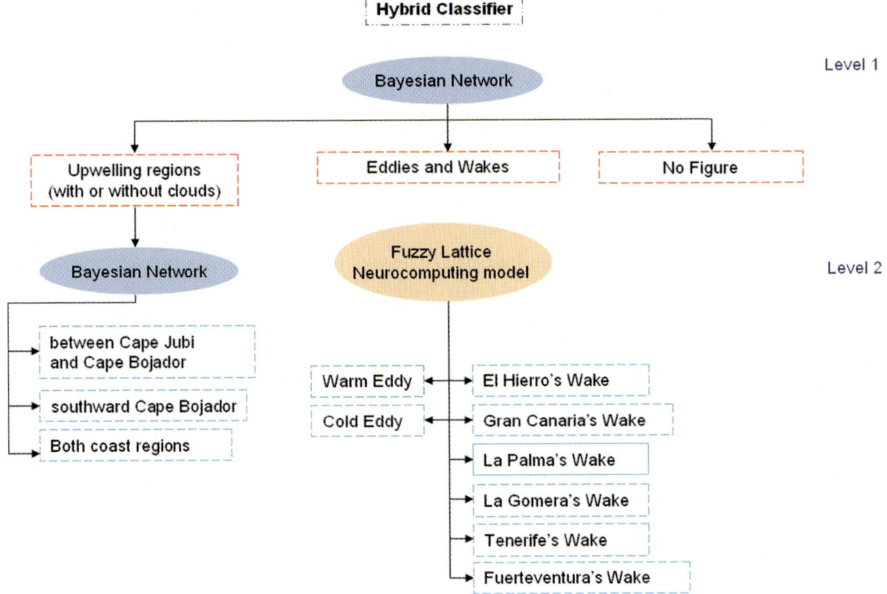

Fig. 11.5. Structure of hybrid hierarchical model

- σ-FLNMAP: 2 classes of eddies (anticyclonic and cyclonic eddies) and 6 classes of wakes (The El Hierro Wake, The Gran Canaria Wake, The La Palma Wake, The La Gomera Wake and The Tenerife Wake)

11.5 Results

All of the classifiers produced good general ocean structure recognition results, as shown in Table 11.1. Classifications were made with 10-fold cross-validation. We have worked with 1000 cases (upwellings: 590 cases, wakes: 180 cases, anti-cyclonic eddies: 10 cases, cyclonic eddies: 40 cases and misclassified regions: 180 cases). 12 classes were used (3 classes of upwellings, 2 classes of eddies, 6 classes of wakes and 1 class of misclassified regions).

The best classifications were provided by NEFCLASS, σ-FLNMAP, GES and the Hybrid Hierarchical Model. Should be emphasized that the neurofuzzy systems and fuzzy lattice neural networks improve the classification results and enable knowledge to be acquired by means of fuzzy rules or fuzzy lattice rules (understandable to the user). Of course, the results of the GES are one of the best, with only 50 rules compared with the 367 of NEFCLASS.

The σ-FLNMAP classifier learns in one run through the training data, furthermore, there is only one parameter to be tuned (the vigilance parameter σ). The vigilance parameter had several values, of which $\sigma = 0.85$ yielded the best results. Classification performance with the dataset was 92.29%.

Table 11.1. Comparative classification results

classifier	no. of Features	no. of Rules	Accuracy rate
NEFCLASS	16	367	96.49
NEFPROX	16	420	89.62
σ-FLNMAP	15	156	92.29
Nave Bayes	16	-	89.68
TAN	14	-	87.08
Hybrid Hierarchical Model	1st level: 11		
	2nd level: 5 and 25	2nd level: 8	94.93
G.E.S.	18	50	95.00

On the first level, hybrid hierarchical model classification performance with the test dataset was 95.79% using 11 features.

On the second level, the classification performance by the Naive Bayes using 5 features was 93.06%. The FLNMAP had an accuracy rate of 95.96% with 25 features. The global classification accuracy rate of the hybrid hierarchical model was 94.93%. This model improves the classification of the rest of the classifiers reducing the number of rules and increasing the accuracy rate. The best classification was by NEFCLASS, but it uses a very large number of fuzzy rules (about 370 rules) and there are about 100 linguistic labels for each feature, whereas the hybrid hierarchical model uses only 8 rules and 8 linguistic labels (one interval for each one of the rules). Therefore, NEFCLASS efficency is 1.5 better than that of the hybrid model but the interpretability is (100 linguistic labels and 370 rules) worse than (8 rules and 8 linguistic labels). The classification performance of GES was 95.00%, which is a good result, but this system was built by human experts, that is, the learning process is manual and not automatic like the neurofuzzy system.

Figure 11.6 and Fig. 11.7 show the original image and the results obtained in the recognition process, where the structures recognized are shown in different colours.

Moreover other important aspects are:

- GES. The learning process is manual and not automatic like the neurofuzzy and hybrid hierarchical systems and does not use imprecise knowledge.
- BNs. The main problem is that they do not use possibilistic knowledge, although the model structure (Naive Bayes) is very easy.
- Neurofuzzy System. In this case the problem is not the use of a lattice model. Oceanographers must understand the variables that are used to calculate the fair value of the oceanic structures. They must also understand how changes in the required variables drive the fair value and the resulting effect. Lattice models involve construction of a binomial tree of different paths with lattice parameters.
- σ-FLNMAP. Knowledge is represented by means of fuzzy lattice rules, which make it possible to find information close to human knowledge.

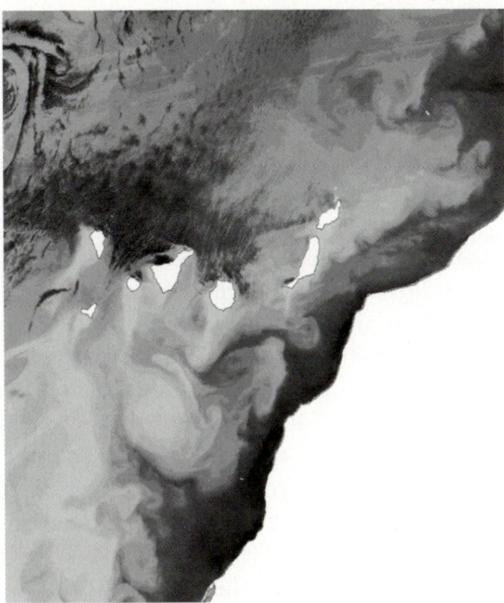

Fig. 11.6. The image aquired on August 10th, 1993

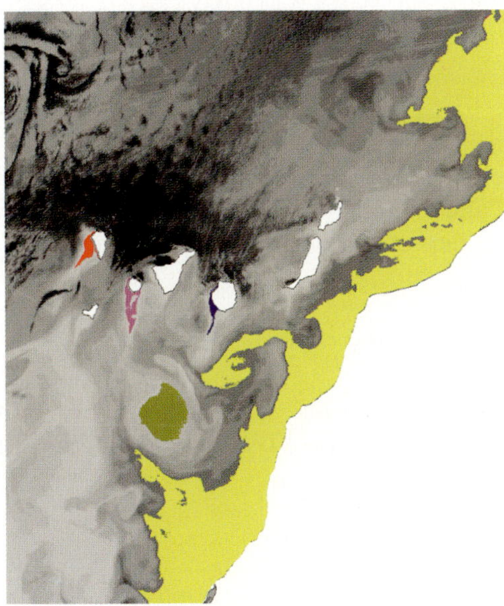

Fig. 11.7. The image obtained by the hybrid automatic interpretation system

- Hybrid hierarchical system. This system improves the accuracy rate and the representation of knowledge. This knowledge is designed by means of possibilistc and probabilistic information on two hierarchical levels.

11.6 Conclusions

The results of this work using classifiers based on neurofuzzy and fuzzy lattice systems improve the results of the other classifiers (GES and BNs) used in previous works and make it possible to manage imprecise knowledge, something impossible with non fuzzy systems. Also, the fuzzy set rules are understandable and reduced. Like in the Hough transform [32] when you divide the parameter space in lattices, the fuzzy lattice group in every box of the hyperspace parameters all the ocean structures that share a common subset of properties and have some similarity. Furthermore, by using fuzzy lattice parameters, it is possible to work with fuzzy intervals instead of fuzzy numbers and this produce a reduced set of fuzzy rules to describe the ocean structures. This allows a clear and simple knowledge representation of the ocean structures. The fuzzy lattice rules can be considered a generalization of fuzzy rules, that is to say, we group the fuzzy features of fuzzy rules into fuzzy lattices.

The main problem with initial classifiers is the number of rules generated, which is often excessive. The hybrid hierarchical model maintains or enhances the accuracy rates in the recognition process, reduces the number of rules generated, thereby increasing the interpretability of the system, and is completely automatic. Moreover, the hybrid hierarchical system is based on probabilistic and possibilistic knowledge through the use of fuzzy lattices.

In future works, we also plan to improve the accuracy rate of the hybrid hierarchical model and the fuzzy lattice systems by including more features, tuning training parameters and optimizing the rule set.

11.7 Acknowledgments

This work was partially supported by the Spanish Commission for Science and Technology (CICYT) through Project 2004-TIN-05346.

References

1. Torres JA, Guindos F, López M, Cantón M (2003) Competitive neural-net-based system for the automatic detection of oceanic mesoscalar structures on AVHRR scenes. IEEE Trans Geoscience and Remote Sensing 41:845–852
2. Piedra JA, Cantón M, Guindos F (2005) Automatic recognition of ocean structures from satellite images by means of hybrid neurofuzzy systems and fuzzy lattice neurocomputing model. In: Proc Image Information Mining - Theory and Application to Earth Observation (ESA-EUSC)

3. Lea SM, Lybanon M (1993) Automated boundary delineation in infrared ocean images. IEEE Transactions on Geoscience and Remote Sensing 31:1256–60

4. Simhadri K, Iyengar S, Holyer RJ, Lybanon M, Zachary JM (1998) Wavelet-based feature extraction from oceanographic images. IEEE Transactions on Geoscience and Remote Sensing 36:767–78

5. Lybanon M (1996) Maltese front variability from satellite observations based on automated detection. IEEE Transactions on Geoscience and Remote Sensing 34:1159–65

6. Thonet H, Lemonnier B, Delmas R (1995) Automatic segmentation of oceanic eddies on AVHRR thermal infrared sea surface images. In: Proc Conf Challenges of Our Changing Global Env (OCEANS) 2:1122–1127

7. García-Weil L, Pacheco M, Rodríguez G, McClimans T (2000) Topographic effects on the mesoscale ocean circulation near Canarian archipelago examined by means of satellite images and laboratory simulations. In: Proc IEEE Int Geoscience and Remote Sensing Symposium (IGARSS) pp 1830–1832

8. Barton ED, Arístegui J, Tett P, Cantón M, García-Braun J, Hernández-León S, Nykjaer L, Almeida C, Almunia J, Ballesteros S, Basterretxea G, Escánez J, García-Weil L, Hernández-Guerra A, López-Laatzen F, Molina R, Montero MF, Navarro-Pérez E, Rodríguez-Pérez JM, Lenning K Van, Vélez H, Wild K (1998) The transition zone of the Canary current upwelling region. Progress in Oceanography 41:455–504

9. García-Weil L (1998) Descripción y Análisis Cuantitativo Mediante Series de Imágenes de Satélite de la Dinámica de las Aguas Superficiales del Noroeste de África. PhD Thesis, Las Palmas de Gran Canaria University, Spain

10. Parada M, Cantón M (1998) The spatial and temporal evolution of thermal structures in the Alboran sea Mediterranean basin. Int Journal of Remote Sensing 19:2119–2131

11. Parada M, Cantón M (1998) Sea surface temperature variability in the Alboran sea from satellite data. Int Journal of Remote Sensing 19:2439–2450

12. Arístegui J, Sangrá P, Hernández-León S, Cantón M, Hernández-Guerra A, Kerling JL (1994) Island-induced eddies in the Canary islands. Deep-Sea Research 41:1509–1525

13. Sangra P, Pelegri JL, Hernández Guerra A, Arregui I, Martín JM, Marrero Díaz A, Martínez A, Ratsimandresy AW, Rodríguez Santana A (2005) Life history of an anticyclonic eddy. Journal of Geophysical Research 110:19

14. Tejera A, García Weil L, Heywood K, Cantón M (2002) Observation of oceanic mesoscale features and variability in the Canary islands area from ERS-1 altimeter, satellite infrared imagery and hydrographic measurements. Int Journal of Remote Sensing 23:4897–4916

15. Torres JA, Guindos F, Peralta M, Cantón M (2003) An automatic cloud-masking system using backpro neural nets for AVHRR scenes. IEEE Trans Geoscience and Remote Sensing 41:826–831

16. Guindos F, Piedra JA, Cantón M (2004) Ocean features recognition in AVHRR images by means of bayesian net and expert system. In: Proc 3rd International Workshop on Pattern Recognition in Remote Sensing 11:1–4

17. Beucher S (1990) Segmentation Tools in Mathematical Morphology. SPIE-1350 Image Algebra and Morphological Image Processing 70–84

18. Canny J (1983) Finding Edges and Lines in Images. Tech Rep 720, Artificial Intelligence Lab, MIT, Cambridge, MA

19. Murphey YL, Hong G (2000) Automatic feature selection - an hybrid statistical approach. In: Proc Int Conf Pattern Recognition pp 3–8
20. Langley P, Sage S (1994) Induction of selective Bayesian classifiers. In: Proc 10th Conf on Uncertainty in Artificial Intelligence pp 399–406. Morgan-Kaufmann
21. Yamagata Y, Oguma H (1997) Bayesian feature selection for classifying multi-temporal SAR and TM Data. In: Proc Int Geoscience and Remote Sensing Symposium (IGARSS) 2: 978–980
22. Hall MA (1999) Correlation-based Feature Selection for Machine Learning. PhD Thesis, Waikato University, Hamilton, NZ
23. Duda RO, Hart PE (2001) Pattern Classification, 2nd ed. John Wiley and Sons, New York
24. Rish Irina (2001) An empirical study of the naive Bayes classifier. In: Proc Workshop on Empirical Methods in Artificial Intelligence (IJCAI)
25. Nauck D, Kruse R (1995) NEFCLASS - A neuro-fuzzy approach for the classification of data. In: Proc ACM Symp Applied Computing pp 461-465. ACM Press, New York
26. Nauck D, Kruse R (1998) A neuro-fuzzy approach to obtain interpretable fuzzy systems for function approximation. In: Proc World Congress on Computational Intelligence (WCCI) 2:1106–1111
27. Kaburlasos VG, Petridis V (2000) Fuzzy lattice neurocomputing (FLN) models. Neural Networks 13:1145–1170
28. Petridis V, Kaburlasos VG (1999) Learning in the framework of fuzzy lattices. IEEE Transactions on Fuzzy Systems 7:422–440
29. Kaburlasos VG, Petridis V (2002) Learning and decision-making in the framework of fuzzy lattices. In: Jain LC, Kacprzyk J (eds) New Learning Paradigms in Soft Computing, ser Studies in Fuzziness and Soft Computing 84:55–96. Physica-Verlag, Heidelberg, Germany
30. Kaburlasos VG (2006) Towards a Unified Modeling and Knowledge-Representation Based on Lattice Theory — Computational Intelligence and Soft Computing Applications, ser Studies in Computational Intelligence 27. Springer, Heidelberg, Germany.
31. Kaburlasos VG, Athanasiadis IN, Mitkas PA (2007) Fuzzy lattice reasoning (FLR) classifier and its application for ambient ozone estimation. Intl J Approximate Reasoning 45(1):152–188
32. Gonzalez RC, Woods RE (2002) Digital Image Processing, 2nd ed. Prentice Hall, Upper Saddle River, New Jersey

Genetically Engineered ART Architectures

Ahmad Al-Daraiseh[1], Assem Kaylani[1], Michael Georgiopoulos[1], Mansooreh Mollaghasemi[1], Annie S. Wu[1], and Georgios Anagnostopoulos[2]

[1] University of Central Florida, Orlando FL, 32816 `michaelg@mail.ucf.edu`
[2] Florida Institute of Technology, Melbourne, FL, 32796 `georgio@fit.edu`

Summary. This chapter focuses on the evolution of ARTMAP architectures, using genetic algorithms, with the objective of improving generalization performance and alleviating the ART category proliferation problem. We refer to the resulting architectures as GFAM, GEAM, and GGAM. We demonstrate through extensive experimentation that evolved ARTMAP architectures exhibit good generalization and are of small size, while consuming reasonable computational effort to produce an optimal or a sub-optimal network. Furthermore, we compare the performance of GFAM, GEAM and GGAM with other competitive ARTMAP structures that have appeared in the literature and addressed the category proliferation problem in ART.

12.1 Introduction

Adaptive resonance theory (ART) was developed by Grossberg (see [18]). Some of the ART architectures that have appeared in the literature include Fuzzy ARTMAP (FAM) (see [10]), Ellipsoidal ARTMAP (EAM) (see [1]), and Gaussian ARTMAP (GAM) (see [36]). All of these ART architectures possess a number of desirable properties, such as they can solve arbitrarily complex classification problems, they converge quickly to a solution (within a few presentations of the list of input/output patterns belonging to the training set), they have the ability to recognize novelty in the input patterns presented to them, they can operate in an on-line fashion (new input patterns can be learned by the ART system without retraining with the old input/output patterns), and they produce answers that can be explained with relative ease.

Since, Fuzzy ARTMAP's inception in 1992, a number of ART related papers have appeared in the neural network literature, some of which (as the ones mentioned above) modified the Fuzzy ARTMAP neural network so as to improve its performance. A related, important contribution in the ART literature is the one contributed by Petridis and Kaburlasos in 1998 ([31]), where they introduced the Fuzzy Lattice Neural Network (FLNN), a cross fertilization of fuzzy set theory and lattice theory, which can handle general

A. Al-Daraiseh: *Genetically Engineered ART Architectures*, Studies in Computational Intelligence (SCI) **67**, 233–262 (2007)
`www.springerlink.com`

types of data in addition to N-dimensional vectors. Most of the lattice work of Kaburlasos is included in a recently published book by Springer Verlag (see [21]) where a unified, cross-fertilizing approach for knowledge representation and modeling based on lattice theory is presented with emphasis on clustering, classification and regression applications. Some of Kaburlasos' recent work related with fuzzy lattice reasoning (FLR), which is the algorithm/software applied on a FLNN, can be found in [22].

The above references paint only an incomplete picture of the work that researchers have contributed into the ART literature, since Fuzzy ARTMAP's inception. However, since our goal in this chapter is to focus on the category proliferation problem in ART we are limiting, from this point on, our referrals to papers that have addressed this ART problem. Quite often the category proliferation problem in ART is connected with the issue of overtraining in ART. Over-training happens when ART is trying to learn the training data perfectly at the expense of degraded generalization performance (i.e., classification accuracy on unseen data) and also at the expense of creating many categories to represent the training data (leading to the category proliferation problem). Categories in ART are formed in order to compress the input data prior to mapping these compressed data to their respective outputs. The categories in Fuzzy ARTMAP are hyperboxes, in Ellipsoidal ARTMAP are ellipsoids, and in Gaussian ARTMAP they are Gaussian multi-dimensional probability distributions represented by their center points and widths (means and standard deviations) across every dimension. A number of authors have tried to address the category proliferation problem in ART. Amongst them we refer to the work by Marriott and Harrison, 1995, (see [28]), where the authors eliminate the match tracking mechanism of Fuzzy ARTMAP when dealing with noisy data; the work by Charalampidis et al., 2001 (see [13]), where the Fuzzy ARTMAP equations are appropriately modified to compensate for noisy data; the work by Verzi, et al., 2001 (see [35]), Anagnostopoulos, et al., 2002 and 2003 (see [2, 3]), and Gomez-Sanchez et al., (see [16, 17]), where different ways are introduced of allowing the ART categories to encode patterns that are not necessarily mapped to the same output (label); the work by Koufakou, et al., 2001, (see [25]), where cross-validation is employed to avoid the overtraining/category proliferation problem in Fuzzy ARTMAP; and the work by Carpenter, et al., 1998 (see [11]), Williamson, 1997 (see [37]), Parrado, et al., 2003 (see [30]), where the ART structure is changed from the winner-take-all to a distributed version and simultaneously slow learning is employed with the intent of creating fewer ART categories and reducing the effect of noisy patterns.

In this paper, we propose the use of genetic algorithms (see [15]) to solve the category proliferation problem in ART architectures, such as FAM, EAM and GAM.

Genetic algorithms (GAs) are a class of population-based stochastic search algorithms that are developed from ideas and principles of natural evolution. An important feature of these algorithms is their population based search strategy. Individuals in a population compete and exchange information with each other in order to perform certain tasks. Genetic algorithms have been extensively used to evolve artificial neural networks. For a thorough exposition of the available research literature on evolving neural networks the interested reader is advised to consult the work of Yao (see [40]), 1999. In [40], the author distinguishes three different strategies in evolving neural networks. The first strategy is the one used to search for the weights of the neural network (see for example [32]). The second one is the one used to design the structure of the network (see for example [27] where only the structures are evolved, and [39], where both the structure and the interconnection weights are evolved), and the third one is the one where the learning rules of the neural network are evolved (see [19]). Furthermore, GAs may also be used to select the features that are input to the neural network. Since the pioneering work by Siedlecki and Sklanski (see [33]), genetic algorithms have been used for many selection problems using neural networks (see [6, 38]), and other classifiers, such as decision trees (see [4]), k-nearest neighbors (see [24]), and Naive Bayes classifiers (see [8, 20]).

To the best of our knowledge, work that utilizes GAs and ART neural network architectures is rather limited. For instance, in Liu, et al., 2003 (see [26]), a GA algorithm was employed to appropriately weigh attributes of input patterns before they were fed into the input layer of Fuzzy ARTMAP. The results reported reveal that this attribute weighting was beneficial because it produced a trained ART architecture of improved generalization. Furthermore, in Burton and Vladimirova, 1997 (see [7]), a Fuzzy ART neural network is employed as a GA fitness function evaluator, however the brevity of the published paper did not allow for the discussion of the details pertinent to this work.

In our work here, we use GAs to evolve simultaneously the weights, as well as the topology of the ART neural networks, such as FAM, EAM or GAM. In contrast to the feed-forward neural networks that have been extensively evolved, the aforementioned ART neural networks have a number of topological constraints, such as (a) they have one hidden layer of nodes, called category representation layer, and (b) every interconnection weight value from every node of the input layer to a node in the hidden layer is important (e.g., in FAM they are representing the minimum or the maximum of the values of input patterns across every dimension that were encoded by this node). Consequently, the only element of the ART's topology that can be evolved is the number of nodes in the category representation layer of the ART network. Furthermore, in our application we start with a population of trained ARTs, whose number of nodes in the hidden layer and the values of the interconnection weights converging to these nodes are fully determined (at the beginning of the evolution) by the ART's training rules. To this initial population of ART networks, GA operators are applied to modify these trained ART architectures (i.e., number of nodes in the hidden layer, and values of the interconnection weights) in a way that encourages better generalization and smaller size architectures.

It is worth reminding the reader that, as with many neural network architectures, the knowledge in ART is stored in its interconnection weights that have a very interesting geometrical interpretation (see [2, 9]. In particular, the interconnection weights in Fuzzy ARTMAP (converging to the nodes in the hidden layer) represent the lower and upper end-points of hyper-rectangles (referred to as categories) that enclose within their boundaries clusters of data that are mapped to the same label. Furthermore, the interconnection weights in Ellipsoidal ARTMAP represent the size and the direction of ellipsoids that enclose within their boundaries clusters of data that are mapped to the same label. Finally, the interconnection weights in Gaussian ARTMAP represent the mean vectors and variance vectors of Gaussianly distributed data that are mapped to the same label. Eventually, the evolution of these trained ARTs produces ART architectures, referred to as GFAM, GEAM or GGAM. GFAM, GEAM and GGAM are extracted from the last generation of the ART evolution, as the Fuzzy ARTMAP, Ellipsoidal ARTMAP, or Gaussian ARTMAP, respectively, that attained the highest fitness value. The fitness value of an ART network is defined in a way that attains higher values for better generalizing and smaller size ART networks.

It is apparent that, in evolving neural network architectures, one has to decide on the genotype representation scheme for the neural network architecture under consideration, on the genetic operators used to evolve these neural network architectures, and on the fitness function used to guide this evolution. In this paper we address these issues in a manner that fits the characteristics of the specific ART neural network, under consideration, and our ultimate objective of reducing category proliferation in ART, while preserving good generalization performance. In addition to successfully addressing the issues related with the evolution of ART structures, mentioned above, we also compare in this paper the final product of ART's evolution with other approaches proposed in the ART literature that also addressed the category proliferation problem in ART. This comparison is based on the accuracy of the architectures and size of the architectures produced by these techniques. This comparison demonstrates that GFAM, GEAM, and GGAM perform well compared to a number of other approaches proposed in the literature that have claimed that they address the ART category proliferation problem.

The organization of this chapter is as follows: In section 12.2 we emphasize some of the basics of the ART architectures (FAM, EAM, GAM). In Sect. 12.3 we also provide the step-by-step description of the evolved ART architectures, such as GFAM, GEAM and GGAM, and we introduce the fitness function used for the evolution of ART architectures. In Sect. 12.4, we describe the experiments and the datasets used to assess the performance of GFAM, GEAM, and GGAM. In particular, in this section we explain the experiments that we conducted to come up with good default parameter values for the probability of category mutation, and for the probability with which a category is deleted in the evolution of ART architectures. Then for these good default parameter values we assess the performance of these genetically engineered architectures, and we offer performance comparisons between the GFAM, GEAM and GGAM and other ART architectures that were proposed as solutions for the category proliferation problem in ART. In Sect. 12.5, we summarize our findings, and we offer some conclusive remarks.

12.2 ART Preliminaries

The Fuzzy ARTMAP (FAM) neural network architecture was introduced by Carpenter and Grossberg in their seminal paper (see [10]). Since its introduction, other ART architectures have been introduced into the literature and the focus on this paper is on Fuzzy ARTMAP and two other ART architectures Ellipsoidal ARTMAP (see [1]) and Gaussian ARTMAP (see [36]). Our objective in this paper is to illustrate how we can design genetically engineered ART architectures from a population of Fuzzy ARTMAPs, or Ellipsoidal ARTMAPs, or Gaussian ARTMAPs. We assume that the reader is familiar with all these ART architectures. In this section we will only describe the specifics of ART architectures that are needed to understand the genetically engineered ART structures, introduced in section 12.3. For simplicity we refer to all these ART architectures as ART and we use their specific name (FAM, or EAM or GAM) only when we want to discriminate one from the other.

The block diagram of an ART architecture is shown in Fig. 12.1 (for FAM) and Fig. 12.2 (for EAM and GAM). Notice that this block diagram is simpler than the block diagram of FAM, reported in Carpenter and Grossberg in

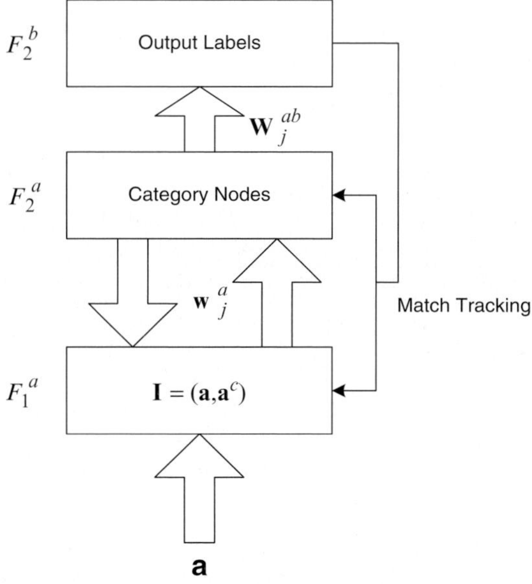

Fig. 12.1. The block diagram of a FAM Architecture

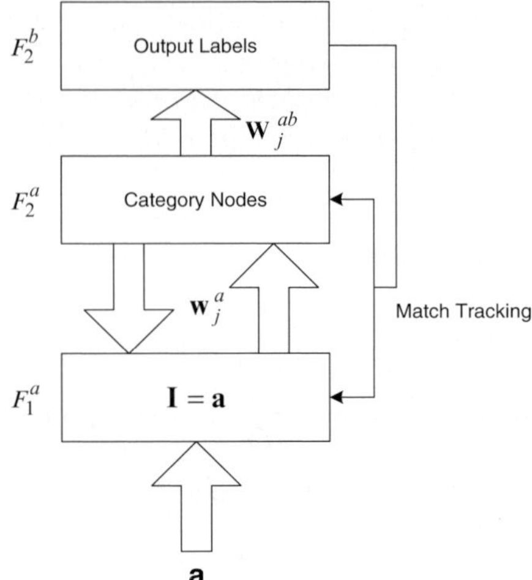

Fig. 12.2. The block diagram of an EAM or GAM Architecture

1992, and it has been adopted by various researchers in the field (e.g., Kasuba, 1993 [23], as well as Taghi, et al., 2003 [34]) because it can completely and succinctly describe the functionality of a variety of ART architectures, dealing with classification problems. As the focus of our paper is on classification problems, we also adopt this simpler ART architecture. The ART architecture, depicted in Fig. 12.1 (FAM) and Fig. 12.2 (EAM or GAM), has three major layers. The input layer F_1^a where the input patterns (designated by $\mathbf{I}$) are presented, the category representation layer F_2^a, where compressed representations of these input patterns (designated as $\mathbf{w}_j^a$), are formed, and the output layer F_2^b that holds the labels of the categories formed in the category representation layer. An additional layer, not shown in Fig. 12.1 and Fig. 12.2, and designated by F_0^a, is a pre-processing layer and its functionality is to pre-process the input patterns, prior to their presentation to ART. The first level of ART pre-processing takes the input vector and normalizes it so that each one of its components lies in the interval [0, 1], and that is the only level of pre-processing needed for EAM and GAM. The second level of pre-processing (needed only for FAM) takes the normalized input vector, referred to as $\mathbf{a}$ and complementary encodes it, by appending to it another vector, referred to as $\mathbf{a}^c$. The complement of vector $\mathbf{a}$ is defined as

$$\mathbf{a}^c = (1 - a(1), 1 - a(2), \ldots, 1 - a(M_a)) \tag{12.1}$$

where

$$\mathbf{a} = (a(1), a(2), \ldots, a(M_a)) \tag{12.2}$$

and M_a, in the above equations, stands for the dimensionality of the input pattern of the pattern classification task under consideration. It is worth mentioning that the complementary encoding of the input patterns is necessary for the successful operation of Fuzzy ARTMAP (for an explanation see [14]), however it is not needed by either EAM or GAM. Therefore, in this paper we assume that the inputs to FAM are normalized and complementary encoded, while the inputs to EAM and GAM are simply normalized (see Fig. 12.1 for FAM, and Fig. 12.2 for EAM and GAM). Note that normalization of inputs prior to their presentation to a neural network is common practice in the neural network literature.

ART can operate in two distinct phases: the training phase and the performance (test) phase. The training phase of ART can be described as follows: Given a set of inputs and associated label pairs, $\mathbf{I}_1, label(\mathbf{I_1}), \mathbf{I}_2, label(\mathbf{I_2}),...,$ $\mathbf{I}_{PT}, label(\mathbf{I_{PT}})$ (called the training set), we want to train ART to map every input pattern of the training set to its corresponding label. To achieve the aforementioned goal we present the training set to the ART architecture repeatedly. That is, we present $\mathbf{I}_1$ to F_1^a, $label(\mathbf{I_1})$ to F_2^b, then $\mathbf{I}_2$ to F_1^a, $label(\mathbf{I_2})$ to F_2^b, and finally, $\mathbf{I}_{PT}$ to F_1^a, $label(\mathbf{I_{PT}})$ to F_2^b. We present the training set to the ART network as many times as it is necessary for ART to correctly classify these input patterns. The task is considered accomplished (i.e., learning is complete) when the weights in ART do not change during a training set presentation, or after a specific number of list presentations is reached. The performance phase of ART works as follows: Given a set of input patterns (referred to as the test set), we want to find the ART output (label) produced when each one of the aforementioned test patterns is presented at its layer. In order to achieve this goal, we present the test set to the trained ART network and we observe the network's output. For the purposes of this paper, we assume that the reader knows the details of the training/performance phases in ART (FAM, EAM or GAM).

As we mentioned above, the weights (templates) in ART create compressed representations of the input patterns presented to the ART network during its training phase. These compressed representations have a geometrical interpretation. In particular, every node (category) in the category representation layer of Fuzzy ARTMAP (FAM) has template weights that completely define the lower and upper endpoints of a hyperbox. This hyperbox includes within its boundaries all the input patterns that chose this category as their representative category in FAM's training phase and were subsequently encoded by this category. In Fig. 12.3, we show the hyperbox of a category in a FAM architecture (2-D example), with lower endpoint $\mathbf{u}_j$, and upper endpoint $\mathbf{v}_j$, and the input patterns (the $\mathbf{I}$'s that it has encoded). Also, every node (category) in the category representation layer of Ellipsoidal ARTMAP (EAM) has template weights that completely define an ellipsoid through its center, direction of major axis, length of the major axis, and ratio of lengths of minor axes to major axis in the ellipsoid. This ellipsoid includes within its boundaries all the input patterns that chose this category as their representative category in EAM's training phase and were subsequently encoded by this category. In Fig. 12.4, we show the ellipsoid of a category in a EAM architecture

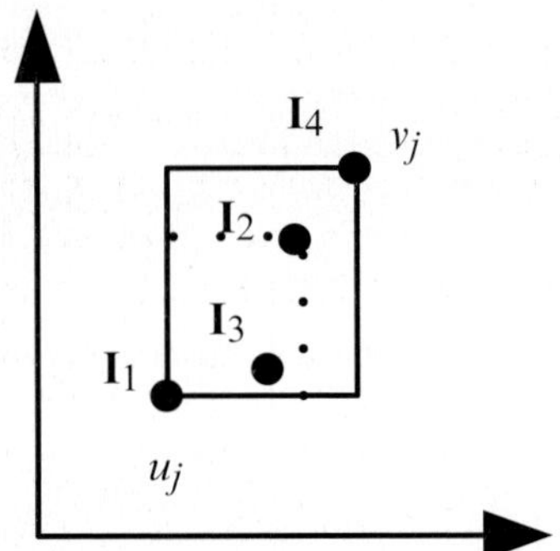

Fig. 12.3. A hyperbox category representation in FAM. This hyperbox has encoded patterns $\mathbf{I}_1, \mathbf{I}_2, \mathbf{I}_3, \mathbf{I}_4$. In the figure, the **a** portion of these input patterns is depicted, as well as the lower end-point $\mathbf{u}_j$ and the upper endpoint $\mathbf{v}_j$ of this hyperbox

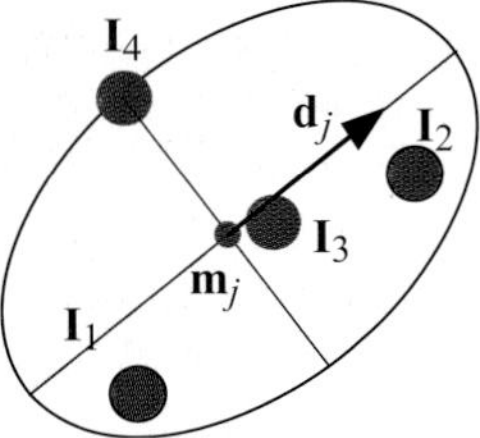

Fig. 12.4. An ellipsoidal category representation in EAM. This ellipsoid has encoded patterns $\mathbf{I}_1, \mathbf{I}_2, \mathbf{I}_3, \mathbf{I}_4$. In the figure, the center point $\mathbf{m}_j$ and the direction vector $\mathbf{d}_j$ are shown, while the radius of the major axis, and the ratio of lengths of minor to major axis are easily deduced from the figure

(2-D example), with center $\mathbf{m}_j$, direction of the major axis $\mathbf{d}_j$, length of the major axis, represented by its radius r_j (implied from the figure), ratio of the lengths of minor axes to major axis μ (implied from the figure), and the input patterns $\mathbf{I}$'s that it has encoded. Finally, every node (category) in the category representation layer of Gaussian ARTMAP has template weights that define the mean vector, the standard deviation vector of a multi-dimensional Gaussian distribution, and the number of points that are associated with this Gaussian distribution. The mean vector of this Gaussian distribution and the standard deviation vector of this Gaussian distribution are defined in terms of the means and the standard deviations (across every dimension) of the points that chose this node (category) as their representative category, while the number of the points associated with this Gaussian distribution are the number of points that chose this category as their representative category. In Fig. 12.5, we show the Gaussian distribution of a category in a GAM architecture (1-D example), with mean $\mathbf{m}_j$, standard deviation σ_j, number of points n_j (in our example $n_j = 4$), and the input patterns (i.e., $\mathbf{I}$'s) that it has encoded.

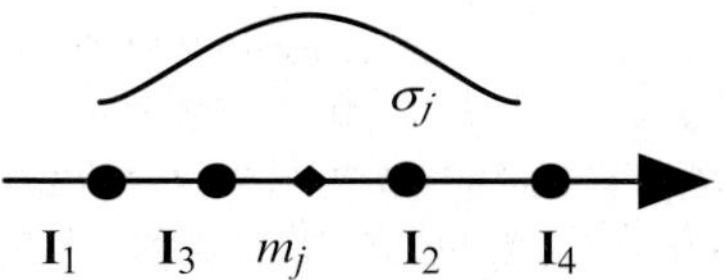

Fig. 12.5. A Gaussian curve category representation in GAM. This Gaussian distribution has encoded patterns $\mathbf{I}_1, \mathbf{I}_2, \mathbf{I}_3, \mathbf{I}_4$. In the figure, the center point $\mathbf{m}_j$ and the standard deviation vector σ_j of the Gaussian curve are shown, while the number of points n_j that this Gaussian curve represents is deduced as being equal to 4

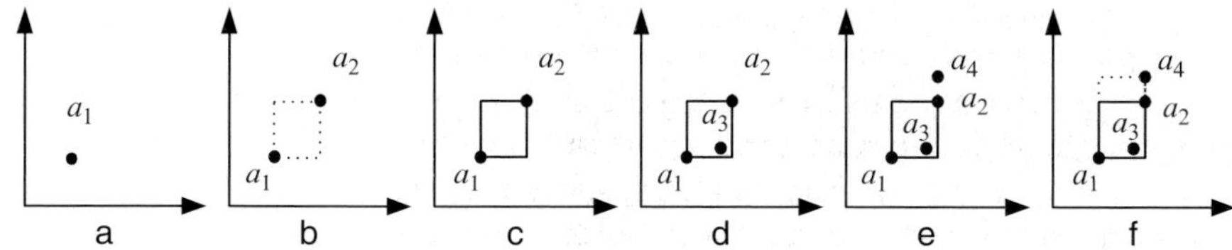

Fig. 12.6. FAM Learning (2-D Example): (a) A category with 0 size; (b) Introducing a new pattern $\mathbf{I}_2$, represented by $\mathbf{a}_2$; (c) The category expands to include $\mathbf{a}_2$; (d) Since new pattern $\mathbf{I}_3$, represented by $\mathbf{a}_3$ is inside the category, it doesn't change its size; (e) New Pattern $\mathbf{I}_4$, represented by $\mathbf{a}_4$ is presented; (f) Since $\mathbf{a}_4$ is outside the category, the category is expanded to include $\mathbf{a}_4$, within its boundaries

In essence, at the beginning of training, every category of FAM starts as a trivial hyperbox (equal to a point) and subsequently it expands to incorporate within its boundaries all the input patterns that in the training phase choose this hyperbox as their representative hyperbox, and are encoded by it (see Fig. 12.6, where the category expansion of FAM is shown for an example dataset). The size of hyperbox is the sum of the lengths of its sides.

Similarly, at the beginning of training, every EAM category starts as a trivial ellipsoid (equal to a point) and subsequently it expands to incorporate within its boundaries all the input patterns that in the training phase chose this ellipsoid as their representative ellipsoid, and are encoded by it (see Fig. 12.7, where the category expansion of EAM is shown for an example dataset). The size of the ellipsoid is measured as the length of the major axis.

Finally, at the beginning of training, every category of GAM starts as a collection of Gaussianly distributed data, with mean equal to the input pattern that was first encoded by this category, and a small standard deviation vector (part a of Fig. 12.8); as training progresses in GAM this GAM category is modified to incorporate the information of the additional input patterns that are encoded by it (see part b of Fig. 12.8 for an illustration of how the GAM category is modified for an example dataset). At any point in time the mean vector of this Gaussian distribution, corresponding to a category, is equal to the mean vector of all the input patterns encoded by the category, and the variance vector of the Gaussian distribution is equal to the variance vector corresponding to the input patterns that were encoded by the category.

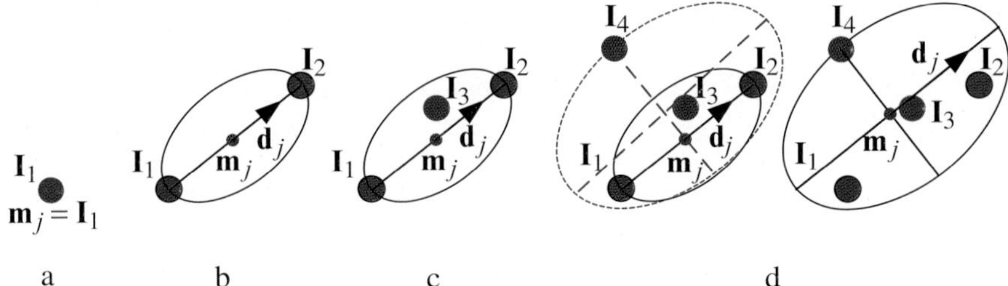

Fig. 12.7. EAM Learning (2-D Example): (a) A category with 0 size; (b) Introducing a new pattern $\mathbf{I}_2$; the category expands to include $\mathbf{I}_2$; (c) Introducing a new pattern $\mathbf{I}_3$; since the category includes $\mathbf{I}_3$, it does not change its size; (d) Pattern $\mathbf{I}_4$ is presented; since this pattern is outside the category, the category is expanded to include $\mathbf{I}_4$ within its boundaries

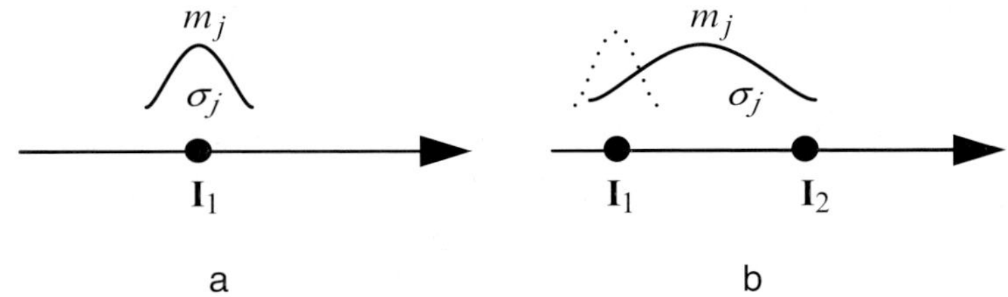

Fig. 12.8. GAM Learning (a) A category with 0 size; (b) Introducing a new pattern $\mathbf{I}_2$; the category characteristics (mean, standard deviation, of the Gaussian curve, as well as number of points encoded by the Gaussian curve) change to include the new knowledge that the new input pattern brings

The ability of the category to encode new input patterns depends on the Mahalanobis distance of an input pattern from the mean/variance vectors of the Gaussian distribution corresponding to the category.

It is also worth mentioning that the categories in FAM, EAM and GAM are allowed to expand up to a point allowed by a threshold, controlled by a network parameter denoted as the baseline vigilance parameter . This parameter assumes values in the interval $[0, 1]$. Small values of this parameter allow the creation of large categories, while large values of this parameter allow the creation of small categories. In the one extreme when $\bar{\rho}_a$ is equal to 0, a FAM or EAM category, equal to the whole input space, could be created, while at the other extreme when $\bar{\rho}_a$ is equal to 1 only point categories are formed. In GAM, small values of this parameter allow more and more patterns to be encoded by a GAM category, while large values of this parameter allow only a few patterns to be encoded by a GAM category. It turns out that this parameter has a significant effect on the number and type of categories formed, and consequently it affects the performance of these networks.

The performance of ART networks is measured in terms of the number of categories created in its training phase (small number of categories is good), and how well it generalizes on unseen data (high generalization accuracy is good). In the process of discovering GFAM, GEAM or GGAM we are starting from a population of trained FAM, EAM or GAM networks that have been trained with different baseline vigilance parameter values, and different orders of pattern presentations of the training data (it has been a known fact that performance in ART is affected by the order according to which training data are presented to an ART architecture).

The performance of Fuzzy ARTMAP (Ellipsoidal ARTMAP) is also affected by another network parameter called choice parameter. In the training of the Fuzzy ARTMAP (Ellipsoidal ARTMAP) networks used in our experiments, we fixed this choice parameter to the value of 0.1 (0.01). The performance of Ellipsoidal ARTMAP is also affected by another network parameter called length of minor to major axis parameter (denoted by the symbol μ), and expressing the ratio of lengths of minor axes versus major axis of the ellipsoid that corresponds to an EAM category. In the training of the Ellipsoidal ARTMAP networks used in our experiments, we fixed this choice parameter to the value of 1. The performance of Gaussian ARTMAP is also affected by another network parameter called initial variance parameter (denoted by the symbol γ), and representing the initial variance of a GAM category (after it has encoded a single input pattern). In the training of our Gaussian ARTMAP networks, we fixed this choice parameter to the value of 0.6.

12.3 Evolution of ART Architectures

In the rest of the paper we assume that for every classification problem we are provided with a training set, a validation test, and a test set. The training set is used for the training of GFAM, GEAM, and GGAM architectures under consideration. The validation set is used to optimize the produced GFAM, GEAM or GGAM network in ways that will become apparent in the following text. Finally, the test set is used to assess the performance of the optimized GFAM, GEAM, or GGAM network created.

GFAM, GEAM, and GGAM are evolved FAM, EAM, GAM networks, respectively, that are produced by applying, repeatedly, genetic operators on an initial population of trained FAM, EAM, or GAM networks. GFAM, GEAM, GGAM use tournament selection with elitism, as well as genetic operators, including crossover and mutation. In addition, GFAM, GEAM and GGAM use a special operator, Cat_{del}; this special operator is needed so that categories could be destroyed in the ART architectures, thus leading us, through evolution, to smaller ART structures.

To better understand how ART (FAM, or EAM, or GAM) is genetically designed, we resort to a step-by-step description of this design. The genetic design of ART can be articulated through a sequence of basic steps, defined succinctly below. Before we proceed, appropriate terminology is needed and is outlined in the Appendix to this chapter.

244   A. Daraiseh et al.

The following pseudo-code shows the basic steps of GFAM, GEAM and GGAM :

begin

1:          Generate-Initial-Population()
2:          Repeat
2.a.   Evaluate-Fitness-And-Sort()
2.b.   Selection()
2.c.   CrossOver()
2.d.   Catdel()
2.e.   Mutate()
            Until [max number of generations reached]

3.   Return bestNetwork
end

Step 1: Generate Initial Population: The algorithm starts by training Pop_{size} ARTMAP networks (FAM, EAM or GAM), each one of them trained with a different value of the baseline vigilance parameter $\bar{\rho}_a$, and order of training pattern presentation. In particular, we first define $\bar{\rho}_a^{inc} = \frac{\bar{\rho}_a^{max} - \bar{\rho}_a^{min}}{Pop_{size}-1}$, and then the baseline vigilance parameter of every network is determined by the equation $\bar{\rho}_a^{min} + i\bar{\rho}_a^{inc}$, where $i \in \{1, 2, ..., Pop_{size} - 1\}$. The choice parameter in a FAM network was chosen to be equal to 0.1. The choice parameter in an EAM network was chosen to be equal to 0.01. The ratio of the length of the minor axes to major axes in EAM was chosen equal to 1. The initial value of the standard deviation γ in a GAM network is chosen to be equal to 0.6. In our experiments with GFAM and GEAM we chose $\bar{\rho}_a^{min} = 0.1$ and $\bar{\rho}_a^{max} = 0.95$, while in our experiments with GGAM we chose $\bar{\rho}_a^{min} = 0.1$ and $\bar{\rho}_a^{max} = 0.75$.

We assume that the reader is familiar with how training of FAM, EAM and GAM networks is accomplished, and thus the details here are omitted. Once the Pop_{size} networks are trained, they need to be converted to chromosomes so that they can be manipulated by the genetic operators. The following is a description of the encoding schemes used in GFAM, GEAM and GGAM:

- GFAM uses a real number representation to encode the networks. Each FAM chromosome consists of two levels, level 1 containing all the categories of the FAM network, and level 2 containing the lower and upper endpoints of the hyperboxes, representing every category in level 1, as well as the label of every category in level 1 (see Fig. 12.9). We denote a category of a trained FAM network with index p, $1 \le p \le Pop_{size}$, by $\mathbf{w}_j^p$, where $\mathbf{w}_j^a(p) = (\mathbf{u}_j^a(p), (\mathbf{v}_j^a(p))^c, l_j(p))$, where $\mathbf{u}_j^a(p)$, and $\mathbf{v}_j^a(p)$ are the lower and

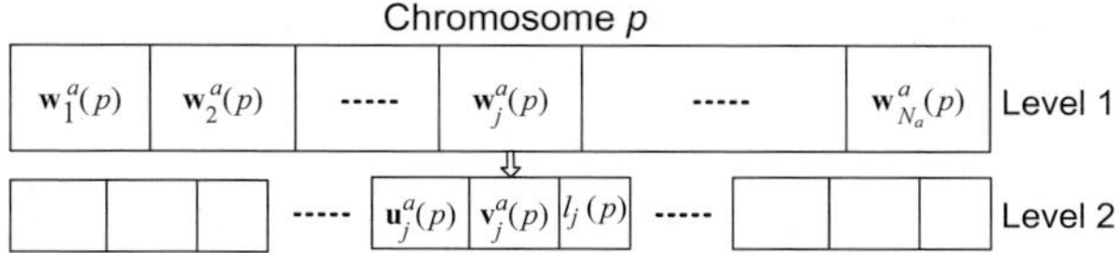

Fig. 12.9. GFAM chromosome structure. At level 2, the category's weight $\mathbf{w}_j^a$ contains the information about the lower end-point, $\mathbf{u}_j^a$, and the upper end-point, $\mathbf{v}_j^a$, of the hyperbox corresponding to the category, as well as the label l_j of the category

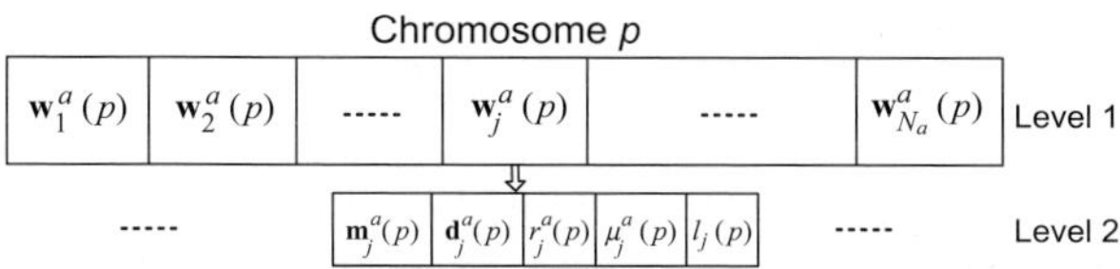

Fig. 12.10. GEAM chromosome structure. At level 2, the category's weight $\mathbf{w}_j^a$ contains the information of the center, $\mathbf{m}_j^a$, the direction vector of the major axis, $\mathbf{d}_j^a$, the radius (half length) of the major axis, r_j, and the ratio of the lengths of the minor axes over the length of the major axis, μ_j, of the ellipsoid corresponding to this category, as well as the label l_j of the category

upper endpoints of the hyperbox corresponding to this category, and $l_j(p)$ is the label of this category.

- GEAM also uses a real number representation to encode the networks. Each EAM chromosome consists of two levels, level 1 containing all the categories of the EAM network, and level 2 containing the center, direction, radius of the major axis, the ratio of the minor axes to major axis of every category (ellipsoid) in level 1, as well as the label of the category (see Fig. 12.10). We denote the category of a trained EAM network with index p, $1 \le p \le Pop_{size}$, by $\mathbf{w}_j^a(p)$, where $\mathbf{w}_j^a(p) = (\mathbf{m}_j^a(p), \mathbf{d}_j^a(p), r_j^a(p), \mu_j^a(p), l_j(p))$. The first four components of this vector are the center, direction of the major axis, half length of the major axis, and ratio of the lengths of minor axes to major axis of the ellipsoid that represents this category, while the fifth component of this vector is the label of this category.

- GGAM also uses a real number representation to encode the networks. Each GAM chromosome consists of two levels, level 1 containing all the categories of the GAM network, and level 2 containing the mean, standard deviation, number of training patterns, and the label of every category (Gaussian curve) in level 1 (see Fig. 12.11). We denote the category of a trained GAM network with index p, $1 \le p \le Pop_{size}$, by $\mathbf{w}_j^a(p)$, where $\mathbf{w}_j^a(p) = (\mathbf{m}_j^a(p), \sigma_j^a(p), n_j^a(p), l_j(p))$. The first three components of this vector are the mean, standard deviation, and number of patterns that chose this category as their representative category, while the fourth component of this vector is the label of the category.

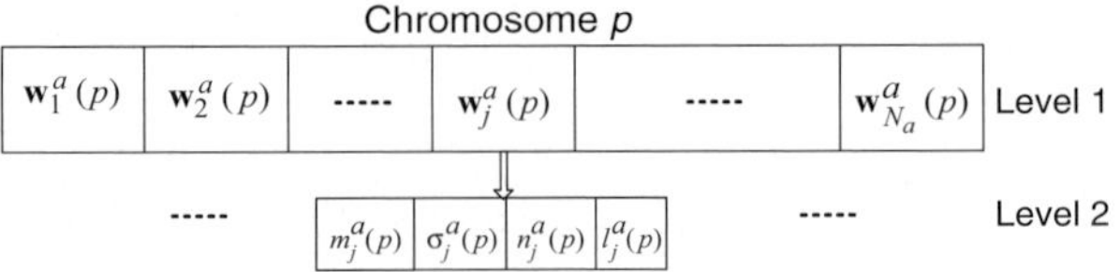

Fig. 12.11. GGAM chromosome structure. At level 2, the category's weight $\mathbf{w}_j^a$ contains the information of the center of the Gaussian curve, $\mathbf{m}_j^a$, the standard deviation vector of the Gaussian curve, σ_j^a, and the number of points represented by the Gaussian curve, n_j, as well as the label l_j of the category

In this step, we eliminate all single-point categories in the trained networks, referred to as cropping the chromosomes. Since our ultimate objective is to design a network that reduces the network size and improves generalization we discourage at this stage the creation of single-point categories. Our experiments have shown that cropping single-point categories is beneficial because it speeds-up the convergence of the GA.

Step 2 (Apply Genetic Operators): In this step a GA is applied to the population of the ART trained networks.

- **Sub-step 2a (Fitness Evaluation):** Calculate the fitness of each ART chromosome (ART trained network). The fitness function for the p-th ART network is denoted by $Fit(p)$, and it depends on the $PCC(p)$ and $N_a(p)$ values of this network. Note that, $PCC(p)$ designates the percentage of correct classification, exhibited by the p-th network, on the validation set, while $N_a(p)$ is the number of categories of the p-th network. The fitness function $Fit(p)$ of the p-th network is defined as follows:

$$Fit(p) = PCC(p) - 0.5(N_a(p) - Cat_{min}) \qquad (12.3)$$

Obviously, this fitness function increases as $PCC(p)$ increases or as $N_a(p)$ decreases. The value of Cat_{min} is chosen to be equal to the number of classes of the classification problem at hand. It is evident from the fitness equation that one percentage of better correct classification of a network, or two categories less of a network, increase the fitness function by the same amount (i.e., by an amount of 1). This is one of the simplest ways of defining a fitness function that depends on two measures (generalization of the network and size of the network) and has been extensively adopted in the classification literature (e.g., see [5]).

- **Sub-step 2b (Selection):** Initialize an empty generation referred to as the *temporary generation*. The algorithm searches for the best NC_{best} chromosomes (i.e., the chromosomes having the NC_{best} highest fitness values) from the current generation and copies them to the temporary generation without change.

- **Sub-step 2c (Cross-Over Operation):** The remaining $Pop_{size}-NC_{best}$ chromosomes in the temporary generation are created by crossing over pairs of parents from the current generation. The parents are chosen using a deterministic tournament selection, as follows: Randomly select two groups of four chromosomes each from the current generation, and use as a parent, from each group, the chromosome with the best fitness value in the group. If it happens that from both groups the same chromosome is chosen then we choose from one of the groups the chromosome with the second best fitness value. If two parents with indices p, p' are crossed over, two random numbers n, n' are generated from the index sets $1, 2, ..., N_a(p)$ and $1, 2, ..., N_a(p')$, respectively. Then, all the categories with index greater than index n' in the chromosome with index p' and all the categories with index less than or equal to index n in the chromosome with index p are moved into an empty chromosome within the temporary generation. Notice that crossover is done at level 1 of the chromosome. This operation is pictorially illustrated in Fig. 12.12.

- **Sub-step 2d (Category Deletion):** The operator Cat_{del} deletes one of the categories of every chromosome (created in Step 2c) with probability $Pr(Cat_{del})$. If a chromosome is chosen to have one of its categories deleted then this category is picked randomly from the collection of the chromosome's categories; however if the number of categories for this chromosome, due to deletion, falls below the designated minimum number of categories Cat_{min} the deletion is prohibited.

- **Sub-Step 2e (Category Mutation):** Every chromosome created by step 2d gets mutated as follows:
  - In GFAM, with probability $Pr(Mut)$ every category is mutated. If a category is chosen to be mutated, either its **u** or **v** endpoint is selected randomly (with 50% probability) and then every component of this selected vector gets mutated by adding to it a small number. This number is drawn from a Gaussian distribution with mean 0 and standard deviation 0.01. If the component of the chosen vector becomes smaller than 0 or greater than 1 (after mutation), it is set back to 0 or 1, respectively.

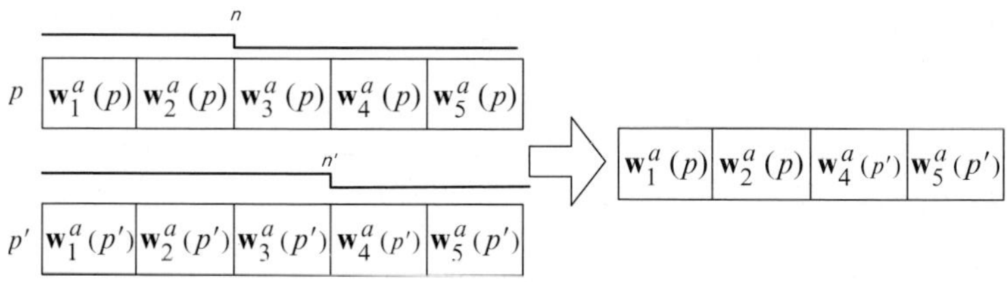

Fig. 12.12. GFAM, GEAM, GGAM Crossover implementation. In crossover the weight vectors of chromosome p, with index smaller than or equal to index n, and the weight vectors of chromosome p' with index larger than n', are combined (concatenated) to produce a new chromosome

- In GEAM, with probability $Pr(Mut)$ every category is mutated. If a category is chosen to be mutated, then every component of the ellipsoidal center $\mathbf{m}$ gets mutated by adding to it a small number. This number is drawn from a Gaussian distribution with mean 0 and standard deviation 0.01. If the component of the chosen vector becomes smaller than 0 or greater than 1 (after mutation), it is set back to 0 or 1, respectively. Furthermore, the mutated category's axis ratio μ or radius r is selected with 50% probability. We add a small number, to the axis ratio or the radius, if they are chosen to be mutated. The small number is drawn from a Gaussian distribution with zero mean and standard deviation 0.01. However if μ, or r, due to mutation, become larger than 1, they are set back to the value of 1, while if they become smaller than zero we set their value to 0.0001.
- In GGAM, with probability $Pr(Mut)$ every category is mutated. If a category is mutated, its mean vector $\mathbf{m}$, or standard deviation vector σ is selected randomly (50% probability). Then every component of this selected vector is mutated by adding to it a small number. This number is drawn from a Gaussian distribution with mean 0 and standard deviation 0.01. If the component of the chosen vector becomes smaller than 0 or greater than 1 (after mutation), it is set back to 0 or 1, respectively.

  Notice that mutation is applied at level 2 of the chromosome structure. The label of the chromosome is not mutated because our initial GA population consists of trained networks, and consequently we have a lot of confidence in the labels of the categories that these trained networks have discovered through the training process.

Step 3: If evolution has not reached the maximum number of iterations, Gen_{max}, replace the current generation with the members of the temporary generation and go to step 2a. Otherwise calculate the performance of the best-fitness network on the test set.

12.4 Experiments and Results

We conducted a number of experiments to assess the performance of the genetically engineered ART architectures. There were two objectives for this experimentation. The first one is to find proper values for the ranges of two of the GA parameters, the probability of deleting a category, $Pr(Cat_{del})$, and the probability of mutating a category, $Pr(mut)$. The second objective is to compare GFAM, GEAM and GGAM performance (for good parameter values) to that of popular ART architectures, such as ssFAM, ssEAM, ssGAM, and micro-ARTMAP.

12.4.1 Databases

We have experimented with 19 databases, 16 simulated databases and 3 real databases. Each dataset in the database was randomly divided into three subsets; training, validation and testing. Each one of these datasets is described, in more detail, in the text that follows.

- **Gaussian Databases: (# 1-12)** These are artificial databases, where we created 2-dimensional data sets, Gaussianly distributed, belonging to 2-class, 4-class, and 6-class problems. In each one of these databases, we varied the amount of overlap of data belonging to different classes. In particular, we considered 5%, 15%, 25%, and 40% overlap. Note that 5% overlap means that the optimal Bayesian Classifier would have 5% misclassification rate on the Gaussianly distributed data. There are a total of $3 \times 4=12$ Gaussian databases. We refer to these databases as **G#c-##** where the first number is the number of classes and the second number is the percentage of class overlap. For example, G2c-05 means that the Gaussian database is a 2-class and 5% overlap database.

- **Structures within a Structure databases:** These are artificial databases that were inspired by the circle (structure) - in the - square (structure) problem. This problem has been extensively examined in the ART, and other than ART neural network literature. Four different datasets were generated by changing the structures (type, number and probability) that we were dealing with. The data-points within each structure of these artificial datasets are uniformly distributed within the structure. The number of points within each structure is chosen in a way that the probability of finding a point within this structure is equal to a pre-specified number. Some of these artificial datasets were also considered in [30] where four different ART architectures were compared, Fuzzy ARTMAP, FasART, distributed Fuzzy ARTMAP, and distributed FasART.
  4Ci/Sq (# 13): This is a four circle in a square problem, a five class classification problem. The probability of finding a data point within a circle or inside the square and outside the circles is equal to 0.2. We refer to this database as **4Ci/Sq**.
  1Ci/Sq (# 14): This is a one circle in a square problem, a two class classification problem. The probability of finding a data point within a circle or inside the square and outside the circle is equal to 1/2. The sizes of the areas in the circle and outside the circle and inside the square are the same. This is the benchmark circle in the square problem. We refer to this database **1Ci/Sq**.
  1Ci/Sq/30:70 (# 15): This is a one circle in a square problem, a two class classification problem. The probability of finding a data point within a circle or inside the square and outside the circle is equal to 0.3 and 0.7, respectively. The sizes of the areas in the circle and outside the circle and inside the square are 0.3 and 0.7, respectively. This is a modified version of the circle in the square problem. We refer to this database as **30:70**.

2Ci/Sq/20:30:50 (# 16): This is two circles in a square problem, a three class classification problem. One of the circles is smaller than the other. The probability of finding a data point within the small circle, the large circle, and outside the circles but inside the square is 0.2, 0.3, and 0.5, respectively. We refer to this database as **20:30:50**.

- **Modified Iris (MOD-IRIS) Database (# 17):** In this database we started from the IRIS dataset (see [29]) of the 150 data-points, 3-class problem. We eliminated the data corresponding to the class that is linearly separable from the other 2 classes. Thus, we ended up with 100 data-points. From the four input attributes of this IRIS dataset we focused on only two attributes (attribute 3 and 4) because they seem to have enough discriminatory power to separate the 2-class data. Finally, in order to create a reasonable size dataset from these 100 points (so we can reliably perform cross-validation to identify the optimal ART, genetically engineered ART networks) we created noisy data around each one of these 100 data-points (the noise was Gaussian of zero mean and small variance) to end up with approximately 10,000 points. We named this database Modified Iris. We refer to this database as **MOD-IRIS**.

- **Modified Abalone (ABALONE) Database (# 18):** This database is originally used for prediction of the age of an abalone (see [29]). It contains 4177 instances, each with 7 numerical attributes, 1 categorical attribute, and 1 numerical target output (age). We discarded the categorical attribute in our experiments, and grouped the target output values into 3 classes: 8 and lower (class 1), 9-10 (class 2), 11 and greater (class 3). This grouping of output values has been reported in the literature before. We refer to this database as **ABALONE**.

- **Page Blocks (PAGE) Database (# 19):** This dataset represents the problem of classifying the blocks of the page layout in a document (see [29]). It contains 5473 examples coming from 54 distinct documents. Each example has 10 numerical attributes (e.g., height of the block, length of the block, eccentricity of the block, etc.,) and one target (output) attribute, representing the type of the block (text, horizontal line, graphic, vertical line, and picture). One of the noteworthy points about this database is that its major class (text) has a high probability of occurring (above 80%). This dataset has five classes, four of them make only 10% of the total instances. We refer to this database as **PAGE**.

The summarized specifics of each one of the aforementioned datasets are depicted in Table 12.1.

12.4.2 Selection of the GA parameters

As we have mentioned above, part the first objective of our experimentation was devoted to the selection of good values for the GA parameters: probability of deleting an ART category, $Pr(Cat_{del})$, and probability of mutating an

Table 12.1. Datasets used for experimentation, and their characteristics

	Database Name	# Training Instances	# Validation Instances	# Test Instances	# Attributes	# Classes	% Major Class
1	G2c-05	500	5000	5000	2	2	50.0
2	G2c-15	500	5000	5000	2	2	50.0
3	G2c-25	500	5000	5000	2	2	50.0
4	G2c-40	500	5000	5000	2	2	50.0
5	G4c-05	500	5000	5000	2	4	25.0
6	G4c-15	500	5000	5000	2	4	25.0
7	G4c-25	500	5000	5000	2	4	25.0
8	G4c-40	500	5000	5000	2	4	25.0
9	G6c-05	504	5004	5004	2	6	16.7
10	G6c-15	504	5004	5004	2	6	16.7
11	G6c-25	504	5004	5004	2	6	16.7
12	G6c-40	504	5004	5004	2	6	16.7
13	4Ci/Sq	2000	5000	3000	2	5	20.0
14	1Ci/Sq	2000	5000	3000	2	2	50.0
15	30:70	2000	5000	3000	2	2	70.0
16	20:30:50	2000	5000	3000	2	3	50.0
17	MOD-IRIS	500	4800	4800	2	2	50.0
18	ABALONE	501	1838	1838	7	3	33.3
19	PAGE	500	2486	2487	10	5	83.2

ART category, $Pr(Mut)$. As it is evident from our prior discussion there are a few other GA parameters that one has to carefully choose, such as Pop_{size}, Gen_{max}, and NC_{best}; we did not perform exhaustive experimentation to decide on the values of these parameters, but limited experimentation with these parameters for some of the above databases showed that reasonable choices for these parameters were: $Pop_{size} = 20$, $Gen_{max} = 500$, $NC_{best} = 3$.

Our approach to select good values for the GA parameters $Pr(Cat_{del})$, and $Pr(Mut)$ consisted of a number of steps delineated below:

- **Select GA Step 1:** We selected four different values for the $Pr(Cat_{del})$ to experiment with. These were: 0.05, 0.1, 0.2, and 0.4. We also selected four different values for the $Pr(Mut)$ to experiment with. These were: 0.0, 0.1, 0.2, and 0.4. This resulted in 16 combinations of parameter settings for $Pr(Cat_{del})$, and $Pr(Mut)$.
- **Select GA Step 2:** For each one of the 16 settings of the $Pr(Cat_{del})$, and $Pr(Mut)$ parameters, and for each of the 19 databases, described in an earlier section, we applied the GA optimization of FAMs, EAMs, and GAMs, as delineated in Sect. 12.2, 10 different times (using a different initial seed for the GA optimization). Consequently, for each database, and each parameter setting, and each of the genetically engineered ART algorithms we obtained 10 PCC and 10 N_a numbers.
- **Select GA Step 3:** For each genetically engineered ART algorithm (i.e., GFAM, GEAM, or GGAM), and each dataset, we chose the best-performing

(with respect to average validation PCC of the 10 experiments) parameter setting. Then, we used ANOVA statistical tests to choose other parameter settings that did not significantly differ (statistically) from the best performing parameter setting. We marked these parameter settings as good settings for this database and the associated GART (GFAM or GEAM or GGAM) algorithm.

- **Select GA Step 3:** After we performed Step 3 for all databases and all genetically engineered ART algorithms we counted the number of databases for each GART algorithm for which a particular parameter setting was deemed as "good" from the Select GA Step 3. Based on these counts we recommended the best parameter setting for each GART algorithm, and a range of acceptable parameter settings.

The following table (Table 12.2) summarizes the results for GFAM. Similar tables have been produced for GEAM and GGAM but are omitted due to lack of space. In Table 12.2 an entry of "1" for a database indicates that the corresponding parameter setting performed well (with respect to the average PCC on the validation set). An underscored "1" entry indicates that the corresponding parameter setting performed the best for this database (with respect to the average PCC on the validation set). In Table 12.2 the "1" entries corresponding to the Number of Categories criterion (actually average number of categories criterion) are omitted to preserve the table's clarity. However an entry of "1" for the PCC resulted also in an entry of "1" for the Number of Categories (not shown in Table 12.2). In Table 12.2 we designated with an asterisk the parameter setting that performed best for this database (with respect to the average Number of Categories criterion). A careful observation of the results shown in Table 12.2 indicate that any value of $Pr(Cat_{del})$ in the interval $[0.2, 0.4]$, and any value of the $Pr(Mut)$ in the interval $[0.05, 0.4]$ gives good results. Also, the results from Table 12.2 indicate that the best performing parameter setting for GFAM is $Pr(Cat_{del}) = 0.1$, and $Pr(Mut) = 0.4$, since for this parameter setting we observe the highest number of good performances (19), and best performances (7) of the associated GFAM (the count of the best performances consider the best observed performances with respect to the average PCC on the validation set or the average number of categories). Finally, we can also deduce from the results of Table 12.2 that a probability of mutation equal to 0 is not recommended, since it always (for all databases) results in a GFAM network that is not performing well.

From similar tables produced for GEAM and GGAM (omitted due to lack of space) we can draw similar conclusions. In particular, a careful observation of the GEAM results indicate that any value of $Pr(Cat_{del})$ in the interval $[0.2, 0.4]$, and any value of the $Pr(Mut)$ in the interval $[0.05, 0.4]$ gives good results for GEAM. Also, the best performing parameter setting for GEAM is $Pr(Cat_{del}) = 0.2$, and $Pr(Mut) = 0.4$, since for this parameter setting we observe the highest number of good performances (19), and best performances (6) of the associated GEAM. Finally, a probability of mutation equal to 0

Table 12.2. Goodness of Parameter Settings for the GA parameters $Pr(Cat_{del})$, and $Pr(Mut)$. An entry of "1" in the table indicates that the corresponding parameter setting is good, while the lack of an entry "1" indicates that the parameter setting is not good. The column before last counts the number of "1" entries for a particular parameter setting that are good, while the last column counts the number of times that a particular parameter setting is the best of all parameter settings (with respect to the average PCC on the validation set or the average number of categories). Entries are depicted for the PCC value and not for the number of categories value, in order to avoid cluttering the table; however whenever there is a "1" or lack of a "1" entry for the PCC value there is also a "1" or lack of a "1" entry for the number of categories value. Underscored "1" entries in this table designate parameter settings for which we obtained the best result related to the average PCC value, while asterisks designate parameter settings for which we obtained the best result with respect to the average number of categories value

Description	Data	g2c-05	g2c-15	g2c-25	g2c-40	g4c-05	g4c-15	g4c-25	g4c-40	g6c-05	g6c-15	g6c-25	g6c-40	4cirinsq	cirInSq	cins-70-30	2cins-50-30-20	Iris	abalone	pageblocks	sum	best
$Pr(Cat_{del}) = 0.05$	PCC	1																			1	
$Pr(Mut) = 0.0$	Num Cats																					
$Pr(Cat_{del}) = 0.05$	PCC	_1_	1	1	_1_	_1_	1	1	_1_	1	1	1		1		1	1	1	1	1	17	4
$Pr(Mut) = 0.1$	Num Cats																					
$Pr(Cat_{del}) = 0.05$	PCC	1	1	1	1	1	1	1	1	1	1	1	1	1	1	1	1	1	1	1	19	
$Pr(Mut) = 0.2$	Num Cats																			*		1
$Pr(Cat_{del}) = 0.05$	PCC	1	1	1	1	1	1	1	1	1	1	1	1	1	1	_1_	_1_	1	1	1	19	2
$Pr(Mut) = 0.4$	Num Cats																					
$Pr(Cat_{del}) = 0.1$	PCC																				0	
$Pr(Mut) = 0.0$	Num Cats																					
$Pr(Cat_{del}) = 0.1$	PCC	1	1	1	1	1	1	1		1	1	1		1		1	1	1	1	1	16	
$Pr(Mut) = 0.1$	Num Cats																					
$Pr(Cat_{del}) = 0.1$	PCC	1	1	1	1	1	1	1	1	1	1	1	1	_1_	1	1	1	1	1	1	19	1
$Pr(Mut) = 0.2$	Num Cats																					
$Pr(Cat_{del}) = 0.1$	PCC	1	1	_1_	1	1	1	_1_	1	1	1	1	_1_	1	_1_	1	1	1	1	_1_	19	
$Pr(Mut) = 0.4$	Num Cats																		*	*		7
$Pr(Cat_{del}) = 0.2$	PCC	1																			1	
$Pr(Mut) = 0.0$	Num Cats																					
$Pr(Cat_{del}) = 0.2$	PCC	1	1	1	1	1	1	1	1	1	1	1	1	1		1	1	_1_		1	17	1
$Pr(Mut) = 0.1$	Num Cats																					
$Pr(Cat_{del}) = 0.2$	PCC	1	_1_	1	1	1	1	1	1	1	1	1	1	1	1	1	1	1	1	1	19	
$Pr(Mut) = 0.2$	Num Cats																					1
$Pr(Cat_{del}) = 0.2$	PCC	1	1	1	1	1	1	1	1	1	_1_	1	1	1	1	1	1	1	1	1	19	1
$Pr(Mut) = 0.4$	Num Cats																					
$Pr(Cat_{del}) = 0.4$	PCC																				0	
$Pr(Mut) = 0.0$	Num Cats																					
$Pr(Cat_{del}) = 0.4$	PCC	1	1	1	1	1	_1_	1	1	1	1	1		1	1		1	1	1	1	17	
$Pr(Mut) = 0.1$	Num Cats																					1
$Pr(Cat_{del}) = 0.4$	PCC	1	1	1	1	1	1	1	1	_1_	1	1	1	1	1	1	1	1	1	1	19	1
$Pr(Mut) = 0.2$	Num Cats																					
$Pr(Cat_{del}) = 0.4$	PCC	1	1	1	1	1	1	1	1	1	1	_1_	1	1	1	1	1	1	1	1	19	
$Pr(Mut) = 0.4$	Num Cats													*								2

is not recommended for GEAM, since it always (for all databases) results in a GEAM that is not performing well. Additionally, a careful observation of the GGAM results indicate that any value of $Pr(Cat_{del})$ in the interval $[0.2, 0.4]$, and any value of the $Pr(Mut)$ in the interval $[0.05, 0.4]$ gives good results for GGAM. Also, the best performing parameter setting for GGAM is $Pr(Cat_{del}) = 0.4$, and $Pr(Mut) = 0.1$, since for this parameter setting we observe the highest number of good performances (19), and best performances (4) of the associated GGAM. Finally, a probability of mutation equal to 0 is not recommended for GGAM, since it always (for all databases) results in a GGAM that is not performing well.

12.4.3 Comparison of GART with other ART architectures

The second objective of our experimentation was to compare GFAM, GEAM, and GGAM (summarily referred to as GART) with other ART architectures that have previously addressed the category proliferation problem in ART. The ART architectures that we chose to compare GFAM, GEAM, GGAM with are: ssFAM, ssEAM, ssGAM, and safe micro-ARTMAP. The first three are based on the principle of semi-supervision, introduced by Anagnostopoulos, et al., 2002, [2], and Verzi, et al., 2001, [35]. Semi-supervision is a term attributed to learning in an ART architecture (FAM, EAM or GAM), where categories in ART are allowed to encode patterns of different labels provided that the percentage of patterns that belong to the plurality label exceed a certain threshold. Safe micro-ARTMAP is a Fuzzy ARTMAP that allows categories in Fuzzy ARTMAP to encode patterns that are mapped to different labels. In safe micro-ARTMAP (see Gomez-Sanchez, et al., 2001, [16]) though the mixture of labels allowed in a category, or in all of the categories is controlled by the entropy of the category or categories.

For each of the ssFAM, ssEAM, ssGAM, and safe micro-ARTMAP networks, and for each of the 19 databases, we performed a number of experiments with different settings of their network parameter values. For each one of these network parameter settings we calculated the resulting network's fitness function (we used the same fitness function as the one utilized for the GART networks (see equation 12.3)). For the training of ssFAM, ssEAM, ssGAM, and safe micro-ARTMAP we used the same training set as the one used for the GART networks, and for the validation of the performance of each of the ssFAM, ssEAM, ssGAM, and Safe micro-ARTMAP networks we used the same validation set as the one used for the GART networks. The parameter setting of the ssFAM, ssEAM, ssGAM, and safe micro-ARTMAP network that maximized the fitness function was chosen as the best parameter setting for the specific database; the number of categories created by the "best" parameter setting network, and its corresponding percentage of correct classification on the test set are reported in Table 12.3.

In particular, the parameter settings that we experimented with ssFAM were: baseline vigilance values ranging from 0 to 0.9 with step size of 0.1, choice parameter values of 0.001 and 0.01, maximum allowable mixture threshold values ranging from 0 to 1 with step size of 0.1, and 100 different orders of pattern presentations of the training data (resulting in 22,000 different parameter settings). Furthermore, the settings for ssEAM were: baseline vigilance values ranging from 0 to 0.9 with step size of 0.1, choice parameter values of 0.001 and 0.01, maximum allowable mixture threshold values ranging from 0 to 1 with step size of 0.1, minimum axes to maximum axis ratio values ranging from 0.1 to 1 with step size of 0.1, and 100 different orders of pattern presentations of the training data (resulting in 220,000 different parameter settings). Also, the settings for ssGAM were: baseline vigilance values ranging from 0 to 0.9 with step size of 0.1, initial standard deviation parameter ranging from 0.1 to 1 with step size of 0.1, maximum allowable mixture threshold values ranging from 0 to 1 with step size of 0.1, and 100 different orders of pattern presentations of the training data (resulting in 110,000 different parameter settings). Finally, the settings for safe micro-ARTMAP were: baseline vigilance values ranging from 0 to 0.4 with step size of 0.2, baseline vigilance parameter values of 0.001 and 0.01, 5 values for the maximum "all-categories" entropy threshold, 6 different ratios of the values of the "categories" entropy threshold to the "all-categories" entropy threshold, three values of the maximum allowable expansion of a category, and 100 different orders of pattern presentations of the training data (resulting in 90,000 different parameter settings).

The best parameter setting, identified in the previous sub-section, for GFAM, GEAM, and GGAM was used for each of the 19 databases. Ten (10) experiments per database were conducted for 10 different initial seeds of the GA optimization process. The network that produced the maximum value of the fitness function, was deemed as "best". The number of categories of the "best" GFAM, GEAM and GGAM for each database and its corresponding performance (PCC) on the test set are reported in Table 12.3. The results shown in Table 12.3 are truncated to one decimal place.

12.4.4 Observations from the results

Some of the conclusions that can be deduced from the comparative results, depicted in Table 12.3, are emphasized below.

- **Observation 1 (Overall Performance of GART networks):** GFAM, GEAM and GGAM attain good performance on all the datasets, and quite often, optimal performance (e.g., see performance of all the networks in the Gaussian databases, and performance of GGAM on the structures-within-structure problems, and on the real databases). The best performing network from the class of GART networks (GFAM, GEAM, and GGAM) is GGAM.
- **Observation 2 (Comparative Performance of GART networks, with respect to each other).** GGAM and GEAM outperform the performance of GFAM on all the structures, within structure problems. For

Table 12.3. Best results obtained from GFAM, GEAM and GGAM compared to best results obtained from Safe μARTMAP, ssFAM, ssEAM and ssGAM

Dataset Name	GFAM PCC	Cat	GEAM PCC	Cat	GGAM PCC	Cat	Safe μAM PCC	Cat	ssFAM PCC	Cat	ssEAM PCC	Cat	ssGAM PCC	Cat
G2c-05	95.4	2	95.3	2	95.3	2	95.2	2	94.7	2	94.6	2	94.6	2
G2c-15	85.3	2	85.2	2	85.2	2	85.0	2	85.5	2	85.2	2	85.5	2
G2c-25	75.2	2	75.2	2	75.2	2	74.9	2	75.0	2	75.1	2	75.0	2
G2c-40	62.0	2	61.8	2	61.7	2	61.4	3	59.5	2	59.5	2	59.5	3
G4c-05	95.1	4	95.0	4	95.0	4	95.0	4	94.5	5	94.9	4	95.5	4
G4c-15	84.7	4	84.6	4	84.7	4	83.2	4	82.7	4	82.0	4	83.4	6
G4c-25	75.0	4	75.1	4	75.3	4	74.5	4	70.3	9	72.9	4	72.3	21
G4c-40	59.9	4	59.8	4	75.3	4	58.9	4	57.8	7	54.7	7	59.5	14
G6c-05	94.8	6	94.7	6	94.8	6	92.3	6	87.2	8	93.4	6	94.6	8
G6c-15	84.8	6	85.1	6	85.2	6	80.9	6	80.5	6	82.0	7	83.4	9
G6c-25	74.3	6	74.1	6	74.4	6	67.9	6	70.2	15	71.4	7	71.2	13
G6c-40	60.1	6	59.9	6	60.0	6	54.0	6	55.1	17	49.3	7	55.1	13
4Ci/Sq	95.0	9	99.1	7	98.9	6	95.4	8	87.2	18	94.6	5	93.4	12
1Ci/Sq	97.7	7	99.6	3	99.8	2	94.7	8	92.9	8	97.0	8	91.0	8
30:70	97.9	6	99.9	2	99.9	2	96.8	8	93.2	8	97.1	8	92.3	8
20:30:50	97.5	5	98.1	3	99.5	3	97.2	6	90.2	12	97.0	3	95.6	9
MOD-IRIS	95.3	2	95.3	2	94.9	2	94.9	2	94.7	8	94.7	2	94.7	2
ABALONE	61.8	3	62.2	3	62.6	3	58.1	3	60.0	6	58.8	3	56.3	2
PAGE	96.7	5	95.0	5	96.2	5	92.9	5	87.9	3	93.8	2	94.3	5

all the other problems the differences between GEAM, and GGAM versus GFAM are not statistically significant.

- **Observation 3 (Comparative Performance of GART networks compared with ssFAM):** ssFAM performs as well as the GART networks for the 2-class Gaussian datasets. For all the other datasets at least one (if not all) the GART networks perform better (achieving higher PCC with fewer ART categories). The largest difference in PCC observed is almost 12% (for the 4 Circle in the Square problem), while the largest ratio of number of ssFAM versus GART categories is for the modified IRIS problem (ratio of 4).

- **Observation 4 (Comparative Performance of GART networks compared with ssEAM):** ssEAM performs as well as the GART networks for the 2-class Gaussian datasets. For all the other datasets at least one (if not all) the GART networks perform better (achieving higher PCC with fewer ART categories). The largest difference in PCC observed is more than 10% (for the 6 class Gaussian problem with 40% overlap), while the largest ratio of number of ssEAM versus GART categories is for Circle in the Square problem (ratio of 4).

- **Observation 5 (Comparative Performance of GART networks compared with ssGAM):** ssGAM performs as well as the GART networks for the 2-class Gaussian datasets. For all the other datasets at least one (if not all) the GART networks perform better (achieving higher PCC with fewer ART categories). The largest difference in PCC observed is more than 8% (for the 1 Circle in the Square problem), while the largest ratio of number of ssGAM versus GART categories is for the four Gaussian dataset with 25% overlap problem (ratio larger than 5).

- **Observation 6 (Comparative Performance of GART networks compared with safe micro-ARTMAP):** Safe micro-ARTMAP performs as well as the GART networks for the 2-class, and 4-class Gaussian datasets. For all the other datasets at least one (if not all) the GART networks perform better (achieving higher PCC with fewer ART categories). The largest difference in PCC observed is more than 6% (for the 6 class Gaussian dataset with 25% overlap), while the largest ratio of number of safe micro-ARTMAP versus GART categories is for the Circle in the Square problem (ratio of 4).

What is also worth pointing out is that the better performance of the GART network is attained with reduced computations as compared with the computations needed by the alternate methods (ssFAM, ssEAM, ssGAM, safe micro-ARTMAP). Specifically, the performance attained by ssFAM, ssEAM, ssGAM and the safe micro-ARTMAP required training these networks for a large number of network parameter settings (at least 22,000 experiments) and then choosing the network that achieved the highest value for the fitness function that we introduced earlier (through cross-validation). In GFAM, GEAM and GGAM cases we trained only a small number of these networks ($Pop_{size} = 20$ of them), compared to the large number of networks trained in the ssFAM, ssEAM, ssGAM or micro-ARTMAP cases (at least 22,000). Furthermore, in GFAM, GEAM and GGAM cases we evolved the trained networks $Gen_{max} = 500$ times, each evolution requiring cross-validating $Pop_{size} = 20$ networks. Hence, the total number of networks cross-validated in the ssFAM, ssEAM, ssGAM and micro-ARTMAP cases were at least 22,000, while in the GFAM, GEAM and GGAM networks were 10,000; furthermore the networks cross-validated in the ssFAM, ssEAM, ssGAM, and micro-ARTMAP cases have higher number of category nodes than the ones cross-validated in the GFAM, GEAM and GGAM cases. As a result, we can argue that the improved performance (smaller number of nodes and better generalization) of GFAM, GEAM, and GGAM, compared with ssFAM, ssEAM, ssGAM, and micro-ARTMAP, is achieved with reduced computational effort.

12.5 Conclusions

Adaptive Resonance Theory (ART) neural networks have been introduced into the literature by Carpenter, Grossberg and their colleagues at Boston University, as well as other researchers in the field. The consensus with ART networks is that they converge fast to a solution for arbitrary classification problems they can provide explanations for the answers that they produce, they can function in an on-line training mode, and they solve effectively a variety of classification problems. However, all these benefits sometimes come at the expense of unnecessarily creating (at times) too many categories to solve the problem at hand, referred to as the category proliferation problem in ART. This problem is more acute when ART is confronted with classification problems that deal with noisy or highly overlapping data. To alleviate this problem a number of researchers have proposed solutions such as ssFAM, ssEAM (see Anagnostopoulos, et al., 2002 and 2003, [2, 3], and Verzi, et al., 2001, [35]), ssGAM (see Chalasani, 2005 [12]), Safe micro-ARTMAP (see Gomez, et al., 2001, [16]), dFAM (Carpenter, et al., 1998, [11]), FasART, and dFasART (Parado-Hernandez, et al., 2003, [30]), to mention only a few.

In this paper, we have introduced yet another method of solving the category proliferation problem in ART. This method relies on evolving a population of trained ART networks, such as FAM, EAM or GAM. The evolution of trained ART networks creates an ART network, referred to as GFAM, or GEAM or GGAM.

We have experimented with a number of databases that helped us identify good default parameter settings for the evolution of FAM, EAM or GAM. We defined a fitness function that gave emphasis to the creation of a small size ART networks which exhibited good generalization. In the evolution of ART trained networks, we used a unique (and needed) category operator, referred to as Cat_{del} operator (this operator allowed us to evolve into ART networks of smaller size). The ART network identified at the end of the evolutionary process (last generation) was the FAM, EAM or GAM network that attained the highest fitness value (referred to as GFAM, GEAM, or GGAM, respectively). Our method for creating GFAM, GEAM and GGAM resulted in a networks that performed well on a number of classification problems, and on a few of them **it performed optimally** (see our observations in earlier sections).

Furthermore, GART networks were found to be superior to a number of other ART networks (ssFAM , ssEAM , ssGAM , safe micro-ARTMAP) that have been introduced into the literature to address the category proliferation problem in ART. More specifically, GFAM , GEAM , and GGAM gave a **better generalization performance** (in almost all problems tested) and a **smaller than, or equal, size** network (in all problems tested), compared to these other ART networks, **requiring reduced computational effort** to achieve these advantages. More specifically, in some instances the difference in classification performance of GFAM, GEAM, and GGAM with these other ART networks quite significant (as high as 12%). Also, in some instances the ratio of the number of categories created by these other ART networks, compared to the categories created by GFAM, GEAM or GGAM was large (as high as 5).

Acknowledgment This work was supported in part by the National Science Foundation (NSF) grant CRCD: 0203446, and the National Foundation grant DUE 05254209. Georgios Anagnostopoulos and Michael Georgiopoulos acknowledge the partial support from the NSF grant CCLI 0341601.

References

1. Anagnostopoulos GC (2001) Novel Approaches in Adaptive Resonance Theory for Machine Learning. PhD Thesis, Univ Central Florida, Orlando
2. Anagnostopoulos GC, Georgiopoulos M, Verzi SJ, Heileman GL (2002) Reducing generalization error and category proliferation in ellipsoid ARTMAP via tunable misclassification error tolerance: boosted ellipsoid ARTMAP. In: Proc IEEE-INNS-ENNS Int J Conf Neural Networks (IJCNN) 3:2650–2655
3. Anagnostopoulos GC, Bharadwaj M, Georgiopoulos M, Verzi SJ, Heileman GL (2003) Exemplar-based pattern recognition via semi-supervised learning. In: Proc IEEE-INNS-ENNS Int J Conf Neural Networks (IJCNN) 4:2782–2787
4. Bala J, De Jong K, Huang J, Vafaie H, Wechsler H, (1996) Using learning to facilitate the evolution of features for recognizing visual concepts. Evolutionary Computation 4:297-311
5. Breiman L, Friedman J, Stone CJ, Olshen RA (1984) Classification and Regression Trees. Wadsworth
6. Brotherton TW, Simpson PK (1995) Dynamic feature set training of neural nets for classification. In: McDonnel JR, Reynolds RG, Fogel DB (eds) Evolutionary Programming IV pp 83–94. MIT Press, Cambridge, MA
7. Burton AR, Vladimirova T (1997) Utilisation of an adaptive resonance theory neural network as a genetic algorithm fitness evaluator. In: Proc IEEE Int Symp Information Theory pp 209
8. Cantu-Paz E (2002). Feature sub-set selection by estimation of distribution algorithms. In: Langdon WB, Cantu-Paz E, Mathias K, Roy R, Davis D, Poli R, Balakrishna K, Honavar V, Rudolph G, Wegener J, Bull L, Potter MA, Schultz AC, Miller JF, Burke E, Jonoska N (eds) Proc Genetic and Evolutionary Computation Conf pp 303–310. Morgan Kaufmann Publishers, San Francisco, CA
9. Carpenter GA, Grossberg S, Reynorlds JH (1991) Fuzzy ART: fast stable learning and categorization of analog patterns by an adaptive resonance system. Neural Networks 4:759–771
10. Carpenter GA, Grossberg S, Markuzon N, Reynolds JH, Rosen DB (1992) Fuzzy ARTMAP: a neural network architecture for incremental supervised learning of analog multidimensional maps. IEEE Transactions on Neural Networks 3:698–713
11. Carpenter GA, Milenova B, Noeske B (1998) dARTMAP: a neural network for fast distributed supervised learning. Neural Networks 11:793–813
12. Chalasani R (2005) Optimization of Network Parameters and Semi-supervision in Gaussian ART Architectures. MS Thesis, Univ Central Florida, Orlando
13. Charalampidis D, Kasparis T, Georgiopoulos M (2001) Classification of noisy signals using fuzzy ARTMAP neural networks. IEEE Transactions on Neural Networks 12:1023–1036
14. Georgiopoulos M, Huang J, Heileman GL (1994) Properties of learning in ARTMAP. Neural Networks 7:495–506

15. Goldberg DE (1989) Genetic Algorithms in Search, Optimization, and Machine Learning. Addison-Wesley, Reading, MA
16. Gomez-SanchezE, Cano-Izquierdo JM, Lopez-Coronado J (2001) Safe-uARTMAP: a new solution for reducing category proliferation in fuzzy ARTMAP. In: Proc Int J Conf on Neural Networks
17. Gomez-Sanchez E, Dimitriadis YA, Cano Izquierdo JM, Lopez-Coronado J (2002) uARTMAP: use of mutual information for category reduction in Fuzzy ARTMAP. IEEE Transactions on Neural Networks 13:58–69
18. Grossberg S (1976) Adaptive pattern classification and universal recoding, II: feedback, expectation, olfaction, and illusions. Biological Cybernetics 23:187–202
19. Hancock PJB (1992). Pruning neural networks by genetic algorithm. In: Aleksander I, Taylor J (eds) Proc Int Conf on Neural Networks 2:991–994. Elsevier Science, Amsterdam, Netherlands
20. Inza I, Larranaga P, Etxeberria R, Sierra B (2000) Feature sub-set selection by Bayesian networks based optimization. Artificial Intelligence, 123:157–184
21. Kaburlasos VG (2006) Towards a Unified Modeling and Knowledge-Representation Based on Lattice Theory — Computational Intelligence and Soft Computing Applications, ser Studies in Computational Intelligence 27. Springer, Heidelberg, Germany
22. Kaburlasos VG, Athanasiadis IN, Mitkas PA (2007) Fuzzy lattice reasoning (FLR) classifier and its application for ambient ozone estimation. Intl J Approximate Reasoning 45(1):152–188
23. Kasuba T (1993) Simplified fuzzy ARTMAP. AI Expert 18–25
24. Kelly J, Davis L (1991) Hybridizing the genetic algorithm and the k-nearest neighbors classification algorithm. In: Belew RK, Booker LB (eds) Proc Fourth Int Conf Genetic Algorithms pp 377–383
25. Koufakou A, Georgiopoulos M, Anagnostopoulos GC, Kasparis T (2001) Cross-validation in fuzzy ARTMAP for large databases. Neural Networks 14:1279–1291
26. Liu H, Liu Y, Liu J, Zhang B, Wu G (2003) Impulse force based ART network with GA optimization. In: Proc Int Conf Neural Networks and Signal Processing 1:499–502
27. Martin FJM, Hernandez FS (1993) Genetic synthesis of discrete-time recurrent neural networks. In: Proc Int Workshop Artificial Neural Networks (IWANN) LNCS 686:179–184. Springer-Verlag, London, UK
28. Marriott S, Harrison RF (1995) A modified fuzzy ARTMAP architecture for the approximation of noisy mappings. Neural Networks 8:619–641
29. Newman DJ, Hettich S, Blake, CL, Merz CJ (1998) UCI Repository of machine learning databases `http://www.ics.uci.edu/ mlearn/MLRepository.html` University of California, Irvine, CA
30. Parrado-Hernadez E, Gomez-Sanchez E, Dimitriadis YA (2003) Study of distributed learning as a solution to category proliferation in fuzzy ARTMAP based neural systems. Neural Networks 16:1039–1057
31. Petridis V, Kaburlasos VG (1998) Fuzzy lattice neural network (FLNN): a hybrid model for learning. IEEE Transactions on Neural Networks 9: 877–890
32. Sexton RS, Dorsey RE, Jonson JD (1998) Toward global optimization of neural networks: a comparison of the genetic approach and back-propagation. Decision Support Systems 22:171–185
33. Siedlecki W, Sklansky J (1989) A note on genetic algorithms for large-scale feature selection. Pattern Recognition Letters 10:335–347

34. Taghi M, Bagmisheh V, Pavesic N (2003) A fast simplified fuzzy ARTMAP network. Neural Processing Letters 17:273–316
35. Verzi SJ, Heileman GL, Georgiopoulos M, Healy, M (2001) Rademacher penalization applied to fuzzy ARTMAP and boosted ARTMAP. In: Proc IEEE-INNS Int J Conf on Neural Networks (IJCNN) pp 1191–1196
36. Williamson JR (1996) Gaussian ARTMAP: a neural network for fast incremental learning of noisy multidimensional maps. Neural Networks 9:881–897
37. Williamson JR (1997) A constructive, incremental-learning network for mixture modeling and classification. Neural Computation 9:1517–1543
38. Yang J, Honavar V (1998). Feature subset selection using genetic algorithms. IEEE Intelligent Systems 13:44–49
39. Yao X, Liu Y (1998) Making use of population information in evolutionary artificial neural networks. IEEE Transactions on Systems, Man, Cybernetics part B 28:417-425
40. Yao X (1999) Evolving artificial neural networks. Proceedings of the IEEE 87(9):1423–1447

Chapter 12 Appendix

GFAM: It denotes the genetically engineered Fuzzy ARTMAP architecture.
GEAM: It denotes the genetically engineered Ellipsoidal ARTMAP architecture.
GGAM: It denotes the genetically engineered Gaussian ARTMAP architecture.
GART: It is a generic name referring to the genetically engineered ART architectures, as a whole, such as GFAM, GEAM, and GGAM.
ssFAM: It denotes the semi-supervised Fuzzy ARTMAP architecture. For more details see [2, 3].
ssEAM: It denotes the semi-supervised Ellipsoidal ARTMAP architecture. For more details see [2, 3].
ssGAM: It denotes the semi-supervised Gaussian ARTMAP architecture. For more details see [12].
Safe micro-ARTMAP: It the ART architecture introduced by Gomez-Sanchez, et al., in 2001 (see [16]).
$\bar{\rho}_a$; **Baseline Vigilance Parameter:** One of the parameters of the Fuzzy ARTMAP (FAM), Ellipsoidal ARTMAP (EAM) and Gaussian ARTMAP (GAM) architectures that controls the size of the categories created in FAM, EAM and GAM.
Choice Parameter: One of the parameters of the Fuzzy ARTMAP (FAM) and Ellipsoidal ARTMAP (EAM) architectures that controls the value of bottom-up input of a category in FAM or EAM.
γ; **Initial Standard Deviation Parameter:** One of the parameters of the Gaussian ARTMAP architecture that controls the initial width of the Gaussian probability distribution of a GAM category.

μ; **Minor Axes to Major Axis Parameter:** One of the parameters of the Ellipsoidal ARTMAP architecture that defines the ratio of the minor axes to major axis of the ellipsoidal categories in the EAM architecture.

m; Center of a Category: One of the parameters that describes a category in EAM or GAM. In EAM it corresponds to the center of the ellipsoidal category, while in GAM it corresponds to the center of the Gaussian probability distribution, representing a GAM category.

d; Direction Vector of a Category: One of the parameters that describes a category in EAM. It corresponds to the direction of the major axis of the ellipsoid.

r; **Radius of a Category:** One of the parameters that describes a category in EAM. It is equal to the half length of the major axis.

σ; **Standard Deviation Vector of a Category:** One of the parameters that describes a category in GAM. The components of this vector define the standard deviation of the Gaussian distribution across every dimension of the input pattern space.

n; **Number of Patterns Encoded by a Category:** One of the parameters that describes a category in GAM.

13

Fuzzy Lattice Reasoning (FLR) Classification Using Similarity Measures

Al Cripps[1] and Nghiep Nguyen[2]

[1] Middle Tennessee State University, Murfreesboro, TN 37132, USA
Dept. of Computer Science
acripps@mtsu.edu

[2] Dept. of Economics and Finance
nguyen@mtsu.edu

Summary. In this work, we show that the underlying inclusion measure used by fuzzy lattice reasoning (FLR) classifiers can be extended to various similarity and distance measures often used in cluster analysis. We show that for the cosine similarity measures, we can weigh the contribution of each attribute found in the data set. Furthermore, we show that evolutionary algorithms such as genetic algorithms, tabu search, particle swarm optimization, and differential evolution can be used to weigh the importance of each attribute and that this weighting can provide additional improvements over simply using the similarity measure. We present experimental evidence that the proposed techniques imply significant improvements.

13.1 Introduction

This work provides a framework to extend fuzzy lattice reasoning (FLR) classifiers [13] to use similarity and distance measures. Furthermore we demonstrate experimentally the effectiveness of using similarity measures and evolutionary algorithms to improve fuzzy lattice classifiers in the Cleveland heart benchmark classification problem.

Stemming from the adaptive resonance theory (ART) neural networks [2, 10], fuzzy lattice reasoning bases both its learning and generalization on the computation of hyperboxes in space R^N. Note that other classifiers including ART, Min-Max neural networks and variations [2, 8, 10] compute hyperboxes in R^N. Nevertheless, a unique advantage of FLR is its applicability in a lattice data domain including the product $\mathsf{L} = \mathsf{L}_1 \times \ldots \times \mathsf{L}_N$ of N *constituent lattices* $\mathsf{L}_1, \ldots, \mathsf{L}_N$ [14].

A practical advantage of a lattice applicability is the capacity to deal with disparate types of data, e.g. vectors of numbers, fuzzy sets, symbols, graphs, etc. in applications [13, 14, 17, 18]. Note that with the proliferation of information technologies, the latter capacity may be useful in dealing as well with non-numeric data. Learning and generalization are effected in a lattice L

A. Cripps and N. Nguyen: *Fuzzy Lattice Reasoning (FLR) Classification Using Similarity Measures*, Studies in Computational Intelligence (SCI) **67**, 263–284 (2007)
www.springerlink.com

by computing lattice L intervals. Apart from the simplicity of the method, this work also shows that the computation of lattice intervals can imply significant improvements in classification problems.

The layout of this paper is as follows. Section 13.2 provides the fuzzy lattice classifier theoretic background. Section 13.3 explains principles of similarity and distance measures while Section 13.4 discusses evolutionary algorithms. Section 13.5 presents our theoretical classification improvements for the classifiers found in [6, 7, 14]. Section 13.6 investigates the characteristics of the weighted cosine similarity measure. Section 13.7 provides empirical results that demonstrate the improvement of our enhanced FLR classifiers when compared to other classification methods for the Cleveland heart data. Finally, section 13.8 summarizes the contribution of this work and delineates future work.

13.2 Fuzzy Lattice Theoretic Background

In this paper we employ the lattice theoretic notation introduced in [11, 13, 14]. More specifically, let L denote a mathematical lattice, then $\tau(\mathsf{L})$ denotes the set of lattice L intervals. For this paper, we deal exclusively with a complete lattice L, where O and I denote, respectively, the least and greatest elements in L. It follows that $\tau(\mathsf{L})$ is, also, a complete lattice [14]. Furthermore, $\alpha(\mathsf{L})$ denotes the set of *atoms* in a lattice L, the latter set $\alpha(\mathsf{L})$ includes all trivial intervals (singletons); hence $\alpha(\mathsf{L}) \subset \tau(\mathsf{L})$.

Note that the conventional space R^N of N-dimensional vectors is a product-lattice of N identical, totally-ordered lattices R. A trivial interval in R^N includes the set of N-dimensional points. Moreover, the unit N-dimensional hypercube I^N is a complete lattice, where I equals the closed interval $\mathsf{I} = [0, 1]$.

An *inclusion measure* is a map $\sigma : \mathsf{L} \times \mathsf{L} \to [0, 1]$, which satisfies the following conditions for $u, w, x \in \mathsf{L}$.

(C0) $\sigma(x, O) = 0, x \neq O$,

(C1) $\sigma(x, x) = 1$,

(C2) $w \wedge u < w \Rightarrow \sigma(w, u) < 1$, and

(C3) $u \leq w \Rightarrow \sigma(x, u) \leq \sigma(x, w)$ - Consistency Property

We remark that $\sigma(x, u)$ denotes the degree of inclusion of lattice element x to lattice element u; hence symbols $\sigma(x, u)$ and $\sigma(x \leq u)$ are used interchangeably.

A *positive valuation* v on a lattice L is a real function $v : \mathsf{L} \to \mathsf{R}$ which satisfies both (1) $v(x) + v(y) = v(x \wedge y) + v(x \vee y)$ and (2) $x < y$ implies $v(x) < v(y)$ for $x, y \in \mathsf{L}$. Given a positive valuation function in a lattice L, one can show that both of the following are inclusion measures: $k(u \leq w) = \frac{v(w)}{v(u \vee w)}$ and $s(u \leq w) = \frac{v(u \wedge w)}{v(u)}$ [11]. Furthermore note that for a positive valuation function v in a lattice L, a metric distance is given in L by $d(x, y) = v(x \vee y) - v(x \wedge y)$ [11]. The work in [14] has shown how a positive valuation function in a lattice L can be extended to the lattice $\tau(\mathsf{L})$ of intervals.

Given a category function $g : \alpha(\mathsf{L}) \to \mathcal{M}$ where $\mathcal{M}$ is a finite set of category labels, then the finite labeled training data set is the pairs $(a_i, g(a_i))$, $i = 1, \ldots, n$ where a_i is an atom in $\alpha(\mathsf{L})$ and $g(a_i)$ is the corresponding category label. For classification problems, the goal is to learn a valid approximation of category function $g : \alpha(\mathsf{L}) \to \mathcal{M}$ so as to achieve an acceptable generalization on the testing data.

Learning from the training data is effected by a fuzzy lattice algorithm by computing *fits*, the latter are intervals of lattice elements computed by the lattice-join of training data in the same category. A *contradiction* occurs when a training datum from one category is included in a fit of a different category. A fit is called *tightest* when the lattice-join with any training datum from the same category (and not already in the fit) causes a contradiction. We are most interested in those category functions that are tightest fits.

In [14], positive valuations of the form $v(x_1, \ldots, x_N) = c_1 x_1 + \ldots + c_N x_N$, where $c_i > 0$ are defined, but are only used for one simple data set, and in that case, a heuristic approach is used to find the constants $c_i > 0$. Almost exclusively in [6, 7, 14] a positive valuation function of the form $v(x_1, \ldots, x_N) = x_1 + \ldots + x_N$ is used, i.e., $c_i = 1$, $\forall i$. We will refer to positive valuations of the form $v(x_1, \ldots, x_N) = c_1 x_1 + \ldots + c_N x_N$, where $c_i > 0$ as *positive hyperplane* valuations and the valuation of the form $v(x_1, \ldots, x_N) = x_1 + \ldots + x_N$ as the *unit* valuation.

In [6, 7, 14] the following five tightest fit classifiers are defined: *FLR tightest fit (FLRtf)*, *FLR first fit (FLRff)*, *FLR maximal tightest fit (FLRmtf)*, *FLR ordered tightest fit (FLRotf)*, and *FLR selective fit (FLRsf)*. In the sections that follow, we focus our attention upon these five classifiers plus the valuations and inclusion measures used by these classifiers.

13.3 Introduction to Similarity and Distance Measures

In section 13.5 we will use similarity and distance measures to extend the FLR classifiers found in [6, 7, 14]. In this section we provide an introduction to these measures. In general, *distance* measures numerically how unlike (different) two datum are where as *similarity* measures numerically how alike two datum are. For a similarity measure, the idea is that a higher value indicates greater similarity where as for a distance measure, the lower (positive) value indicates greater similarity. In concept, a similarity measure is the converse of a distance measure. A formal definition of distance measure is given below, however, we find no formal definition found for similarity measures in the literature. There are many popular measures defined in the literature (e.g. L_p norm, Squared chord, squared Chi-squared, Canberra measure, Czekanowski coefficient, cosine, and correlation coefficient) for which all can be thought of as calculating a correlation between two data items (one of which is usually a group centroid). Generally, for these measures it is necessary that each of the two data items be expressed by a real-valued array of feature attributes.

In section 13.5 we extend the FLR classifiers to use distance measures. For our experimental application, we are mostly interested in the cosine similarity measure often used in cluster analysis. For the cosine similarity measure, if two data items $\mathbf{x}$ and $\mathbf{y}$ have K attributes each, then $\mathbf{x}$ and $\mathbf{y}$ can be thought of as two vectors. The cosine similarity measure is then the cosine of the angle between the two vectors x and y and is described by the formula [21]

$$CS(\mathbf{x}, \mathbf{y}) = \frac{\sum_1^K x_i \, y_i}{\sqrt{\sum_1^K x_i^2 \sum_1^K y_i^2}}$$

The *cosine distance metric* is defined as $CD(\mathbf{x}, \mathbf{y}) = 1 - CS(\mathbf{x}, \mathbf{y})$.

In section 13.5 we also use distance measures to extend the FLR classifiers. Typically, a distance measure falls into one of two groups: *metric* and *semi-metric*. To be classified as *metric*, a non-negative distance between two vectors x and y must obey: 1) $d(x, y) = 0 \Rightarrow x = y$, 2) $d(x, y) = d(y, x)$, 3) $d(x, x) = 0$, and 4) when considering three objects, x, y and z, $d(x, y) \leq d(x, z) + d(z, y)$. Distance measures that obey the last three rules, but fail to obey rule 1 are referred to as *semi-metric*. The L_p norm is a well known group of distance measures as defined by the following. The L_p norm between two data items $\mathbf{x}$ and $\mathbf{y}$ each of which has K attributes is given by the formula

$$L_p(\mathbf{x}, \mathbf{y}) = \left[\sum_{i=1}^K |x_i - y_i|^p \right]^{1/p}$$

If $p = 1$ then this distance is known as the *Manhattan* distance; if $p = 2$ then the formula produces the well-known Euclidean distance. If $p = \infty$, then the distance is called the Chebychev distance.

13.4 Introduction to Evolutionary Algorithms

Often times when a problem has no known algorithmic solution, i.e., not known to be a tractable problem, then some type of search technique can be used to arrive at a reasonable solution. In section 13.5 we are interested in finding coefficients for hyperplane valuations associated with FLR classifiers for which no known algorithmic solution exists. In section 13.7 we employ a group of search techniques generally categorized as evolutionary algorithms to find reasonable coefficients for hyperplane valuations associated with our experimental data set. In general, evolutionary algorithms are based upon concepts found in the study of population genetics. More specifically, the algorithms in this category maintain a population (set) of candidate solutions (individuals) to a problem. The fitness of a candidate solution is determined by evaluating how well it does on the problem, and the most fit candidate solution has some type of effect on the overall makeup of the candidate population. Thus, the population of candidate solutions evolves over time, hopefully resulting in improved solutions for the problem.

In section 13.7 we employ four different Evolutionary Algorithms: Genetic Algorithms [23], Tabu Search [9], Particle Swarm Optimization [15], and Differential Evolution [19] to find coefficients associated with hyperplane valuations for FLR classifiers. The remainder of this section is devoted to an introductory description of the four search techniques.

Genetic Algorithms [23] were introduced in the 1970's by John Holland. For this search technique, an initial population of candidate solutions is established (either randomly or with some known constraints), on which a genetic algorithm performs the following four steps until some stopping criterion is met: 1. Establish the fitness of each candidate solution in the population from time t; 2. Clone individuals from population t for a new population for time $t+1$ using the fitness distribution of the population from time t; 3. Perform crossover in population $t+1$; 4. Perform mutation in population $t+1$. The set of steps $1-4$ is generally called a generation. In many cases, the stopping criteria is merely a set number of generations. The crossover (step 3) combines two chromosomes from the parents to produce a new offspring chromosome with the desired affect being the offspring is better than the parents. Mutation (step 4) is a random change of a parameter value somewhere in the population, i.e., an alteration of one or more gene values in a chromosome. While it is possible to use a high mutation rate, this effectively produces random search. Most researchers use a very low mutation rate. Steps 2 and 3 are a bit more complicated than the other steps.

In step 2, individuals from population t are selected to appear in population $t+1$ based on their fitness relative to the rest of the population. The selection is generally either roulette wheel selection (where the entire set of fitnesses is used to establish the composition of the roulette wheel), or else the selection is a binary tournament selection (where pairs of individuals are chosen randomly and the one with the best fitness is copied into the new population).

In step 3, a percentage of the population is chosen in pairs to create new candidate solutions. The chosen pairs are called parents, and the new individuals are called children. Children are created by choosing some parameters from one parent and other parameters from the other parent. Typically the children replace the parents in the population. There are three common forms of crossover: one point, two point, and uniform. One point crossover randomly selects a crossover point and everything to the left of that point comes from one parent, while everything to the right of that point comes from the other parent. Two point crossover randomly selects two points and everything between the two points comes from one parent, while everything outside the two points comes from the other parent. Uniform crossover randomly chooses the parent for each parameter.

Tabu Search [9] is generally attributed to Fred Glover and like Genetic Algorithms originated in the 1970's. Unlike Genetic Algorithms, Tabu Search works with a population of a single candidate solution and can be considered a neighborhood search method. In addition to the single candidate solution population, there is a list of individuals that are neighbors of the candidate solution and a historical memory of previously encountered candidate solutions. The historical memory is called the Tabu list (there can be more than one Tabu list). The first individual may be generated randomly or based on some type of constraint. Once the individual's fitness is determined, the Tabu search algorithm will perform the following four steps until some stopping criterion is met: 1. create a neighborhood of individuals for the current candidate solution each of which is slightly different than the current candidate solution; 2. determine which of the newly created individuals has the best fitness; 3. if the new best fitness is better than the current candidate solution's fitness, then keep the new individual and place the old individual on the Tabu list; 4. if the new fitness is worse than the previous fitness but it is not on the Tabu list and it is only slightly worse than the current candidate solution, then keep the new individual and place the old individual on the Tabu list. As with Genetic Algorithms, Tabu Search has several parameters that must be set: the size of the neighborhood, the size of the Tabu list, the number of Tabu lists, the definition of "slightly worse", and the stopping criterion.

Particle Swarm Optimization [15], which was introduced by Russell C. Eberhart and James Kennedy in the 1990's, is both a global and a neighborhood optimization technique. The concept comes from observing the behavior of groups of organisms in nature, e.g., schools of fish and flocks of birds. Individuals within the groups have a position and a velocity which change over time relative to other individuals in the group. With this behavior in mind, Particle Swarm Optimization maintains a population of candidate solutions each of which has a position, x (i.e. a point in the search space) and a velocity v (i.e. the distance to move from the current position in the next time step). After initializing the population, the algorithm performs the following five steps until some stopping criterion is met: 1. Calculate each candidate solution's new position $x = x + v$; 2. Calculate the fitnesses of the new positions; 3. If a new fitness is better than the best fitness the individual has seen, χ, then $\chi = x$; 4. If a new fitness is better than the best fitness χ_g the swarm has seen then $\chi_g = x$; 5. Calculate the velocities for the new individuals using the formula $v = wv + c_1 r_1(\chi - x) + c_2 r_2(\chi_g - x)$.

Differential Evolution [19] was developed by Kenneth Price in the 1990's. Differential Evolution is very similar to genetic algorithms without the cloning step or the mutation step. Instead, the algorithm performs the following four steps for each individual (target) in the population: 1. Choose two random population members and calculate their weighted sum; 2. Add the calculated answer to a third randomly chosen individual; 3. Create a child by performing crossover with the answer from step 2 and the target individual; 4. If the child's fitness is better than the target's fitness, then replace the target with the child.

13.5 Classification Enhancements

This section describes enhancements to the inclusion measure used in [6, 7, 14]. Recall that each of the classifiers described and applied in [6, 7, 14] is based upon an inclusion measure given by the function $\sigma(u \leq w) = \frac{v(w)}{v(u \vee w)}$, where v is a positive valuation function. We will refer to this inclusion measure as the σ-inclusion measure. Generally, for the experimental results found in [6, 7, 14] the unit valuation function of the form $v(x_1, \ldots, x_N) = x_1 + \ldots + x_N$ is used. As noted in [14], we can also *weigh* each of the attributes $x_1, \ldots, x_N$ via a positive hyperplane valuation function $v(x_1, \ldots, x_N) = c_1 x_1 + \ldots + c_N x_N$, where $c_i > 0$.

In the following, we investigate the effect of using the combination of hyperplane valuations and the σ-inclusion measure. First, consider a point $C = (c_1, \ldots, c_N)$ where each $c_i > 0$ represents the coefficients used for the positive hyperplane valuation, i.e. $v(x_1, \ldots, x_N) = c_1 x_1 + \ldots + c_N x_N$. Since the point C is in R^N, we can translate C into polar coordinates as follows:

Let $\rho = \sqrt{c_1^2 + \ldots + c_N^2}$, $\theta_1, \ldots, \theta_{N-2}$ be the polar angles and let θ_{N-1} be the azimuthal angle, then

$c_1 = \rho \, cos\theta_1$

$c_2 = \rho \, sin\theta_1 \, cos\theta_2$

$c_3 = \rho \, sin\theta_1 \, sin\theta_2 \, cos\theta_3$

$\ldots$

$c_{N-1} = \rho \, sin\theta_1 \, sin\theta_2 \, \ldots \, sin\theta_{N-2} \, cos\theta_{N-1}$

$c_N = \rho \, sin\theta_1 \, sin\theta_2 \, \ldots \, sin\theta_{N-2} \, sin\theta_{N-1}$

Consider the evaluation of the positive hyperplane valuation function $v(x)$,

$$v(x) = v(x_1, \ldots, x_N) = c_1 x_1 + \ldots + c_N x_N$$
$$= \rho \, cos\theta_1 x_1 + \rho \, sin\theta_1 \, cos\theta_2 x_2 + \ldots + \rho \, sin\theta_1 \, \ldots \, sin\theta_{N-1} x_N$$
$$= \rho(cos\theta_1 x_1 + sin\theta_1 \, cos\theta_2 x_2 + \ldots + sin\theta_1 \ldots \, sin\theta_{N-1} x_N)$$
$$= \rho(b_1 x_1 + b_2 x_2 + \ldots + b_N x_N)$$
$$= \rho \, v'(x),$$

where the point $B = (b_1, b_2, \ldots, b_N)$ is the point on the (unit) hypersphere corresponding to the point C in R^N, i.e. $1 = \sqrt{b_1^2 + \ldots + b_N^2}$ with $b_i > 0$. Hence for the σ-inclusion measure with hyperplane valuation is given by

$$\sigma(u \leq w) = \frac{v(w)}{v(u \vee w)} = \frac{\rho v'(w)}{\rho v'(u \vee w)} = \frac{v'(w)}{v'(u \vee w)}$$

Thus, the σ-inclusion measure defined in [14] using hyperplane valuations is equivalent to using (unit) *hypersphere* valuations. In particular, this means that we can limit our search space from the first quadrant of R^N to the (unit) hypersphere in the first quadrant of R^N when trying to find reasonable coefficients for a hyperplane valuation.

Next, we consider the relationship of the following three entities: 1) σ-inclusion measure, 2) the distance (metric) $d(u, w) = v(u \vee w) - v(u \wedge w)$ defined in [3] where v is a valuation, and 3) the $L_1(u, w)$ norm. For simplicity, we use the unit hypercube I^N in R^N, the unit valuation, and a point, say w in I^N; we also use *complement coding* (in regards to the set A below) to represent a data vector as detailed in [14]. With these conditions, we have the following three sets being identical: $A = \{u \in I^N : \sigma(u \leq w) = \frac{N}{N+m}\}$, $B = \{u \in I^N : d(u, w) = m\}$, and $C = \{u \in I^N : L_1(u, w) = m\}$. Using the relationship between A, B, and C plus the fact that in I^N, $\sigma(u \leq w) = \frac{N}{N+m} = \frac{v(w)}{v(w)+L_1(u,w)}$ we can see that the inclusion measure defined in [14] is based upon the L_1 norm. Furthermore, this relationship provides a general technique to define a measure based upon distance measures, i.e. $\delta(u, w) \equiv \frac{v(w)}{v(w)+d(u,w)}$, where d is a distance measure. Likewise, if s is a similarity measure such that $0 \leq s(u, w) \leq 1$ then we can define a measure as follows: $\delta(u, w) \equiv \frac{v(w)}{v(w)+(1-s(u,w))}$. We will refer to measures defined in this way (either using a distance measure or similarity measure) as δ-measures.

Now consider the training data given in Fig. 13.1 which consists of 12 training data and 3 test data. We are interested in a solution with 100% accuracy for the test data that uses the fewest number of hyperboxes possible. Using the geometric interpretations described in [14] where the data are grouped via hyperboxes, we quickly conclude there is no solution using the unit valuation that is 100% accurate for the test data. In other words, one cannot place point g in a training data rectangle without that rectangle also including point 7 or 8. This can be verified via experiments as well. To solve this problem, we turn our attention to hyperplane valuations and σ-inclusion measure. Since we have shown that we may limit our search for coefficients of the hyper-

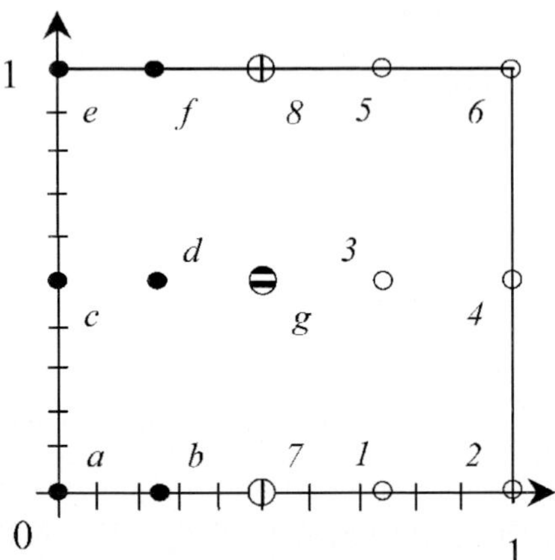

Fig. 13.1. Fifteen training data are presented in the order $a(0,0)$, $b(.25,0)$, $c(0,.5)$, $d(.25,.5)$, $e(0,1)$, $f(.25,1)$ (category "•"), and $1(.75,0)$, $2(1,1)$, $3(.75,.5)$, $4(1,.5)$, $5(.75,1)$, $6(1,1)$ (category "o"). Three test data are $g(.5,.5)$ from category "•" and $7(.5,0)$, $8(.5,1)$ from category "o". The problem is to find a tightest fit category function that uses the fewest hyperboxes and has 100% test data accuracy

plane valuation to the first quadrant hypersphere, an exhaustive search is not unreasonable, i.e. we can restrict our search to the circumference of the unit circle in the first quadrant. Via experiments, we found that none of the five classifiers *FLRtf, FLRff, FLRmtf, FLRotf,* and*FLRsf* can group the training data with 100% test data accuracy using σ-inclusion measure with hyperplane valuation. Actually for the given data found in Fig. 13.1, each of the five classifiers *FLRtf, FLRff, FLRmtf, FLRotf,* and *FLRsf* provide the same test data accuracy results (one data point incorrect using two hyperboxes) whether a unit or hyperplane valuation is used.

In an effort to find better inclusion measures, we have conducted numerous experiments on different data sets using different δ-measures. We tried well known distance and/or similarity measures such as the L_p norms, cosine, squared chord, squared Chi-squared, Canberra measure, Czekanowski coefficient, and correlation coefficient. However, none of these δ-measures consistently performed as well as the σ-inclusion measure defined in [14] when using a unit valuation nor are these δ-measures able to solve the problem noted in Fig. 13.1 *when using a unit valuation.*

In order to solve the problem in Fig. 13.1, we now turn our focus to a δ-measure that uses weighted attributes. As previously noted, the σ-inclusion measure using a hyperplane valuation is equivalent to a σ-inclusion measure $\frac{v'(w)}{v'(u\vee w)}$ where the coefficients for v' are taken from the unit hypersphere. This equivalency is not true, however, of all δ-measures. Consider the weighted (attribute) cosine similarity measure

$$ACS(u,w) = \frac{\sum\limits_{i} c_i u_i w_i}{\sqrt{(\sum\limits_{i} c_i u_i^2)(\sum\limits_{i} c_i w_i^2)}}$$

and the following four δ measures with ACD representing the attribute cosine distance metric

$$\delta 1(u,w) = \frac{v(w)}{v(w)+ACD(u,w)},$$

$$\delta 2(u,w) = \frac{v(w)}{v(w)+ACD(u\vee w,w)},$$

$$\delta 3(u,w) = \frac{v(w)}{v(w\vee u)+ACD(u\vee w,w)}, \text{ and}$$

$$\delta 4(u,w) = \frac{v(w)-(1-ACD(u\wedge w,w))}{v(w)}.$$

Note that for each measure, constant vector $C = (c_1,\ldots,c_N)$ is used in both the valuation function $v(x)$ and the attribute cosine distance metric ACD.

We now reconsider the problem described in Fig. 13.1 and conduct experiments using the five classifiers *FLRtf, FLRff, FLRmtf, FLRotf,* and *FLRsf* using each of the four weighted cosine δ-measures: $\delta 1$, $\delta 2$, $\delta 3$, and $\delta 4$. For each experiment, no FLR training parameters (such as epsilon for improved accuracy) are used. Hence, this requires the coefficients used in the weighted cosine δ-measure to completely compensate for all FLR parameters. Thus for our experiments, each FLR classifier is simply used to compute the number of hyperboxes needed by the training data, and in return, the number of correctly classified test data. The goal in each experiment is to find coefficients for the weighted cosine δ-measure such that the number of hyperboxes is the fewest possible and 100% test data accuracy. Recall for the weighted cosine δ-measure, our search space is not restricted to the unit circle. Hence an exhaustive search of the first quadrant in R^2 is used. When using the four weighted cosine δ-measures, each of the five classifiers are able to solve the problem in Fig. 13.1 using three hyperboxes and 100% test data accuracy. The graph of the solution space for the classifier *FLRotf* using the δ2-measure is given in Fig. 13.2. Table 13.1 provides a single solution (a pair of coefficients) for each of the modified five classifiers (there are infinitely many solutions) using the δ2-measure. As we previously stated, none of the five classifiers *FLRtf*, *FLRff*, *FLRmtf*, *FLRotf*, and *FLRsf* can solve the problem given in Fig. 13.1 using the original inclusion measure given in [14] and hyperplane valuations. It is only when one combines both the weighted cosine δ-measure and hyperplane valuations are the modified five classifiers *FLRtf*, *FLRff*, *FLRmtf*, *FLRotf*, and *FLRsf* able to solve the problem in Fig. 13.1. Via experiments, we found that the δ2-measure is most successful (of the four δ1, δ2, δ3, and δ4) in solving the problem in Fig. 13.1 and problems similar to those given in Fig. 13.8, Fig. 13.9, and Fig. 13.10 below. Because of the success of δ2, future unqualified references to *weighted cosine δ-measure* are assumed to be a reference to the δ2 definition only.

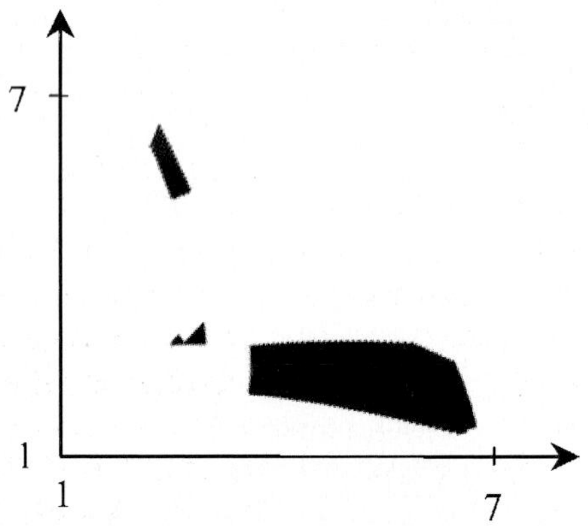

Fig. 13.2. FLRotf weighted cosine solution for the problem described in Fig. 13.1

Table 13.1. Coefficients for weighted cosine δ measure associated with problem described in Fig. 13.1

FLR classifier	C1 coefficients	C2 coefficients
FLRtf	6.391859	2.853784
FLRff	2.488825	5.349365
FLRmtf	6.391859	2.853784
FLRotf	2.761679	2.893634
FLRsf	2.761679	2.893634

Let's revisit the definition of an inclusion measure given above, namely σ is a map $\sigma : \mathsf{L} \times \mathsf{L} \to [0,1]$, which satisfies the following conditions

(C0) $\sigma(x, O) = 0, x \neq O$,

(C1) $\sigma(x, x) = 1$,

(C2) $w \wedge u < w \Rightarrow \sigma(w, u) < 1$, and

(C3) $u \leq w \Rightarrow \sigma(x, u) \leq \sigma(x, w)$ - Consistency Property

and our definition of the weighted cosine δ-measure given above. First, consider conditions (C0) and (C1), the unit hypercube, and the inclusion measure defined by $\sigma(u, w) = \frac{v(w)}{v(u \vee w)}$ where v is a positive valuation. Under these circumstances, in order to satisfy condition (C1), $\sigma(O, O)$ is defined to be 1 [11]. This definition is needed since one has division by zero whenever $v(u \vee w)$ is zero in $\frac{v(w)}{v(u \vee w)}$. Note that σ is not continous at O since as u approaches O, $\sigma(u, O)$ is zero and $\sigma(u, u)$ is one. For the weighted cosine δ-measure, a similar problem exists in that one has a division by zero whenever $ACD(u \vee w, w)$ is undefined or equivalently whenever $v(u \vee w) = 0$ or $v(w) = 0$ or equivalently whenever $v(u) = 0$ or $v(w) = 0$. If one defines $ACD(u, w)$ to be some value geater than zero, say 1, whenever $v(u) = 0$ or $v(w) = 0$, then another problem occurs when one evaluates $\delta(O, O)$, i.e. the value of $\delta(O, O)$ evaluates to 0 and in conflict with condition (C1). One solution is to define $ACD(u, w)$ to be zero whenever either $v(u) = 0$ or $v(w) = 0$ and define $\delta(O, O) = 1$. Hence, by defining special case values, we can satisfy conditions (C0) and (C1). An alternative would be to modify (C1) to read as $\sigma(x, x) = 1$, for all $x \neq O$ and to define $ACD(u, w) = 1$ whenever $v(u) = 0$ or $v(w) = 0$. Now, let's consider condition (C2). In general, condition (C2) is not true even with the restriction to positive valuations and special definitions for ACD. As an example, consider the unit hypercube I^N in R^N and two points (vectors) u and w. Now if the vectors u and w have the same direction (when viewed as vectors) and $u < w$ in I^N, then $w \wedge u = u < w$ and $\delta(w, u) = \frac{v(u)}{v(u)+1-ACS(w \vee u, u)} = \frac{v(u)}{v(u)+1-1} = 1$. To show that $ACS(w \vee u, u)$ is one, consider the following: 1) since w and u have the same direction and $w \wedge u = u < w$, $w \vee u = w$ has the same direction as u, 2) since $ACS(w \vee u, u)$ is the cosine of the angle formed between the vector u and $w \vee u$, the $ACS(w \vee u, u)$ is just the cosine of the angle zero since u and $w \vee u$ have the same direction. This is in direct conflict with condition (C2). An alternative for (C2) is $w \wedge u \leq w \Rightarrow \sigma(w, u) \leq 1$. In regards to condition (C3), the weighted cosine δ measure does not meet this condition. To demonstrate this, consider the unit hypercube I^N in R^N and three points (vectors) x, u and w with 1) the vectors x and u having the same direction, 2) $u < x$ and $u < w$ in I^N, 3) $u \neq O$, and 4) u and w do not have the same direction (see Fig. 13.3 for an example in I^2). Then

$$\delta(x, u) = \frac{v(u)}{v(u)+1-ACS(x \vee u, u)} = \frac{v(u)}{v(u)+1-ACS(x, u)} = \frac{v(u)}{v(u)+1-1} = 1, \text{ and}$$

$$\delta(x, w) = \frac{v(w)}{v(w)+1-ACS(x \vee w, w)} = \frac{v(w)}{v(w)+1-m} < 1,$$

where i) $ACS(x, u)$ is one since the angle between x and u is zero, ii) $ACS(x \vee w, w)$ is some constant m, $0 < m < 1$, i.e. the angle between $x \vee w$ and w, θ,

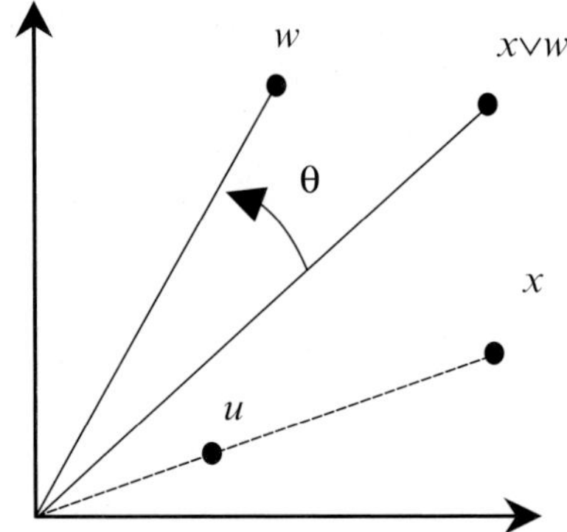

Fig. 13.3. For the given points, we have both $\delta(x, u) = 1$ and $\delta(x, w) < 1$

is between 0 and $\pi/2$, exclusive. Thus the weighted cosine δ measure fails to meet condition (C3) and no alternative for (C3) is offered. Although no proof is given here, note that $\delta 3$ as defined above is an inclusion measure.

13.6 Investigating Further the Weighted Cosine Measure

Let's further consider the *FLRotf* solution noted in Fig. 13.2 and the weighted cosine δ-measure. All of the coefficient solutions given in Fig. 13.2 give the same set of hyperboxes, i.e. if we consider a specific coefficient solution, say $(5, 2)$, and apply the *FLRotf* classifier, then the resulting solution consists of exactly three hyperboxes: one hyperbox containing the single point $(.75, 0)$, one hyperbox containing the single point $(.75, 1)$, and one hyperbox containing the single point $(.25, .5)$. We will refer to these hyperboxes as A, B, and C respectively. See Fig. 13.4. Is this solution a tightest fit solution, i.e. can either of the hyperboxes A, B, and C be expanded to include more training data without incurring a contradiction? Although visually it appears that one can expand each of the hyperboxes, the fit is indeed (based upon the computations) a tightest fit.

Although we cannot graph the weighted cosine δ-measure for the above problem $\{z : z = \delta(u, w),$ where u and w are members of $I^2 \times$ (dual of) $I^2\}$ since the input range is 8 dimensions, we can, however, graphically view the how the weighted cosine δ-measure behaves for specific coefficients. To do this, we consider a pair of coefficients from each of the blocks in the solution space identified in Fig. 13.2: namely the coefficient pairs $(5, 2)$, $(2.75, 3)$, and $(2.3, 5.75)$ giving one pair from each region. To graphically view the measure, we use three data sets: the training and test sets associated with Fig. 13.1 and a third data set (validation) defined to be $\{(x, y, 1) : (x, y) \in I^2$ and 1 represents the category for the pair$\}$. The training and test sets are used to obtain the hyperboxes identified in Fig. 13.4 (initial training) while the third data set is

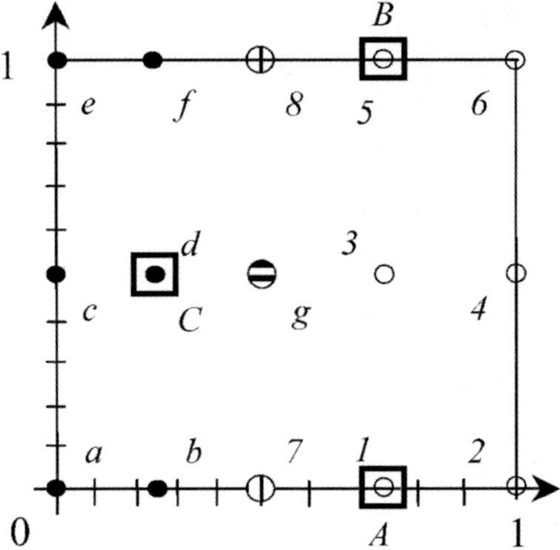

Fig. 13.4. Hyperbox A contains the single point $(.75, 0)$, hyperbox B contains the single point $(.75, 1)$, and hyperbox C contains the single point $(.25, .5)$

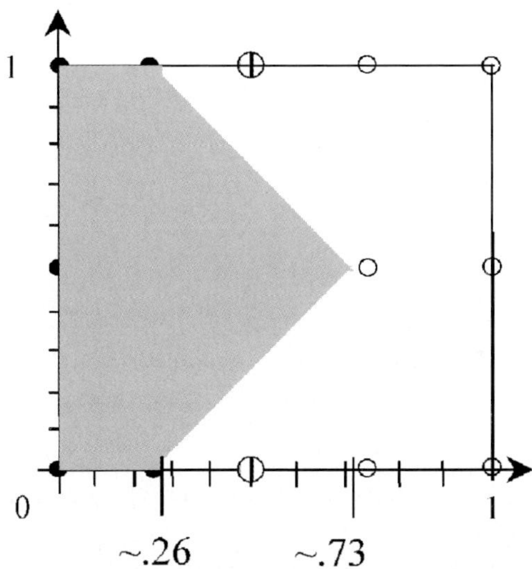

Fig. 13.5. Coefficients $(2.75, 3)$ are employed in this figure

used to locate the boundary between the region correctly classified and the region misclassified. We shall use this technique for multiple experiments that follow. Note that we not using the validation set in the traditional sense, but only to gather information about the boundary of the category regions given since we are no longer dealing with traditional hyperboxes. Hence, for the problem stated in Fig. 13.1 and coefficient solution space given in Fig. 13.2, the category regions for each pair of coefficients are given in Fig. 13.5, Fig. 13.6, and Fig. 13.7. For each the three figures, the category "•" is the shaded

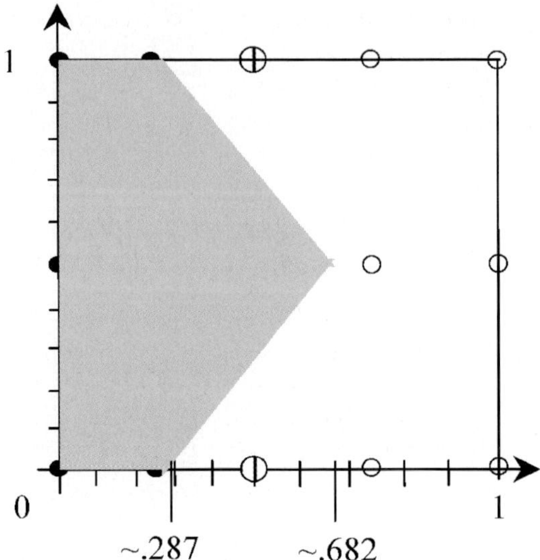

Fig. 13.6. Coefficients $(2.3, 5.75)$ are employed in this figure

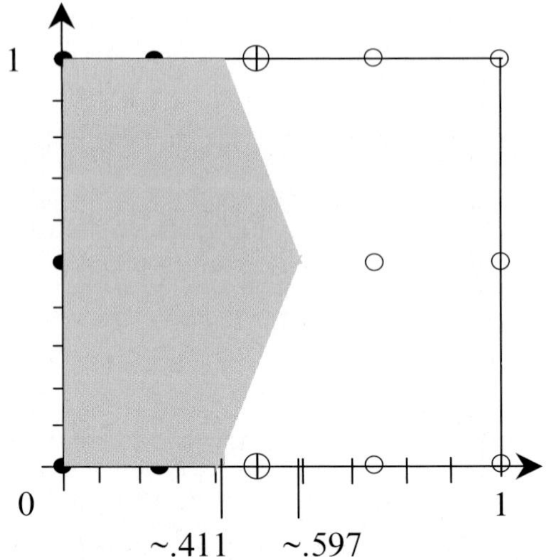

Fig. 13.7. Coefficients $(5, 2)$ are employed in this figure

portion while the category "o" is represented by the unshaded region. As demonstrated in Fig. 13.5, Fig. 13.6, and Fig. 13.7, the shaded triangular region grows and shrinks based upon the coefficients used in the weighted cosine δ-measure. For the coefficients $(1, 1)$, the triangular region disappears with the category "•" being the rectangular region $[0, .5] \times [0, 1]$ and category "o" being the remaining area. This results in two misclassifications for the test data. In other words, without the "weighting" the cosine δ-measure achieves the same results as the traditional *FLNotf* classifier (for this problem). It is also interesting to note that for other coefficient pairs such as $(8, 2)$, the triangular region protrudes to the left instead of to the right, i.e. a mirror image of the ones given in Fig. 13.5, Fig. 13.6, and Fig. 13.7.

Next, we investigate how the weighted cosine δ-measure adapts to a slightly different test set (the training set remains as found in Fig. 13.1). Namely, the test data set: $g(.5, .75)$, $h(.5, .25)$ from category "•" and $7(.5, 0)$, $8(.5, 1)$, $9(.5, .5)$ from category "o". We can perform an exhaustive search and determine that in deed the weighted cosine δ-measure can solve this modified problem as well. For one solution to this problem using the weighted cosine δ-measure with coefficients $(1.71, 2.1)$, see Fig. 13.8. In Fig. 13.8, the curve line near the vertical line $x = 0.5$ represents the boundary between the two categories "•" and "o" and is generated by the weighted cosine δ-measure using coefficients $(1.71, 2.1)$. When using the given coefficients, the *FLNotf* with weighted cosine δ-measure is able to train with 100% accuracy and 100% correct for the test set. Also in Fig. 13.8, the resulting trained hyperboxes are labeled A, B, C, D, E, and F, i.e. there are 6 hyperboxes. Again, this solution is a tightest fit. Visually, it appears that one could form one hyperbox containing both the hyperbox A and hyperbox B, but in doing so, the single box would cause miss classifications along the vertical line at $x = 0.5$.

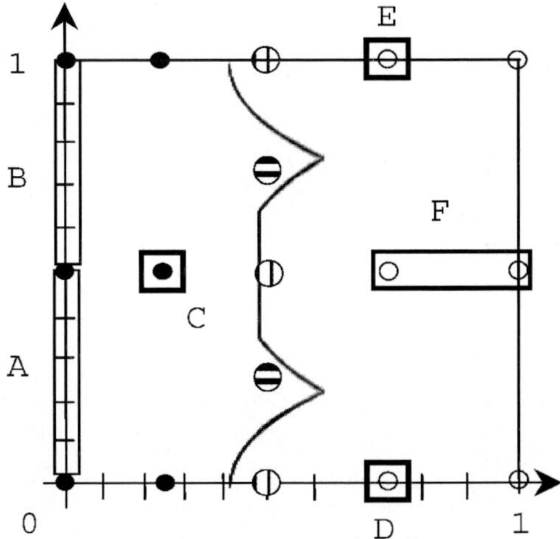

Fig. 13.8. The training set consists of points $a(0, 0)$, $b(.25, 0)$, $c(0, .5)$, $d(.25, .5)$, $e(0, 1)$, $f(.25, 1)$ (category "•") and $1(.75, 0)$, $2(1, 1)$, $3(.75, .5)$, $4(1, .5)$, $5(.75, 1)$, $6(1, 1)$ (category "o") plus the test data set: $g(.5, .75)$, $h(.5, .25)$ from category "•" and $7(.5, 0)$, $8(.5, 1)$, $9(.5, .5)$ from category "o". To produce the curve boundary between the two categories, the weighted cosine δ-measure uses coefficients $(1.71, 2.1)$. The six hyperboxes A, B, C, D, E, and F represent the training results for *FLNotf* and form a tightest fit

Next, we experiment by tilting the series of test points found near $x = .5$ for Fig. 13.8 while the training set remains the same. We tilt the test points to the left, i.e. negative slope, with a pivot point at $(.5, .5)$. Via experimentation, we determine that *FLNotf* is successful in achieving 100% accuracy for the test data with a slope range from vertical down to -7 (actually somewhere between -6 and -7) with each line passing through the pivot point $(.5, .5)$. Thus for a slope of -7, the given line is $y = -7x + 4$ and corresponding points $g(.4643, .75)$, $h(.5357, .25)$ from category "$\bullet$" and $7(.5714, 0)$, $8(.4246, 1)$, $9(.5, .5)$ from category "o" are given in Fig. 13.9. Although the *FLNotf* achieves success in adapting to the tilting, it does so by modifying the curvature of the boundary (see Fig. 13.9) and not by tilting of the curve. Hence the axis remains vertical. Again for these experiments, an exhaustive search is first done to determine possible coefficients for the weighted cosine δ-measure followed by the use of a validation set to determine the boundary between the correctly classified sets.

Finally, we investigate how the weighted cosine δ-measure adapts to a slightly different training and test set. Namely, a training set of $a(0, 0)$, $b(.25, 0)$, $c(0, .5)$, $d(.25, .5)$, $e(0, 1)$, and $f(.25, 1)$ in category "$\bullet$" and $1(.75, 0)$, $2(1, 1)$, $3(.75, 0.5)$, $4(1, 0.5)$, $5(.75, 1)$, and $6(1, 1)$ in category "o" plus the test data set: $(.5, 0.125)$, $(.5, 0.375)$, $(.5, 0.625)$, $(.5, 0.875)$ from category "$\bullet$" and $(.5, 0)$, $(.5, 0.25)$, $(.5, 0.5)$, $(.5, 0.75)$, and $(.5, 1)$ from category "o". We again perform an exhaustive search and determine that in deed the weighted cosine

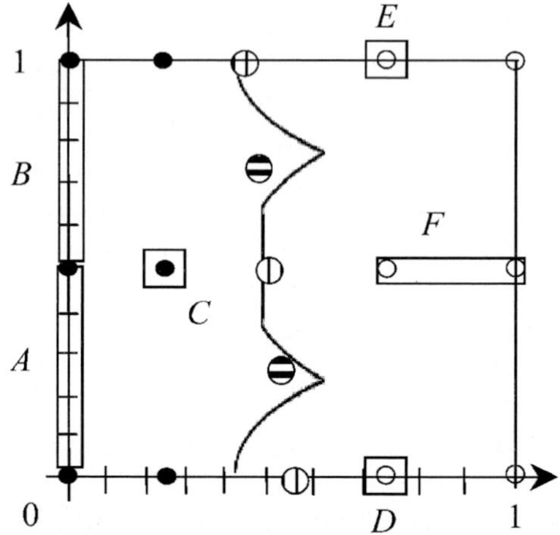

Fig. 13.9. The training set consists of points $a(0, 0)$, $b(.25, 0)$, $c(0, .5)$, $d(.25, .5)$, $e(0, 1)$, $f(.25, 1)$ (category "$\bullet$") and $1(.75, 0)$, $2(1, 1)$, $3(.75, .5)$, $4(1, .5)$, $5(.75, 1)$, $6(1, 1)$ (category "o") plus the test data set: $g(.4643, .75)$, $h(.5357, .25)$ from category "$\bullet$" and $7(.5714, 0)$, $8(.4246, 1)$, $9(.5, .5)$ from category "o". In this case the test data are tilted to the left and are obtained from the line $y = -7x + 4$. To produce the curve boundary between the two categories, the weighted cosine δ-measure uses coefficients $(0.600012, 2.42178)$. Six hyperboxes represent the training results for *FLNotf* (there are two hyperboxes along the y axis). These form a tightest fit

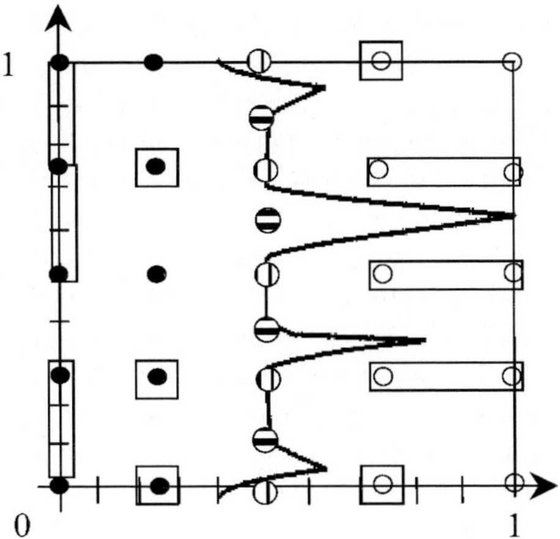

Fig. 13.10. The training set consists of points $a(0,0)$, $b(.25,0)$, $c(0,.5)$, $d(.25,.5)$, $e(0,1)$, $f(.25,1)$ (category "•") and $1(.75,0)$, $2(1,1)$, $3(.75,.5)$, $4(1,.5)$, $5(.75,1)$, $6(1,1)$ (category "o") plus the test data set: $(0.125, 0.5)$, $(0.375, 0.5)$, $(0.625, 0.5)$, $(0.875, 0.5)$ from category "•" and $(.5,0)$, $(.5,0.25)$, $(.5,0.5)$, $(.5,0.75)$, $(.5,1)$ from category "o". To produce the curve boundary between the two category regions, the weighted cosine δ-measure uses coefficients $(0.004, 0.255272)$. When using the given coefficients, the *FLNotf* with weighted cosine δ-measure is able to train with 100% accuracy and 100% correct for the test set. There are 11 hyperboxes in the training results for *FLNotf*

δ-measure can solve this modified problem. For one solution to this problem using the weighted cosine δ-measure with coefficients $(0.004, 0.255272)$, see Fig. 13.10. In Fig. 13.10, the curve line near the vertical line $x = 0.5$ represents the boundary between the solution regions for the two categories "•" and "o". When using the given coefficients, the *FLNotf* with weighted cosine δ-measure is able to train with 100% accuracy and 100% correct for the test set. As noted in Fig. 13.10, the resulting 11 trained hyperboxes are given. It is somewhat disappointing that this boundary region is not symmetrical, i.e. the spikes near $y = 0.375$ and $y = 0.625$ are not the same. This observation tends to imply that if the following two data points were added to the training set: $(.75, 0.625)$, $(1, 0.625)$, then certainly this solution would not work. When these additional two points are added, we are not able via experimentation using an exhaustive search for the weighted cosine δ-measure coefficients determine a solution that is 100% accurate for the test data.

13.7 Experiments and Results

The general learning capacity of the classifiers *FLRff*, *FLRotf*, and *FLRsf* with the weighted cosine δ-measure is demonstrated in this section by considering the Cleveland heart benchmark classification problem from the UCI repository of machine learning data sets [4].

The Cleveland heart data set involves numeric data of various sizes including missing attribute values. A missing attribute value is dealt with by replacing it with the least element in the corresponding constituent lattice as explained in [14]. Thus each data vector is represented in the lattice R^N. *Complement coding* is used to represent a data vector, in particular instead of $(x_1, \ldots, x_N)$ the vector $(1 - x_1, x_1, \ldots, 1 - x_N, x_N)$ is used [12, 14].

For the Cleveland heart data, the problem is to diagnose heart disease in a patient from a 14-attribute vector. This benchmark consists of 303 data vectors. The severity of heart disease is indicated by an integer ranging from 0 (no heart disease) to 4. By collapsing the classes into two, i.e. absence versus presence of heart disease, the problem becomes a "2-categories" problem as opposed to the original "5-categories" problem. Because no training and testing data sets are given explicitly, a *keep-250-in* series of 100 experiments is carried out such that in each experiment 250 randomly selected data are used for training and the remaining 53 data are left out for testing. Also, since we are wanting to compare the weighted cosine δ-measure results with existing published results for other classifiers. Since our comparison is to published works that did not use a validation set in their experiments, neither will our experiments use a validation set. Again, we are primarily interested in showing the malleability of the weighted cosine δ-measure via our experiments.

Table 13.2 shows the results by various methods from the literature in the 2-categories problem. Table 13.3 shows results for the 5-category problem using various FLR classifiers as reported in [6, 7, 14]. In Table 13.3, the average and the corresponding standard deviation are shown for both the classification accuracy and the number of rules in a series of experiments. The same 100 randomly generated data sets are used in all 5-category problems listed in Table 13.3.

For our experiments involving the Cleveland heart data, the same 100 randomly generated data sets are used in all 5-category problems. For each of the three classifiers *FLRff*, *FLRotf*, and *FLRsf*, four evolutionary algorithms are employed to find reasonable coefficients for the weighted cosine δ-measure. All total, we conducted 1200 experiments (100 data sets with three classifiers

Table 13.2. Performance of various methods from the literature in classifying Cleveland's 2-categories benchmark "heart problem"

Pattern classification method	Classification accuracy (%)
Probability analysis	79.0
Conceptual clustering (CLASSIT)	78.9
ARTMAP-IC (10 voters)	78.0
Discriminant analysis	77.0
Instance based prediction (NTgrowth)	77.0
Instance based prediction (C4)	74.8
Fuzzy ARTMAP (10 voters)	74.0
KNN (10 neighbors)	67.0

Table 13.3. Performance of classifiers *FLRtf*, *FLRff*, *FLRotf*, *FLRmtf*, and *FLRsf* in the Cleveland's benchmark "5-categories" heart problem. Results are from [7]

FLR classifier	Classification Accuracy		Number of rules	
	Average (%)	std. deviation	Average (%)	std. deviation
FLRff	66.74	4.96	53.47	10.30
FLRotf	66.60	5.52	51.18	11.00
FLRsf	66.49	5.59	49.47	9.80
FLRmtf	65.38	5.00	61.24	7.00
FLRtf	56.74	7.23	86.81	4.67

and four evolutionary algorithms per classifier). For each of the four evolutionary algorithms used, the same parameters are used on all 100 data sets and for each classifier. Each FLR classifier is merely used to determine the fitness of an evolutionary algorithm's solution as noted in Sect. 13.4. We use the Genetic Algorithm Utility Library (GAUL) software [1] for genetic algorithms, tabu search, and differential evolution portion of our experiments. For particle swarm optimization, we adapted software originally written by Clerc [5].

For our experiments, we modify each FLR classifier *FLRff*, *FLRotf*, and *FLRsf* to use the weighted cosine δ-measure in place of the traditional σ-inclusion measure. For the problem given in Fig. 13.1, it is reasonable to use an exhaustive search to find the hyperplane valuation for the weighted cosine since the search space is the first quadrant of $\mathbb{R}^2$. However for larger problem sets (with more attributes) such as the Cleveland heart data, an exhaustive search is not realistic. For larger problem sets we employ evolutionary algorithms to find a hyperplane valuation for the weighted cosine δ-measure.

In particular, we use evolutionary algorithms to provide a set of coefficients for the weighted cosine δ-measure. In each experiment, the initial coefficient set is randomly generated. Each of the FLR classifiers are then used to determine the fitness of the coefficients provided by the evolutionary algorithm, i.e. to determine the number of test data correctly identified and the number of hyperboxes used. Using the fitness information provided by the FLR classifier, the evolutionary algorithm updates its candidate solution and in return provides an additional set of coefficients for the weighted cosine δ-measure. For each evolutionary algorithm and all data sets, the evolutionary algorithm parameters remain fixed.

The parameter settings for each evolutionary algorithm follows. In all cases, the stopping criterion for each evolutionary algorithm is a fixed number of generations. For the genetic algorithm, a set of coefficients for the weighted cosine δ-measure represent a single chromosome. We use a population size of 500 which is initially randomly generated/seeded within the range 3.0 to 44.0. The stopping criterion is set to 200 generations; our mutation rate is 0.001 (i.e. one out of every 1000 parameters in a population was altered). For selection we use binary tournament selection where the elitist parents survive from one generation to the next. For crossover, we use a two point crossover in which two crossover points are randomly selected, and we set the percentage of the population that undergo crossover at 60%. We experimented with multiple techniques of selection and mutation. For example, we experimented with directional adjustments to the chromosome similar to what one would do using a "hill climbing" technique, i.e. we iterated through the genes of a chromosome making small adjustments and testing for improvement. Unfortunately, none of techniques showed improvement over a mutation rate of 0.001 and binary tournament selection. For tabu search, we used a population size of 500 which was initially randomly generated within the range of 0.3 and 50. The stopping criterion was set to 300 generations. We used a single tabu list size of 400 with a neighborhood size of 100. For mutation, we randomly chose whether to swap or shift a list item. A member was tagged as tabu if it differed no more than 0.1 from other members. For differential evolution, we used a population size of 500 which was initially randomly generated within the range 0.3 and 50. The stopping criterion was set to 100 generations. We chose to use the strategy "best", crossover "exponential", perbations "1", weighting factor 0.5, and crossover 0.8. For particle swarm optimization, we used a swarm size of 500, confidence coefficient of 1.42694, 100 informers, and 15 executions as the stopping criterion.

Table 13.4 shows the results for the classifiers *FLRff*, *FLRotf*, and *FLRsf* with the weighted cosine δ-measure for the 5-category problem. The use of the weighted cosine δ-measure shows considerable improvement (approximately 10%) in the classification accuracy for this problem as compared to the same classifiers for the same data sets using σ-inclusion measure. In fact, the 5-category results are comparable to most other methods when they are applied to the 2-category problem. See Table 13.2. When differential evolution is used to find the coefficients of the weighted cosine δ-measure, each of three classifiers performed better than the original classifier in every case, i.e. in each of the 100 data sets. Overall, the genetic algorithm and differential evolu-

Table 13.4. Both classification accuracy and no. rules statistics for *FLRff*, *FLRotf*, and *FLRsf* in the Cleveland "heart problem" using weighted cosine δ-measure

| | FLRff | | | | | FLRotf | | | | | FLRsf | | | | |
| | % clas accu | | | no. rules | | % clas accu | | | no. rules | | % clas accu | | | no. rules | |
Method	av	sd	1-1*	av	sd	av	sd	1-1*	av	sd	av	sd	1-1*	av	sd
GA	75.2	2.7	100	49.0	3.6	75.3	2.5	99	49.7	3.6	75.3	2.6	99	48.9	3.5
Tabu	72.8	2.7	97	49.9	3.1	73.2	2.7	95	49.8	3.3	72.5	2.8	90	49.8	2.9
PSO	70.5	2.6	92	49.9	3.2	70.5	2.6	89	49.8	3.2	70.5	2.6	84	49.9	3.2
DE	75.5	2.3	100	50.8	3.8	75.2	2.5	100	51.2	4.3	75.5	2.3	100	50.9	3.8
Original	66.0	2.6	–	53.5	10.3	66.6	2.9	–	51.2	11.0	66.5	3.0	–	49.5	9.8

* For 100 data sets, a 1-to-1 comparison was made between an enhanced classifier and the original one. For example, entry 97 for Tabu and *FLRff* indicates that of the 100 data sets tested, the enhanced *FLRff* classifier outperformed the traditional *FLRff* on 97 out of 100 data sets.

tion search techniques provided better results than tabu and particle swarm optimization search techniques. Over fitting the training data is ruled out by construction, i.e. no FLR training parameters are used for the FLR classifiers and a fixed number of generations are used for each evolutionary algorithm — independent of the data set used.

13.8 Discussion and Conclusion

This work builds upon previous work found in [6, 7, 14] by modifying the definition of an inclusion measure used in those articles to a measure that we refer to as a δ measure. Although this measure is no longer an inclusion measure in the strictest sense, it can still be used in the traditional FLR tightest fit classifiers. We have also shown that the traditional σ-inclusion measure based upon hyperplane valuations (used in previous articles) is actually limited to the unit hypersphere valuation. It is also shown that the σ-inclusion measure is related to the L_1 norm (in I^N). We have also extended the σ-inclusion measure to a δ measure that is based upon distance and similarity measures. This extension opens a corridor between FLR and cluster analysis research, i.e. the similarity measures of cluster analysis can be used in FLR, and in return FLR classifiers can be more readily applied to cluster analysis. We have shown that FLR classifiers using the weighted cosine measure are able to solve problems for which σ-inclusion hyperplane measures could not. We have used the weighted cosine δ-measure in FLR classifiers plus evolutionary algorithms to find the weights associated with the cosine δ-measure. An application of three FLR classifiers using δ-measure to the Cleveland heart data 5-category problem shows significant improvements over alternative classification methods from the literature. In fact, our results for the 5-category problem are comparable to alternative classification methods reported in the literature for the 2-category problem.

Future work includes a comparison of the weighted cosine δ-measure and σ-inclusion measure on other data sets. Additional work needs to be done in search of similarity measures now used in cluster analysis that may be used as δ measures. Additional research needs to be done to determine if there is a more efficient way to find coefficients used in the weighted cosine δ-measure for a given problem set. Open questions: is there a methodology that can be used to find an optimal or near optimal set of coefficients for different problem sets; are there other measures that can generally perform better than the weighted cosine δ-measure.

References

1. Adcock S, Genetic Algorithm Utility Library, http://gaul.sourceforge.net/, version devel-0.1849
2. Anagnostopoulos GC, Georgiopoulos M (2001) New geometric concepts in fuzzy-ART and fuzzy-ARTMAP: category regions. In: Proc Intl J Conf Neural Networks (IJCNN) 1·32–37

3. Birkhoff G (1967) Lattice Theory. American Math Society, Colloquium Publications XXV
4. Blake CL, Merz CJ (1998) UCI repository of machine learning data-bases `http://www.ics.uci.edu/ mlearn/MLRepository.html` Univ California, Irvine
5. Clerc M, Basic particle swarm optimization software `http://clerc.maurice.free.fr/pso/PSO_basic/zo`
6. Cripps A, Kaburlasos VG, Nguyen N, Papadakis SE (2003) Improved experimental results using fuzzy lattice neurocomputing (FLN) classifiers. In: Proc Intl Conf Machine Learning, Models, Technologies and Applications (MLMTA) pp 161–166
7. Cripps A, Nguyen N, Kaburlasos VG (2003) Three improved fuzzy lattice neurocomputing (FLN) classifiers. In: Proc Intl J Conf Neural Networks (IJCNN) 3:1957–1962
8. Gabrys B (2002) Combining neuro-fuzzy classifiers for improved generalisation and reliability. In: Proc Intl J Conf Neural Networks (IJCNN) 3:2410–2415
9. Glover F, Laguna M (1997) Tabu Search. Kluwer, Norwell, MA
10. Granger E, Rubin MA, Grossberg S, Lavoie P (2002) Classification of incomplete data using the fuzzy ARTMAP neural network. In: Proc Intl J Conf Neural Networks (IJCNN) 6:35–40
11. Kaburlasos VG (2006) Towards a Unified Modeling and Knowledge Representation Based on Lattice Theory — Computational Intelligence and Soft Computing Applications, ser Studies in Computational Intelligence 27. Springer, Heidelberg, Germany
12. Kaburlasos VG (2007) Granular enhancement of fuzzy-ART/SOM neural classifiers based on lattice theory. This volume, Chapter 1
13. Kaburlasos VG, Athanasiadis IN, Mitkas PA (2007) Fuzzy lattice reasoning (FLR) classifier and its application for ambient ozone estimation. Intl J Approximate Reasoning 45(1):152–188
14. Kaburlasos VG, Petridis V (2000) Fuzzy lattice neurocomputing (FLN) models. Neural Networks 13(10):1145–1170
15. Kennedy J, Eberhart RC (1995) Particle swarm optimization. In: Proc IEEE Intl Conf Neural Networks IV:1942–1948
16. Mitchell TM (1997) Machine Learning. McGraw-Hill, New York, NY
17. Petridis V, Kaburlasos VG (1998) Fuzzy lattice neural network (FLNN): a hybrid model for learning. IEEE Trans Neural Networks 9(5):877–890
18. Petridis V, Kaburlasos VG (2001) Clustering and classification in structured data domains using fuzzy lattice neurocomputing (FLN). IEEE Trans Knowledge Data Engineering 13(2):245–260
19. Price K, Storn R, Lampinen J (2006) Differential Evolution — A Practical Approach to Global Optimization. Springer-Verlag, New York, NY
20. Quinlan JR (1992) C4.5: Programs for Machine Learning. Morgan Kaufman, San Mateo, CA
21. Rasmussen E (1992) Clustering algorithms. In: Information Retrieval: Data Structures and Algorithms, Frakes WB, Baeza-Yates R (eds) pp 419-442. Prentice Hall, Englewood Cliffs, NJ
22. Vapnik VN (1998) Statistical Learning Theory. Wiley, Chichester, UK
23. Vose MD (1999) The Simple Genetic Algorithm: Foundations and Theory. MIT Press, Cambridge, MA

Part IV

Logic and Inference

Fuzzy Prolog: Default Values to Represent Missing Information

Susana Munoz-Hernandez[1] and Claudio Vaucheret[2]

[1] Departamento de Lenguajes, Sistemas de la Información e Ingeniería del Software, Facultad de Informática - Universidad Politécnica de Madrid, Campus de Montegancedo 28660 Madrid, Spain
susana@fi.upm.es
[2] Departamento de Ciencias de la Computación, Facultad de Economía y Administración, Universidad Nacional del Comahue, Universidad Politécnica de Madrid, Buenos Aires 1400 (8300), Neuquen, Argentina
vaucheret@gmail.com

Summary. Incomplete information is a problem in many aspects of actual environments. Furthermore, in many scenarios the knowledge is not represented in a crisp way. It is common to find fuzzy concepts or problems with some level of uncertainty. There are not many practical systems which handle fuzziness and uncertainty and the few examples that we can find are used by a minority. To extend a popular system (which many programmers are using) with the ability of combining crisp and fuzzy knowledge representations seems to be an interesting issue.

Fuzzy Prolog [5] is a language that models fuzziness and uncertainty. In this chapter we enhance Fuzzy Prolog by using default knowledge to represent incomplete information in Logic Programming. We also provide the implementation of this new framework. This new release of Fuzzy Prolog handles incomplete information, it has a complete semantics (the previous one was incomplete as Prolog) and moreover it is able to combine crisp and fuzzy logic in Prolog programs. Therefore, new Fuzzy Prolog is more expressive to represent real world.

Fuzzy Prolog inherited from Prolog its incompleteness. The incorporation of default reasoning to Fuzzy Prolog removes this problem and requires a richer semantics which it is discussed.

14.1 Introduction

World information is not represented in a crisp way. Its representation is imperfect, fuzzy, etc., so that the management of uncertainty is very important

This research was partly supported by the Spanish MCYT project TIC2003-01036.

S. Munoz-Hernandez and C. Vaucheret: *Fuzzy Prolog: Default Values to Represent Missing Information*, Studies in Computational Intelligence (SCI) **67**, 287–308 (2007)
www.springerlink.com   © Springer-Verlag Berlin Heidelberg 2007

in knowledge representation. There are multiple frameworks for incorporating uncertainty in logic programming:

- fuzzy set theory
- probability theory
- multi-valued logic
- possibilistic logic

In [10] a general framework was proposed that generalizes many of the previous approaches. At the same time an analogous theoretical framework was provided and a prototype for Prolog was implemented [21]. Basically, a rule is of the form $A \leftarrow B_1, \ldots, B_n$, where the assignment I of certainties is taken from a certainty lattice, to the B_is. The certainty of A is computed by taking the set of the certainties $I(B_i)$ and then they are propagated using the function F that is an aggregation operator. This is a very flexible approach and in [5, 22] practical examples in a Prolog framework are presented.

In this work we extend the approach of [5] with arbitrary assignments of default certainty values (non-uniform default assumptions). The usual semantics of logic programs can be obtained through a unique computation method, but using different assumptions in a uniform way to assign the same default truth-value to all the atoms. The most well known assumptions are:

- the *Closed World Assumption* (CWA), which asserts that any atom whose truth-value cannot be inferred from the facts and clauses of the program is supposed to be false (i.e. certainty 0). It is used in stable models [2, 3] and well-founded semantics [12, 13, 15],
- the *Open World Assumption* (OWA), which asserts that any atom whose truth-value cannot be inferred from the facts and clauses of the program is supposed to be undefined or unknown (i.e. certainty in $[0, 1]$). It is used in [10].

There are also some approaches [23, 24] where both assumptions can be combined and some atoms can be interpreted assuming CWA while others follows OWA. Anyway, what seems really interesting is not only to combine both assumptions but to generalize the use of a default value. The aim is working with incomplete information with more guarantees.

The rest of the paper is organized as follows. Section 14.2 introduces the Fuzzy Prolog language. A complete description of the new semantics of Fuzzy Prolog is provided in Section 14.3. Section 14.4 completes the details about the improved implementation using $CLP(\mathcal{R})$ with the extension to handle default knowledge. Some illustrating examples are provided in section 14.5. Finally, we conclude and discuss some future work in section 14.6.

14.2 Fuzzy Prolog

In this section we are going to summarize the main characteristics of the Fuzzy Prolog that we proposed in [5] and that is the basis of the work presented here.

Fuzzy Prolog is more general than previous approaches to introduce fuzziness in Prolog in some respects:

1. A truth value will be a finite union of closed sub-intervals on $[0, 1]$. This is represented by Borel algebra, $\mathcal{B}([0, 1])$, while the algebra $\mathcal{E}([0, 1])$ only considers intervals. A single interval is a special case of union of intervals with only one element, and a unique truth value is a particular case of having an interval with only one element.

2. A truth value will be propagated through the rules by means of an *aggregation operator*. The definition of *aggregation operator* is general in the sense that it subsumes conjunctive operators (triangular norms [9] like min, prod, etc.), disjunctive operators [19] (triangular co-norms, like max, sum, etc.), average operators (like arithmetic average, quasi-linear average, etc) and hybrid operators (combinations of the above operators [17]). In [11][3] a resolution-Like Strategy based on a Lattice-Value Logic is proposed, as in our approach, although it is limited to the Lukasiewicz's implication operator.

3. The declarative and procedural semantics for Fuzzy Logic programs are given and their equivalence is proved.

4. An implementation of the proposed language is presented. A *fuzzy program* is a finite set of
   - *fuzzy facts* ($A \leftarrow v$, where A is an atom and v, a truth value, is an element in $\mathcal{B}([0, 1])$ expressed as constraints over the domain $[0, 1]$), and
   - *fuzzy clauses* ($A \leftarrow_F B_1, \ldots, B_n$, where $A, B_1, \ldots, B_n$ are atoms, and F is an interval-aggregation operator, which induces a union-aggregation, as by Definition 14.2, $\mathcal{F}$ of truth values in $\mathcal{B}([0, 1])$ represented as constraints over the domain $[0, 1]$).

   We obtain information from the program through *fuzzy queries or fuzzy goals* ($v \leftarrow A$? where A is an atom, and v is a variable, possibly instantiated, that represents a truth value in $\mathcal{B}([0, 1])$).

Programs are defined as usual but handling truth values in $\mathcal{B}([0, 1])$ (the Borel algebra over the real interval $[0, 1]$ that deals with unions of intervals) represented as constraints. We refer, for example, to expressions as: ($v \geq 0.5 \wedge v \leq 0.7$) $\vee$ ($v \geq 0.8 \wedge v \leq 0.9$) to represent a truth value in $[0.5, 0.7] \bigcup [0.8, 0.9]$.

A lot of everyday situations can only be represented by this general representation of truth value. There are some examples in [5].

The truth value of a goal will depend on the truth value of the subgoals which are in the body of the clauses of its definition. Fuzzy Prolog [5] uses *aggregation operators* [20] in order to propagate the truth value by means

---

³ This work discusses the resolution based on a Lattice-Valued Logic for the Prolog language at theoretical level. In our approach we provide also an operational semantics and an implementation using an extended Prolog with constraints.

of the fuzzy rules. Fuzzy sets *aggregation* is done using the application of a numeric operator of the form $f : [0,1]^n \rightarrow [0,1]$. An *aggregation operator* must verify $f(0,\ldots,0) = 0$ and $f(1,\ldots,1) = 1$, and in addition it should be monotonic and continuous. If we deal with the definition of fuzzy sets as intervals it is necessary to generalize from *aggregation operators* of numbers to *aggregation operators* of intervals. Following the theorem proved by Nguyen and Walker in [16] to extend T-norms and T-conorms to intervals, we propose the following definitions.

Definition 14.1 (interval-aggregation) *Given an aggregation $f : [0,1]^n \rightarrow [0,1]$, an interval-aggregation $F : \mathcal{E}([0,1])^n \rightarrow \mathcal{E}([0,1])$ is defined as follows:*

$$F([x_1^l, x_1^u], ..., [x_n^l, x_n^u]) = [f(x_1^l, ..., x_n^l), f(x_1^u, ..., x_n^u)].$$

Actually, we work with union of intervals and propose the definition:

Definition 14.2 (union-aggregation) *Given an interval-aggregation $F : \mathcal{E}([0,1])^n \rightarrow \mathcal{E}([0,1])$ defined over intervals, a union-aggregation $\mathcal{F} : \mathcal{B}([0,1])^n \rightarrow \mathcal{B}([0,1])$ is defined over union of intervals as follows:*

$$\mathcal{F}(B_1, \ldots, B_n) = \cup\{F(\mathcal{E}_1, ..., \mathcal{E}_n) \mid \mathcal{E}_i \in B_i\}.$$

In the presentation of the theory of possibility [25], Zadeh considers that fuzzy sets act as an elastic constraint on the values of a variable and fuzzy inference as constraint propagation.

In [5] (and furthermore in the extension that we presented in this paper), truth values and the result of aggregations are represented by constraints. A constraint is a Σ-*formula* where Σ is a signature that contains the real numbers, the binary function symbols $+$ and $*$, and the binary predicate symbols $=$, $<$ and $\leq$. If the constraint c has solution in the domain of real numbers in the interval $[0,1]$ then c is *consistent*, and is denoted as *solvable*(c).

14.3 Semantics

This section contains a reformulation of the semantics of Fuzzy Prolog. This new semantics is complete thanks to the inclusion of default value.

14.3.1 Least model semantics

The *Herbrand universe U* is the set of all ground *terms*, which can be made up with the constants and function symbols of a program, and the *Herbrand base B* is the set of all ground atoms which can be formed by using the predicate symbols of the program with ground *terms* (of the *Herbrand universe*) as arguments.

Definition 14.3 (default value) *We assume there is a function* default *which implement the Default Knowledge Assumptions. It assigns an element of* $\mathcal{B}([0,1])$ *to each element of the Herbrand Base. If the Closed World Assumption is used, then* $default(A) = [0,0]$ *for all A in Herbrand Base. If Open World Assumption is used instead,* $default(A) = [0,1]$ *for all A in Herbrand Base.*

Definition 14.4 (interpretation) *An* interpretation $I = \langle B_I, V_I \rangle$ *consists of the following:*

1. *a subset B_I of the* Herbrand Base,
2. *a mapping V_I, to assign*
   a) *a truth value, in $\mathcal{B}([0,1])$, to each element of B_I, or*
   b) *$default(A)$, if A does not belong to B_I.*

Definition 14.5 (interval inclusion $\subseteq_{II}$) *Given two intervals $I_1 = [a,b]$, $I_2 = [c,d]$ in $\mathcal{E}([0,1])$, $I_1 \subseteq_{II} I_2$ if and only if $c \leq a$ and $b \leq d$.*

Definition 14.6 (Borel inclusion $\subseteq_{BI}$) *Given two unions of intervals $U = I_1 \cup \cdots \cup I_N$, $U' = I'_1 \cup \cdots \cup I'_M$ in $\mathcal{B}([0,1])$, $U \subseteq_{BI} U'$ if and only if $\forall I_i \in U$, $i \in 1..N$, I_i can be partitioned in to intervals $J_{i1}, ..., J_{iL}$, i.e. $J_{i1} \cup ... \cup J_{iL} = I_i$, $J_{i1} \cap ... \cap J_{iL}$ is the set of the border elements of the intervals (except the lower limit of J_{i1} and the upper limit of J_{iL}) and for all $k \in 1..L$, $\exists J'_{jk} \in U'$. $J_{ik} \subseteq_{II} J'_{jk}$ where $jk \in 1..M$.*

The Borel algebra $\mathcal{B}([0,1])$ is a complete lattice under $\subseteq_{BI}$ (Borel inclusion), and the Herbrand base is a complete lattice under $\subseteq$ (set inclusion) and so the set of all *interpretations* forms a complete lattice under the relation $\sqsubseteq$ defined as follows.

Notice that we have redefined interpretation and Borel inclusion with respect to the definitions in [5]. We will also redefine the operational semantics and therefore the internal implementation of the Fuzzy Prolog library. Sections below are completely new too. For uniformity reasons we have kept the same syntax that was used in [5] in fuzzy programs.

Definition 14.7 (interpretation inclusion $\sqsubseteq$) *Let $I = \langle B_I, V_I \rangle$ and $I' = \langle B_{I'}, V_{I'} \rangle$ be interpretations. $I \sqsubseteq I'$ if and only if $B_I \subseteq B_{I'}$ and for all $B \in B_I$, $V_I(B) \subseteq_{BI} V_{I'}(B)$.*

Definition 14.8 (valuation) *A valuation σ of an atom A is an assignment of elements of U to variables of A. So $\sigma(A) \subset B$ is a ground atom.*

In the Herbrand context, a valuation is the same as a substitution.

Definition 14.9 (model) *Given an interpretation $I = \langle B_I, V_I \rangle$,*

- *I is a* model *for a fuzzy fact $A \leftarrow v$, if for all valuations σ, $\sigma(A) \in B_I$ and $v \subseteq_{BI} V_I(\sigma(A))$.*

- *I is a model for a clause $A \leftarrow_F B_1, \ldots, B_n$ when the following holds: for all valuations σ, $\sigma(A) \in B_I$ and $v \subseteq_{BI} V_I(\sigma(A))$, where $v = \mathcal{F}(V_I(\sigma(B_1)), \ldots, V_I(\sigma(B_n)))$ and $\mathcal{F}$ is the union aggregation obtained from F.*
- *I is a model of a fuzzy program, if it is a model for the facts and clauses of the program.*

Every program has a least model which is usually regarded as the intended interpretation of the program since it is the most conservative model. Let $\cap$ (that appears in the following theorem) be the meet operator on the lattice of interpretations $(I, \sqsubseteq)$. We can prove the following result.

Theorem 14.1 (model intersection property) *Let $I_1 = \langle B_{I_1}, V_{I_1} \rangle$, $I_2 = \langle B_{I_1}, V_{I_1} \rangle$ be models of a fuzzy program P. Then $I_1 \cap I_2$ is a model of P.*

Proof. Let $M = \langle B_M, V_M \rangle = I_1 \cap I_2$. Since I_1 and I_2 are models of P, they are models for each fact and clause of P. Then for all valuations σ we have

- for all facts $A \leftarrow v$ in P,
  - $\sigma(A) \subseteq B_{I_1}$ and $\sigma(A) \in B_{I_2}$, and so $\sigma(A) \in B_{I_1} \cap B_{I_2} = B_M$,
  - $v \subseteq_{BI} V_{I_1}(\sigma(A))$ and $v \subseteq_{BI} V_{I_2}(\sigma(A))$, and so hence
  $v \subseteq_{BI} V_{I_1}(\sigma(A)) \cap V_{I_2}(\sigma(A)) = V_M(\sigma(A))$
  therefore M is a model for $A \leftarrow v$
- and for all clauses $A \leftarrow_F B_1, \ldots, B_n$ in P
  - since $\sigma(A) \in B_{I_1}$ and $\sigma(A) \in B_{I_2}$, hence $\sigma(A) \in B_{I_1} \cap B_{I_2} = B_M$.
  - if $v = \mathcal{F}(V_M(\sigma(B_1)), \ldots, V_M(\sigma(B_n)))$, since F is monotonic, $v \subseteq_{BI} V_{I_1}(\sigma(A))$ and $v \subseteq_{BI} V_{I_2}(\sigma(A))$, hence $v \subseteq_{BI} V_{I_1}(\sigma(A)) \cap V_{I_2}(\sigma(A)) = V_M(\sigma(A))$
  therefore M is a model for $A \leftarrow_F B_1, \ldots, B_n$

and M is model of P. $\square$

Remark 14.1 (Least model semantic). If we let $\mathbf{M}$ be the set of all models of a program P, the intersection of all of these models, $\bigcap \mathbf{M}$, is a model and it is the least model of P. We denote the least model of a program P by $lm(P)$.

Example 14.1 *Let's see an example (from [5]). Suppose we have the following program P:*

$$tall(peter) \leftarrow [0.6, 0.7] \vee 0.8$$
$$tall(john) \leftarrow 0.7$$
$$swift(john) \leftarrow [0.6, 0.8]$$
$$good\_player(X) \leftarrow_{luka} tall(X), swift(X)$$

Here, we have two facts, $tall(john)$ and $swift(john)$ whose truth values are the unitary interval $[0.7, 0.7]$ and the interval $[0.6, 0.8]$, respectively, and a clause for the good_player predicate whose aggregation operator is the Lukasiewicz T-norm.

The following interpretation $I = \langle B, V \rangle$ is a model for P, where
$B = \{tall(john), tall(peter), swift(john),$
    $good\_player(john), good\_player(peter)\}$ *and*

$$V(tall(john)) = [0.7, 1]$$
$$V(swift(john)) = [0.5, 0.8]$$
$$V(tall(peter)) = [0.6, 0.7] \vee [0.8, 0.8]$$
$$V(good\_player(john)) = [0.2, 0.9]$$
$$V(good\_player(peter)) = [0.5, 0.9]$$

note that for instance if $V(good\_player(john)) = [0.2, 0.5]$ $I = \langle B, V \rangle$ cannot be a model of P, the reason is that $v = luka([0.7, 1], [0.5, 0.8]) = [0.7 + 0.5 - 1, 1 + 0.8 - 1] = [0.2, 0.8] \not\subseteq_{II} [0.2, 0.5]$.
The least model of P is the intersection of all models of P which is $M = \langle B_M, V_M \rangle$ where
$B_M = \{tall(john), tall(peter), swift(john),$
$good\_player(john)\}$ *and*

$$V_M(tall(john)) = [0.7, 0.7]$$
$$V_M(swift(john)) = [0.6, 0.8]$$
$$V_M(tall(peter)) = [0.6, 0.7] \vee [0.8, 0.8]$$
$$V_M(good\_player(john)) = [0.3, 0.5].$$

14.3.2 Fixed-point semantics

The fixed-point semantics we present is based on a one-step consequence operator T_P. The least fixed-point $lfp(T_P) = I$ (i.e. $T_P(I) = I$) is the declarative meaning of the program P, so is equal to $lm(P)$. We include it here for clarity reasons although it is the same that in [5].

Let P be a fuzzy program and B_P the Herbrand base of P; then the *mapping T_P* over *interpretations* is defined as follows:

Let $I = \langle B_I, V_I \rangle$ be a fuzzy *interpretation*, then $T_P(I) = I'$, $I' = \langle B_{I'}, V_{I'} \rangle$, $B_{I'} = \{A \in B_P \mid Cond\}, V_{I'}(A) = \bigcup\{v \in \mathcal{B}([0, 1]) \mid Cond\}$
where

$$Cond = (A \leftarrow v \text{ is a ground instance of a fact in } P \text{ and}$$
$$solvable(v)) \text{ or}$$
$$(A \leftarrow_F A_1, \ldots, A_n \text{ is a ground instance of a}$$
$$clause \text{ in } P, \text{ and}$$
$$solvable(v), v = \mathcal{F}(V_I(A_1), \ldots, V_I(A_n))).$$

Note that since I' must be an interpretation, $V_{I'}(A) = default(A)$ for all $A \notin B_{I'}$.

The set of interpretations forms a complete lattice, so that T_P it is continuous.

Recall (from [5]) the definition of the *ordinal powers* of a function G over a complete lattice X:

$$G \uparrow \alpha = \begin{cases} \bigcup \{G \uparrow \alpha' \mid \alpha' < \alpha\} & \text{if } \alpha \text{ is a limit ordinal,} \\ G(G \uparrow (\alpha - 1)) & \text{if } \alpha \text{ is a successor ordinal,} \end{cases}$$

and dually,

$$G \downarrow \alpha = \begin{cases} \bigcap \{G \downarrow \alpha' \mid \alpha' < \alpha\} & \text{if } \alpha \text{ is a limit ordinal,} \\ G(G \downarrow (\alpha - 1)) & \text{if } \alpha \text{ is a successor ordinal,} \end{cases}$$

Since the first limit ordinal is 0, it follows that $G \uparrow 0 = \bot_X$ (the bottom element of the lattice X) and $G \downarrow 0 = \top_X$ (the top element). From Kleene's fixed point theorem we know that the least fixed-point of any continuous operator is reached at the first infinite ordinal ω. Hence $lfp(T_P) = T_P \uparrow \omega$.

Example 14.2 *Consider the same program P of the example 14.1 (from [5]), the ordinal powers of T_P are*
$T_P \uparrow 0 = \{\}$
$T_P \uparrow 1 = \{tall(john), swift(john), tall(peter)\}$ *and*

$$\begin{aligned} V(tall(john)) &= [0.7, 0.7] \\ V(swift(john)) &= [0.6, 0.8] \\ V(tall(peter)) &= [0.6, 0.7] \vee [0.8, 0.8] \end{aligned}$$

$T_P \uparrow 2 = \{tall(john), swift(john), tall(peter), good\_player(john)\}$ *and*

$$\begin{aligned} V(tall(john)) &= [0.7, 0.7] \\ V(swift(john)) &= [0.6, 0.8] \\ V(tall(peter)) &= [0.6, 0.7] \vee [0.8, 0.8] \\ V(good\_player(john)) &= [0.3, 0.5] \end{aligned}$$

$T_P \uparrow 3 = T_P \uparrow 2$.

Lemma 14.1 *Let P a fuzzy program. Then M is a model of P if and only if M is a pre-fixpoint of T_P, that is $T_P(M) \sqsubseteq M$.*

Proof. Let $M = \langle B_M, V_M \rangle$ and $T_P(M) = \langle B_{T_P}, V_{T_P} \rangle$.

We first prove the "only if" ($\rightarrow$) direction. Let A be an element of Herbrand Base, if $A \in B_{T_P}$, then by definition of T_P there exists a ground instance of a fact of P, $A \leftarrow v$, or a ground instance of a clause of P, $A \leftarrow_F A_1, \ldots, A_n$ where $\{A_1, \ldots, A_n\} \subseteq B_M$ and $v = \mathcal{F}(V_M(A_1), \ldots, V_M(A_n))$. Since M is a model of P, $A \in B_M$, and each $v \subseteq_{BI} V_M(A)$, then $V_{T_P}(A) \subseteq_{BI} V_M(A)$ and then $T_P(M) \sqsubseteq M$. If $A \notin B_{T_P}$ then $V_{T_P}(A) = default(A) \subseteq_{BI} V_M(A)$.

Analogously, for the "if" ($\leftarrow$) direction, for each ground instance $v = \mathcal{F}(V_M(A_1), \ldots, V_M(A_n))$, $A \in B_{T_P}$ and $v \subseteq_{BI} V_{T_P}(A)$, but as $T_P(M) \sqsubseteq M$, $B_{T_P} \subseteq B_M$ and $V_{T_P}(A) \subseteq_{BI} V_M(A)$. Then $A \in B_M$ and $v \subseteq_{BI} V_M(A)$ therefore M is a model of P. $\square$

Given this relationship, it is straightforward to prove that the least model of a program P is also the least fixed-point of T_P.

Theorem 14.2 *Let P be a fuzzy program. Then $lm(P) = lfp(T_P)$.*

Proof.

$$\begin{aligned}
lm(P) &= \bigcap\{M \mid M \text{ is a model of } P\}\\
&= \bigcap\{M \mid M \text{ is a pre-fixpoint of } P\}\\
&\quad \text{from lemma 14.1}\\
&= lfp(T_P)\\
&\quad \text{by the Knaster-Tarski}\\
&\quad \text{Fixpoint Theorem [18].}\square
\end{aligned}$$

14.3.3 Operational semantics

The improvement of Fuzzy Prolog is remarkable in its new procedural semantics that is interpreted as a sequence of transitions between different states of a system. We represent the state of a *transition system* in a computation as a tuple $\langle A, \sigma, S \rangle$ where A is the goal, σ is a substitution representing the instantiation of variables needed to get to this state from the initial one and S is a constraint that represents the truth value of the goal at this state.

When computation starts, A is the initial goal, $\sigma = \emptyset$ and S is true (if there are neither previous instantiations nor initial constraints). When we get to a state where the first argument is empty then we have finished the computation and the other two arguments represent the answer.

Definition 14.10 (Transition) *A* transition *in the* transition system *is defined as:*

1. $\langle A \cup a, \sigma, S \rangle \rightarrow \langle A\theta, \sigma \cdot \theta, S \wedge \mu_a = v \rangle$
   if $h \leftarrow v$ is a fact of the program P, θ is the mgu of a and h, μ_a is the truth value for a and solvable$(S \wedge \mu_a = v)$.
2. $\langle A \cup a, \sigma, S \rangle \rightarrow \langle (A \cup B)\theta, \sigma \cdot \theta, S \wedge c \rangle$
   if $h \leftarrow_F B$ is a rule of the program P, θ is the mgu of a and h, c is the constraint that represents the truth value obtained applying the union-aggregation $\mathcal{F}$ to the truth values of B, and solvable$(S \wedge c)$.
3. $\langle A \cup a, \sigma, S \rangle \rightarrow \langle A, \sigma, S \wedge \mu_a = v \rangle$
   if none of the above are applicable and solvable$(S \wedge \mu_a = v)$ where $\mu_a = $ default(a).

Definition 14.11 (Success set) *The success set $SS(P)$ collects the answers to simple goals $p(\widehat{x})$. It is defined as follows:* $SS(P) = \langle B, V \rangle$
where $B = \{p(\widehat{x})\sigma \mid \langle p(\widehat{x}), \emptyset, true \rangle \rightarrow^* \langle \emptyset, \sigma, S \rangle\}$ is the set of elements of the Herbrand Base that are instantiated and that have succeeded; and $V(p(\widehat{x})) = \cup\{v \mid \langle p(\widehat{x}), \emptyset, true \rangle \rightarrow^* \langle \emptyset, \sigma, S \rangle, \text{ and } v \text{ is the solution of } S\}$ is the set of truth values of the elements of B that is the union (got by backtracking) of truth values that are obtained from the set of constraints provided by the program P while query $p(\widehat{x})$ is computed.

Example 14.3 *Let P be the program of example 14.1 (from [5]). Consider the fuzzy goal*

$$\mu \leftarrow good\_player(X) \quad ?$$

the first transition in the computation is

$$\langle\{(good\_player(X)\}, \epsilon, true\rangle \rightarrow$$
$$\langle\{tall(X), swift(X)\}, \epsilon, \mu = max(0, \mu_{tall} + \mu_{swift} - 1)\rangle$$

unifying the goal with the clause and adding the constraint corresponding to Lukasiewicz T-norm. The next transition leads to the state:

$$\langle\{swift(X)\}, \{X = john\}, \mu = max(0, \mu_{tall} + \mu_{swift} - 1) \wedge \mu_{tall} = 0.7\rangle$$

after unifying $tall(X)$ with $tall(john)$ and adding the constraint regarding the truth value of the fact. The computation ends with:

$$\langle\{\}, \{X = john\}, \mu = max(0, \mu_{tall} + \mu_{swift} - 1) \wedge \mu_{tall} = 0.7 \wedge 0.6 \leq \mu_{swift} \wedge$$
$$\mu_{swift} \leq 0.8\rangle$$

As $\mu = max(0, \mu_{tall} + \mu_{swift} - 1) \wedge \mu_{tall} = 0.7 \wedge 0.6 \leq \mu_{swift} \wedge \mu_{swift} \leq 0.8$ entails $\mu \in [0.3, 0.5]$, the answer to the query $good\_player(X)$ is $X = john$ with truth value the interval $[0.3, 0.5]$.

In order to prove the equivalence between operational semantic and fixed-point semantic, it is useful to introduce a type of canonical top-down evaluation strategy. In this strategy all literals are reduced at each step in a derivation. For obvious reasons, such a derivation is called *breadth-first*.

Definition 14.12 (Breadth-first transition) *Given the following set of valid transitions:*

$$\langle\{A_1, \ldots, A_n\}, \sigma, S\rangle \rightarrow \langle\{A_2, \ldots, A_n\} \cup B_1, \sigma \cdot \theta_1, S \wedge c_1\rangle$$
$$\langle\{A_1, \ldots, A_n\}, \sigma, S\rangle \rightarrow \langle\{A_1, A_3 \ldots, A_n\} \cup B_2, \sigma \cdot \theta_2, S \wedge c_2\rangle$$
$$\ldots$$
$$\langle\{A_1, \ldots, A_n\}, \sigma, S\rangle \rightarrow \langle\{A_1, \ldots, A_{n-1}\} \cup B_n, \sigma \cdot \theta_n, S \wedge c_n\rangle$$

a breadth-first transition is defined as
$$\langle\{A_1, \ldots, A_n\}, \sigma, S\rangle \rightarrow_{BF} \langle B_1 \cup \ldots \cup B_n, \sigma \cdot \theta_1 \cdot \ldots \cdot \theta_n, S \wedge c_1 \wedge \ldots \wedge c_n\rangle$$
in which all literals are reduced at one step.

Theorem 14.3 *Given an ordinal number n and $T_P \uparrow n = \langle B_{T_{P_n}}, V_{T_{P_n}}\rangle$. There is a successful breadth-first derivation of length less or equal to $n + 1$ for a program P, $\langle\{A_1, \ldots, A_k\}, \sigma, S_1\rangle \rightarrow^*_{BF} \langle\emptyset, \theta, S_2\rangle$ iff $A_i\theta \in B_{T_{P_n}}$ and solvable($S \wedge \mu_{A_i} = v_i$) and $v_i \subseteq_{BI} V_{T_{P_n}}(A_i\theta)$.*

Proof. The proof is by induction on n. For the base case, all the literals are reduced using the first type of transitions or the last one, that is, for each literal A_i, it exits a fact $h_i \leftarrow v_i$ such that θ_i is the mgu of A_i and h_i, and μ_{A_i} is the truth variable for A_i, and $solvable(S_1 \wedge \mu_{A_i} = v_i)$ or $\mu_{A_i} = default(A_i)$. By definition of T_P, each $v_i \subseteq_{BI} V_{T_{P_1}}(A_i\theta)$ where $\langle B_{T_{P_1}}, V_{T_{P_1}} \rangle = T_P \uparrow 1$.

For the general case, consider the successful derivation,
$$\langle \{A_1, \ldots, A_k\}, \sigma_1, S_1 \rangle \rightarrow_{BF} \langle B, \sigma_2, S_2 \rangle \rightarrow_{BF} \ldots \rightarrow_{BF} \langle \emptyset, \sigma_n, S_n \rangle$$
the transition $\langle \{A_1, \ldots, A_k\}, \sigma_1, S_1 \rangle \rightarrow_{BF} \langle B, \sigma_2, S_2 \rangle$

When a literal A_i is reduced using a fact or there is not rule for A_i, the result is the same as in the base case. Otherwise there is a clause $h_i \leftarrow_F B_{1_i}, \ldots, B_{m_i}$ in P such that θ_i is the mgu of A_i and $h_i \in B\sigma_2$ and $B_{j_i}\theta_i \in B\sigma_2$, by the induction hypothesis $B\sigma_2 \subseteq B_{T_{P_{n-1}}}$ and $solvable(S_2 \wedge \mu_{B_{j_i}} = v_{j_i})$ and $v_{j_i} \subseteq_{BI} V_{T_{P_{n-1}}}(B_{j_i}\sigma_2)$ then $B_{j_i}\theta_i \subseteq B_{T_{P_{n-1}}}$ and by definition of T_P, $A_i\theta_i \in B_{T_{P_n}}$ and $solvable(S_1 \wedge \mu_{A_i} = v_i)$ and $v_i = \subseteq_{BI} V_{T_{P_n}}(A_i\sigma_1)$. $\square$

Theorem 14.4 *For a program P there is a successful derivation*

$$\langle p(\widehat{x}), \emptyset, true \rangle \rightarrow^* \langle \emptyset, \sigma, S \rangle$$

iff $p(\widehat{x})\sigma \in B$ and v is the solution of S and $v \subseteq_{BI} V(p(\widehat{x})\sigma)$ where $lfp(T_P) = \langle B, V \rangle$.

Proof. It follows from the fact that $lfp(T_P) = T_P \uparrow \omega$ and from the Theorem 14.3. $\square$

Theorem 14.5 *For a fuzzy program P the three semantics are equivalent, i.e.*

$$SS(P) = lfp(TP) = lm(P).$$

Proof. The first equivalence follows from Theorem 14.4 and the second from Theorem 14.2. $\square$

14.4 Implementation and Syntax

14.4.1 CLP($\mathcal{R}$)

Constraint Logic Programming [7] began as a natural merging of two declarative paradigms: constraint solving and logic programming. This combination helps make CLP programs both expressive and flexible, and in some cases, more efficient than other kinds of logic programs. CLP($\mathcal{R}$) [8] has linear arithmetic constraints and computes over the real numbers.

Fuzzy Prolog was implemented in [5] as a syntactic extension of a CLP($\mathcal{R}$) system. CLP($\mathcal{R}$) was incorporated as a library in the Ciao Prolog system[4].

Ciao Prolog is a next-generation logic programming system which, among other features, has been designed with modular incremental compilation in

---

[4] The Ciao system [1] including our Fuzzy Prolog implementation can be downloaded from `http://www.clip.dia.fi.upm.es/Software/Ciao`

mind. Its module system [1] permits having classical modules and fuzzy modules in the same program and it incorporates CLP($\mathcal{R}$).

Many Prolog systems have included the possibility of changing or expanding the syntax of the source code. One way is using the `op/3` built-in and another is defining *expansions* of the source code by allowing the user to define a predicate typically called `term_expansion/2`. Ciao has redesigned these features so that it is possible to define source translations and operators that are local to the module or user file defining them. Another advantage of the module system of Ciao is that it allows separating code that will be used at compilation time from code which will be used at run-time.

The *fuzzy* library (or *package* in the Ciao Prolog terminology) which implements the interpreter of our Fuzzy Prolog language [5] has been modified to handle default reasoning.

14.4.2 Syntax

Let us recall, from [5], the syntax of Fuzzy Prolog. Each Fuzzy Prolog *clause* has an additional argument in the head which represents its truth value in terms of the truth values of the subgoals of the body of the clause. A *fact* $A \leftarrow v$ is represented by a Fuzzy Prolog fact that describes the range of values of v with a union of intervals (which can be only an interval or even a real number in particular cases). The following examples illustrate the concrete syntax of programs:

$$youth(45) \leftarrow [0.2, 0.5] \bigcup [0.8, 1]$$
$$tall(john) \leftarrow 0.7$$
$$swift(john) \leftarrow [0.6, 0.8]$$
$$good\_player(X) \leftarrow_{min} tall(X),$$
$$swift(X)$$

```
youth(45):~ [0.2,0.5]v[0.8,1]
tall(john):~ 0.7
swift(john):~ [0.6,0.8]
good_player(X):~min tall(X),
                swift(X)
```

These clauses are expanded at compilation time to constrained clauses that are managed by CLP($\mathcal{R}$) at run-time. Predicates $. = ./2$, $. < ./2$, $. <= ./2$, $. > ./2$ and $. >= ./2$ are the Ciao CLP($\mathcal{R}$) operators for representing constraint inequalities, we will use them in the code of predicates definitions (while we will use the common operators $=, <, \leq, >, \geq$ for theoretical definitions). For example the first fuzzy fact is expanded to these Prolog clauses with constraints

```
youth(45,V):-    V .>=. 0.2, V .<=. 0.5.
youth(45,V):-    V .>=. 0.8, V .<.  1.
```

And the fuzzy clause

```
good_player(X) :~ min tall(X),swift(X).
```

is expanded to

```
good_player(X,Vp) :- tall(X,Vq),
                     swift(X,Vr),
                     minim([Vq,Vr],Vp),
                     Vp .>=. 0,
                     Vp .=<. 1.
```

The predicate `minim/2` is included as run-time code by the library. Its function is adding constraints to the truth value variables in order to implement the T-norm *min*.

```
minim([],_).
minim([X],X).
minim([X,Y|Rest],Min):-
             min(X,Y,M),
             minim([M|Rest],Min).

min(X,Y,Z):- X .=<. Y , Z .=. X.
min(X,Y,Z):- X .>. Y, Z .=. Y .
```

We have implemented several *aggregation operators* as `prod`, `max`, `luka` (Lukasiewicz operator), etc. and in a similar way any other operator can be added to the system without any effort. The system is extensible by the user simply adding the code for new *aggregation operators* to the library.

14.5 Combining Crisp and Fuzzy Logic

14.5.1 Example: Teenager student

In order to use definitions of fuzzy predicates that include crisp subgoals we must define properly their semantics with respect to the Prolog Close World Assumption (CWA) [4]. We will present a motivating example from [5].

Fuzzy clauses usually use crisp predicate calls as requirements that data have to satisfy to verify the definition in a level superior to 0, i.e. crisp predicates are usually tests that data should satisfy in the body of fuzzy clauses. For example, if we can say that a teenager student is a student whose age is about 15 then we can define the fuzzy predicate *teenager_student/2* in Fuzzy Prolog as

```
teenager_student(X,V):~
                 student(X),
                 age_about_15(X,V2).
```

In this example we pretend the goal $teenager\_student(X, V)$ provides:

- $V = 0$ if the value of X is not the name of a student.
- The corresponding truth value V if the value of X is the name of a student and we know that his age is about 15 in a certain level.
- Unknown if the value of X is the name of a student but we do not know anything about his/her age.

Note that we can face the risk of unsoundness unless the semantics of crisp and fuzzy predicates is properly defined. CWA means that all non-explicit information is false. E.g., if we have the predicate definition of $student/1$ as

```
student(john).
student(peter).
```

then we have that the goal $student(X)$ succeeds with $X = john$ or with $X = peter$ but fails with any other value different from these; i.e:

```
?- student(john).
yes
```

```
?- student(nick).
no
```

which means that $john$ is a student and $nick$ is not. This is the semantics of Prolog and it is the one we are going to adopt for crisp predicates because we want our system to be compatible with conventional Prolog reasoning. But what about fuzzy predicates? According to human reasoning we should assume OWA (non explicit information in unknown). Consider the following definition of $age\_about\_15/2$

```
age_about_15(john,1):~ .
age_about_15(susan,0.7):~ .
```

The goal $age\_about\_15(X, V)$ succeeds with $X = john$ and $V = 1$ or with $X = susan$ and $V = 0.7$. If we want to work with the CWA, like crisp predicates do, then we will obtain $V = 0$ for any other value of X different from $john$ and $susan$. The meaning is that the predicate is defined for all values and the membership value will be 0 if the predicate is not explicitly defined with other value. In this example we know that the age of $john$ is 15 and $susan$'s age is about 15 and with CWA we are also saying that the rest of the people are not about 15. This is the equivalent semantics to the one in crisp definitions but we think that we usually prefer to mean something different, i.e. in this case we can mean that we know that $john$ and $susan$ are about 15 and that we have no information about the age of the rest of people.

Therefore we do not know if the age of $peter$ is about 15 or not; and we know that $nick$'s age is definitely not about 15. We can explicitly declare

```
age_about_15(nick,0):~ .
```

We are going to work with this semantics for fuzzy predicates because we think it is the most alike to human reasoning. So a fuzzy goal can be true (value 1), false (value 0) or having other membership value. We have added the concept of **unknown** to represent no explicit knowledge in fuzzy definitions. We understand that if we don't have any information about a truth value V then its value is something (a value, an interval or a union of intervals) in the interval $\{0, 1\}$, so the most general assumption is the whole interval $[0, 1]$. The interval is represented by its corresponding constraints $V \geq 0$ and $V \leq 1$.

Our way to introduce crisp subgoals into the body of fuzzy clauses is translating the crisp predicate into the respective fuzzy predicate. In the example the way to obtain it is by overcoming the CWA behavior of the crisp predicate $student/1$ to obtain the truth value 0 for $student(susan)$. The solution is to fuzzify crisp predicates when they are in the body of fuzzy clauses.

For each crisp predicate in the definition of fuzzy predicate, the compiler will generate a fuzzy version to replace the original one in the body of the clause. For the example above of crisp predicate $student/1$, the compiler will produce the predicate $f\_student/2$ that is an equivalent fuzzy predicate to the crisp one. For our example we obtain the following Prolog definition of $teenager\_student/2$.

```
teenager_student(X,V):~
              f_student(X,V1),
              age_about_15(X,V2).
```

Where the default truth value of a crisp predicate is 0.

```
f_student(X,1):- student(X).
:-default(f_student/2,0).
```

Nevertheless, we consider for $age\_about\_15/2$ and $teenager\_student/2$ that the default value is unknown (the whole interval $[0, 1]$).

```
:-default(age_about_15/2,[0,1]).
:-default(teenager_student/2,[0,1]).
```

Observe the following consults:

```
?- age_about_15(john,X).
X = 1
```

```
?- age_about_15(nick,X).
X = 0
```

```
?- age_about_15(peter,X).
X .>=. 0, X .<=. 1
```

This means *john*'s age is about 15, *nick*'s age is not about 15 and we have no data about *peter*'s age.

We expect the same behavior with the fuzzy predicate *teenager_student/2*, i.e.:

```
?- teenager_student(john,V).
V .=. 1


?- teenager_student(susan,V).
V .=. 0


?- teenager_student(peter,V).
V .>=. 0, V .<=. 1
```

as *john* is a "teenager student" (he is a student and his age is about 15), *susan* is not a "teenager student" (she is not a student) and we do not know the value of maturity of *peter* as student because although he is a student, we do not know if his age is about 15.

Now the internal fuzzy resolution is simple, sound and very homogeneous because we only consider fuzzy subgoals in the body of the clause.

14.5.2 Example: Timetable compatibility

Another real example could be the problem of compatibility of a couple of shifts in a work place. For example teachers that work in different class timetables, telephone operators, etc. Imagine a company where the work is divided in shifts of 4 hours per week. Many workers have to combine a couple of shifts in the same week and a predicate *compatible/2* is necessary to check if two shifts are compatible or to obtain which couples of shifts are compatible. Two shifts are compatible when both are correct (working days from Monday to Friday, hours between 8 a.m. and 18 p.m. and there are no repetitions of the same hour in a shift) and in addition when the shifts are disjoint.

```
compatible(T1,T2):-
            correct_shift(T1),
            correct_shift(T2),
            disjoint(T1,T2).
```

But there are so many compatible combinations of shifts that it would be useful to define the concept of compatibility in a fuzzy way instead of in the crisp way it is defined above. It would express that two shifts could be incompatible if one of them is not correct or if they are not disjoint but when they are compatible, they can be more or less compatible. They can have a level of compatibility. Two shifts will be more compatible if the working hours are concentrated (the employee has to go to work few days during the week). Also, two shifts will be more compatible if there are few free hours between the busy hours of the working days of the timetable.

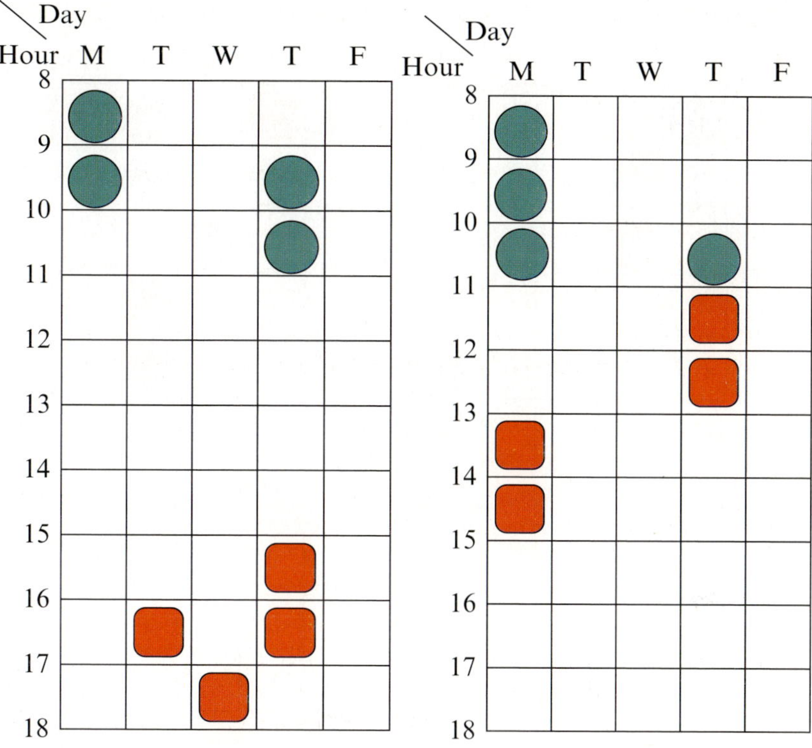

Fig. 14.1. Timetable 1 and 2

Therefore, we are handling crisp concepts (*correct_shift*/1, *disjoint*/2) besides fuzzy concepts (*without_gaps*/2, *few_days*/2). Their definitions, represented in Fig. 14.3 and Fig. 14.4, are expressed in our language in this simple way (using the operator ": #" for function definitions and the reserved word "*fuzzy_predicate*"):

```
few_days :# fuzzy_predicate([(0,1),
                (1,0.8),(2,0.6),
                (3,0.4),(4,0.2),
                (5,0)]).
```

```
without_gaps :# fuzzy_predicate([(0,1),
                (1,0.8),(5,0.3),
                (7,0.1),(8,0)]).
```

A simple implementation in Fuzzy Prolog combining both types of predicates could be:

```
compatible(T1,T2,V):~ min
```

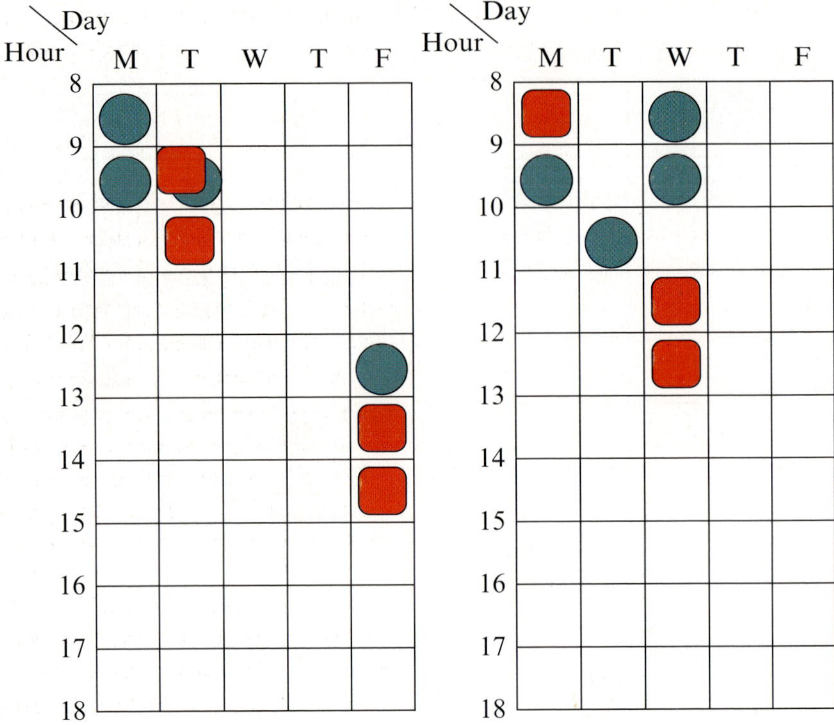

Fig. 14.2. Timetable 3 and 4

```
f_correct_shift(T1,V1),
f_correct_shift(T2,V2),
f_disjoint(T1,T2,V3),
f_append(T1,T2,T,V4),
f_number_of_days(T,D,V5),
few_days(D,V6),
f_number_of_free_hours(T,H,V7),
without_gaps(H,V8).
```

Here *append*/3 gives the total weekly timetable of 8 hours from joining two shifts, *number_of_days*/3 obtains the total number of working days of a weekly timetable and *number_of_free_hours*/2 returns the number of free one-hour gaps that the weekly timetable has during the working days. The **f_***predicates* are the corresponding fuzzified crisp predicates. The aggregation operator *min* will aggregate the value of V from $V6$ and $V8$ checking that $V1$, $V2$, $V3$, $V4$, $V5$ and $V7$ are equal to 1, otherwise it fails. Observe the timetables in Fig. 14.1 and Fig. 14.2. We can obtain the compatibility between the couple of shifts, T1 and T2, represented in each timetable asking the

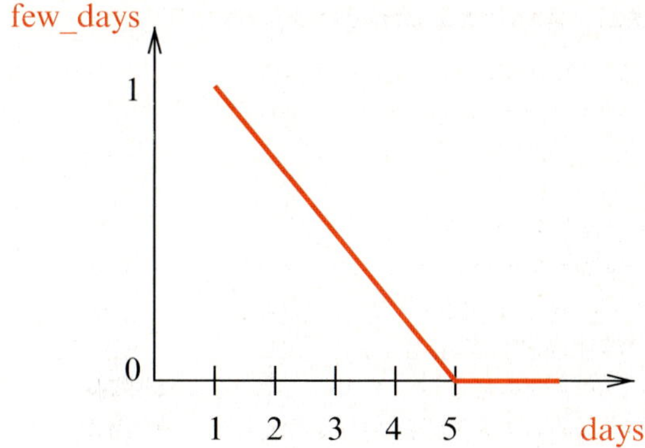

Fig. 14.3. Fuzzy predicate few_days/2

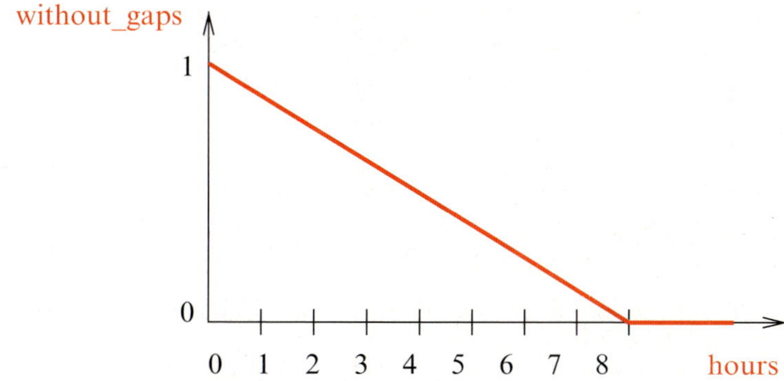

Fig. 14.4. Fuzzy predicate without_gaps/2

subgoal *compatible* $(T1, T2, V)$. The result is $V = 0.2$ for the timetable 1, $V = 0.6$ for the timetable 2, and $V = 0$ for the timetable 3 (because the shifts are incompatible).

Regarding compatibility of shifts in a weekly timetable, we are going to ask some questions about the shifts T1 and T2 of timetable 4 of Fig. 14.2. One hour of T2 is not fixed yet.

We can note: the days of the week as *mo, tu, we, th* and *fr*; the slice of time of one hour as the time of its beginning from 8 a.m. till 17 p.m.; one hour of the week timetable as a pair of day and hour and one shift as a list of 4 hours of the week.

If we want to fix the free hour of T2 in the slice 10-11 a.m. but with a compatibility not null, we obtain that only Tuesday is not compatible.:

```
?- compatible(
      [(mo,9), (tu,10), (we,8), (we,9)],
```

```
        [(mo,8), (we,11), (we,12), (D,10)], V),
    V .>. 0 .
```

```
  D =/= tu
```

If we want to know how to complete the shift T2 given a level of compatibility higher than 70 %, we obtain the slice from 10 to 11 p.m. at Wednesday or Monday morning.

```
  ?- compatible(
        [(mo,9), (tu,10), (we,8), (we,9)],
        [(mo,8), (we,11), (we,12), (D,H)],
        V),
    V .>. 0.7 .


  V = 0.9,  D = we, H = 10 ? ;
  V = 0.75, D = mo, H = 10 ? ;
  no
```

14.6 Conclusions and Future Work

Extending the expressivity of programming systems is very important for knowledge representation. We have chosen a practical and extended language for knowledge representation: Prolog.

Fuzzy Prolog presented in [5] is implemented over Prolog instead of implementing a new resolution system. This gives it a good potential for efficiency, more simplicity and flexibility. For example *aggregation operators* can be added with almost no effort. This extension to Prolog is realized by interpreting fuzzy reasoning as a set of constraints [25], and after that, translating fuzzy predicates into $CLP(\mathcal{R})$ clauses. The rest of the computation is resolved by the compiler.

In this paper we propose to enrich Prolog with more expressivity by adding default reasoning and therefore the possibility of handling incomplete information that is one of the most worrying characteristics of data (i.e. all information that we need usually is not available but only one part of the information is available) and anyway searches, calculations, etc. should be done just with the information that we had.

We have developed a complete and sound semantics for handling incomplete fuzzy information and we have also provided a real implementation based in our former Fuzzy Prolog approach.

We have managed to combine crisp information (CWA) and fuzzy information (OWA or default) in the same program. This is a great advantage because it lets us model many problems using fuzzy programs. So we have extended the expressivity of the language and the possibility of applying it to solve real problems in which the information can be defined, fuzzy or incomplete.

Presently we are working in several related issues:

- Obtaining constructive answers to negative goals.
- Constructing the syntax to work with discrete fuzzy sets and its applications (recently published in [14]).
- Implementing a representation model using unions instead of using backtracking.
- Introducing domains of fuzzy sets using types. This seems to be an easy task considering that we are using a modern Prolog [6] where types are available.
- Implementing the expansion over other systems. We are studying now the advantages of an implementation in XSB system where tabling is used.
- Using our approach for the engine of robots in a RoboCup league in a joint project between our universities.

References

1. Cabeza D, Hermenegildo M (2000) A new module system for Prolog. LNAI 1861:131–148. Springer-Verlag
2. Gelfond M, Lifschitz V (1988) The stable model semantics for logic programming. In: Proc Fifth Intl Conf Symposium on Logic Programming pp 1070–1080
3. Gelfond M, Lifschitz V (1990) Logic programs with classical negation. In: Proc seventh Intl Conf on Logic Programming pp 579–597. MIT Press — Complete version in: New Generation Computing (1991) 9:365–387
4. Clark KL (1978) Negation as failure. In: Gallaire H, Minker J (eds) Logic and Data Bases pp 293–322. Plenum Press, New York, NY
5. Guadarrama S, Munoz-Hernandez S, Vaucheret C (2004) Fuzzy Prolog: a new approach using soft constraints propagation. Fuzzy Sets and Systems 144(1):127–150
6. Hermenegildo M, Bueno F, Cabeza D, Carro M, García de la Banda M, López-García P, Puebla G (1999) The CIAO Multi-Dialect Compiler and System: An Experimentation Workbench for Future (C)LP Systems. Parallelism and Implementation of Logic and Constraint Logic Programming pp 65–85. Nova Science Commack, NY
7. Jaffar J, Lassez JL (1987) Constraint logic programming. In: ACM Symp Principles of Programming Languages pp 111–119
8. Jaffar J, Michaylov S, Stuckey PJ, Yap RHC (1992) The clp(r) language and system. ACM Trans Programming Languages and Systems 14(3):339–395
9. Klement E, Mesiar R, Pap E (2000) Triangular Norms. Kluwer Academic Publishers
10. Lakshmanan L, Shiri N (2001) A parametric approach to deductive databases with uncertainty. IEEE Trans Knowledge and Data Engineering 13(4):554–570
11. Liu J, Ruan D, Xu Y, Song Z (2003) A resolution-like strategy based on a lattice-valued logic. IEEE Trans Fuzzy Systems 11(4):560–567
12. Loyer Y, Straccia U (2002) Uncertainty and partial non-uniform assumptions in parametric deductive databases. In: Proc JELIA, LNCS 2424:271–282

13. Lukasiewicz T (2001) Fixpoint characterizations for many-valued disjunctive logic programs with probabilistic semantics. In: Proc LPNMR 2173:336–350
14. Munoz-Hernandez S, Pérez JG (2005) Solving collaborative fuzzy agents problems with clp(fd). In: Hermenegildo M, Cabeza D (eds) Proc Intl Symp Practical Aspects of Declarative Languages (PADL) LNCS 3350:187–202
15. Ng R, Subrahmanian V (1991) Stable model semantics for probabilistic deductive databases. In: Proc ISMIS, LNCS 542:163–171
16. Nguyen HT, Walker EA (2000) A First Course in Fuzzy Logic. Chapman & Hall/CRC
17. Pradera A, Trillas E, Calvo T (2002) A general class of triangular norm-based aggregation operators: quasi-linear t-s operators. International Journal of Approximate Reasoning 30(1):57–72
18. Tarski A (1955) A lattice-theoretical fixpoint theorem and its applications. Pacific Journal of Mathematics 5:285–309
19. Trillas E, Cubillo S, Castro JL (1995) Conjunction and disjunction on ([0,1],<=). Fuzzy Sets and Systems 72:155–165
20. Trillas E, Pradera A, Cubillo S (1999) A mathematical model for fuzzy connectives and its application to operators behavioural study. Information, uncertainty and fusion. In: Bouchon-Meunier B, Yager R, Zadeh L (eds) ser The Kluwer International Series in Engineering and Computer Sciences 516:307–318
21. Vaucheret C, Guadarrama S, Munoz-Hernandez S (2001) Fuzzy Prolog: a simple implementation using clp(r). Constraints and uncertainty. http://www.clip.dia.fi.upm.es/clip/papers/fuzzy-lpar02.ps
22. Vaucheret C, Guadarrama S, Munoz-Hernandez S (2002) Fuzzy Prolog: a simple general implementation using clp(r). In: Baaz M, Voronkov A (eds) Logic for Programming, Artificial Intelligence, and Reasoning (LPAR) LNAI 2514:450–463
23. Wagner G (1997) A logical reconstruction of fuzzy inference in databases and logic programs. In: Proc IFSA
24. Wagner G (1998) Negation in fuzzy and possibilistic logic programs. Logic programming and soft computing. Research Studies Press
25. Zadeh L (1978) Fuzzy sets as a basis for a theory of possibility. Fuzzy Sets and Systems 1(1):3–28

Valuations on Lattices: Fuzzification and its Implications

Kevin H. Knuth

Department of Physics, University at Albany, State University of New York, Albany, NY 12222 USA
kknuth@albany.edu

Summary. The notion of ordering is perhaps one of the most fundamental of abstract concepts. The zeta function is used to algebraically describe the ordering of elements in a lattice. An appropriate generalization of the zeta function generalizes the concept of inclusion to degrees of inclusion. However, the lattice structure imposes strong constraints on the values that these degrees can take. Here we review our previous work [1] [1] in studying these degrees of inclusion and relate these notions to the fuzzification of the lattice (independently introduced by Vassilis Kaburlasos). We show that an inclusion measure on the Boolean lattice of logical statements leads to Bayesian probability theory, which suggests a fundamental relationship between fuzzification of a Boolean lattice and Bayesian probability theory.

15.1 Introduction

In recent years, Lattice Theory, also known as Order Theory, has become widely recognized (especially in computer science) as a basic mathematical field that introduces a new perspective on the concept of an algebra. An algebra considers a set of elements along with a set of operations. These operations simply map elements of the algebra onto one another. However, a lattice takes a different perspective and considers a set of elements and a binary ordering relation, called a partial order. Instead of studying the mappings of the algebraic operations, one studies how the set of objects are ordered. As we will demonstrate, such changes in perspective often provide unique advantages.

This partial order, denoted by $x \leq y$ and read "y includes x", describes the situation where one element of the set "includes" a second element as defined by the ordering relation. This notion of inclusion can be extremely diverse encompassing the usual "lesser than" relation between real numbers, set inclusion, as well as implication among logical statements. A set of elements

---

[1] This contribution is an expanded version of our paper presented at the 2006 IEEE World Congress on Computational Intelligence (WCCI), Vancouver, BC, Canada.

K.H. Knuth: *Valuations on Lattices: Fuzzification and its Implications*, Studies in Computational Intelligence (SCI) **67**, 309–324 (2007)
www.springerlink.com

and an ordering relation give rise to what is called a partially ordered set, or *poset*. In the event that $x \leq y$ and there does not exist an element z in the set such that $x \leq z \leq y$, it is said that "y covers x." This concept of covering can be used to make diagrams such as those shown in the figure in this paper where y would appear above x in the diagram, and they would be connect by a line. It is in the special case where each pair of elements in a poset possesses a unique least upper bound, denoted $x \vee y$, and a unique greatest lower bound, denoted $x \wedge y$, that the poset is called a *lattice*. For more information on posets and lattices, we refer the reader to the introductory text by Davey and Priestley [2] or the more advanced text by Birkhoff [3].

This notion of inclusion, which is central to the partially ordered set, is encoded neatly by a function called the zeta function $\zeta(\cdot, \cdot)$, defined as

$$\zeta(x, y) = \begin{cases} 1 \text{ if } x \leq y \\ 0 \text{ if } x \nleq y \end{cases} \tag{15.1}$$

This function belongs to an important class of real-valued functions of two variables defined on the poset called the *incidence algebra* [4]. These functions $f(x, y)$ are non-zero only when $x \leq y$, and can be multiplied by performing a convolution over the interval of elements z in the poset where $x \leq z \leq y$

$$h(x, y) = \sum_{x \leq z \leq y} f(x, z)g(z, y).$$

The inversion of the zeta function relies on the Möbius function, $\mu(x, y)$, which is the inverse of the zeta function [4, 5, 6] so that

$$\delta(x, y) = \sum_{x \leq z \leq y} \zeta(x, z)\mu(z, y),$$

where $\delta(x, y)$ is the Kronecker delta function. As one might expect, these functions are indeed related to the more familiar Riemann zeta function of number theory, which originates from the partially ordered set of integers ordered by the relation "divides". These functions play an important role in order theory, and will play an even greater role in the generalizations that we will discuss here.

In this paper we will consider a particular class of lattices called distributive lattices, which include Boolean lattices. In Sect. 15.2 and Sect. 15.3 we review our previous work, which extends algebraic results by Richard Cox [7, 8], to derive Bayesian probability theory from generalizing the dual of the zeta function on the Boolean lattice of logical statements [9]. Last, in Sect. 15.4 we summarize the results and discuss future directions.

15.2 Valuations and Distributive Lattices

In this section, we explore generalizations of the zeta function that enable us to quantify degrees of inclusion. One of the benefits of this methodology is that the *meaning* of the binary ordering relation, which is encoded into the zeta function, is maintained during the process of generalization and carried over into the valuations that we derive.

15.2.1 Lattices and algebras

It is clear from the definition of the zeta function that the values that this function takes depend on the structure of the lattice. All lattices obey the following algebraic properties common to all posets

$$P1.\ \text{For all }\ a,\ \ a \leq a \qquad\qquad (Reflexivity)$$
$$P2.\ \text{If }\ a \leq b \ \text{ and }\ b \leq a, \ \text{ then }\ a = b\ (Antisymmetry)$$
$$P3.\ \text{If }\ a \leq b \ \text{ and }\ b \leq c, \ \text{ then }\ a \leq c\ (Transitivity)$$

as well as the following properties common to all lattices: *Idempotency* (L1), *Commutativity* (L2), *Associativity* (L3), and *Absorption* (L4)

$$L1.\ \ x \vee x = x, \quad x \wedge x = x$$
$$L2.\ \ x \vee y = y \vee x, \quad x \wedge y = y \wedge x$$
$$L3.\ \ x \vee (y \vee z) = (x \vee y) \vee z$$
$$x \wedge (y \wedge z) = (x \wedge y) \wedge z$$
$$L4.\ \ x \vee (x \wedge y) = x \wedge (x \vee y) = x$$

The greatest lower bound $x \wedge y$ can be viewed as an algebraic operation called the *meet*, and the greatest upper bound $x \vee y$ can be viewed as an operation called the *join*. Note that these relations are symmetric with respect to interchange of the meet and the join.

The relationship between the algebraic operations meet and join, and the ordering relation of the lattice can be illustrated by what is called the *consistency relation*

$$x \leq y \quad \Leftrightarrow \quad \begin{matrix} x \wedge y = x \\ x \vee y = y \end{matrix} \qquad (\textit{Consistency Relations})$$

It should be noted that the meaning of the join and the meet depend on both the set and the chosen ordering relation. For example, when the elements are logical statements and the ordering relation is logical implication, the join is the logical OR and the meet is the logical AND. When the elements are sets and the ordering relation is set inclusion $\subseteq$, the join is the set union $\cup$ and the meet is the set intersection $\cap$. As a final example, when the elements are the set of positive integers and the ordering relation is divides, the join is the least common multiple and the meet is the greatest common divisor.

Complete lattices have a top element called the *top*, denoted by $\top$, so that $x \vee \top = \top$ for all $x \in \mathcal{L}$, and dually an element called the *bottom*, denoted by $\bot$ where $x \wedge \bot = \bot$ for all $x \in \mathcal{L}$.

Lattices can have elements that cannot be written as the join of any two elements, so that for such elements $z \in \mathcal{L}$ there do not exist an $x \in \mathcal{L}$ and a

$y \in \mathcal{L}$, with $x \neq y$, such that $z = x \vee y$. Such elements, called *join-irreducible elements*, play an important role in lattice theory. Join-irreducible elements also play an important role in min-max neural networks [11].

15.2.2 Measures and valuations

A measure m typically refers to a function on a Boolean lattice $\mathcal{B}$ that takes a lattice element to a real number so that given $x \in \mathcal{B}$, $m : x \rightarrow \mathbb{R}$. The concept of a valuation is more general.

Definition 15.1 (Valuation) *A valuation v is a function that takes a lattice element to a real number. For all $x, y \in \mathcal{L}$, $v : x \rightarrow \mathbb{R}$ such that*

$$
\begin{aligned}
&1. \quad x \leq y \Rightarrow v(x) \leq v(y) \\
&2. \quad v(x) + v(y) = v(x \wedge y) + v(x \vee y)
\end{aligned}
\tag{15.2}
$$

The first condition above is described as 'v is increasing'. For this reason, it is often convenient in complete lattices to define the valuation such that the bottom element of the lattice has a valuation of zero: $v(\bot) = 0$.

In our previous work [9, 10], we have shown that the second condition need not be included in the definition, as it is a consequence of associativity of the lattice and applies to valuations on all distributive lattices. We begin by considering a valuation where $v(\bot) = 0$. Note that if $v(\bot)$ is not equal to zero, we can define v' so as to shift the origin to make this the case. Now, if the constraint equation we seek is to hold in general, then it had better hold in special cases. This is known as *Skilling's Principle of Induction* after John Skilling [16]. Consider the special case where the meet of any pair of the set of three elements x, y and z is the bottom:

$$
\begin{aligned}
x \wedge y &= \bot \\
y \wedge z &= \bot \\
z \wedge x &= \bot
\end{aligned}
$$

If these lattice elements have any relation between them at all, we would expect that $x \vee y$ would have something to do with x, y, and perhaps even $x \wedge y$. Furthermore, their valuations must somehow be related. So that in general consistency requires that

$$
v(x \vee y) = F[v(x), v(y), v(x \wedge y)],
\tag{15.3}
$$

where $F[\cdot]$ is an unknown function. Since we are looking at the special case where $x \wedge y = \bot$ and $v(\bot) = 0$, we can simplify our expectation of consistency by expressing the relation between $v(x)$, $v(y)$, and $v(x \vee y)$ in terms of a different, but related, unknown function $S[\cdot]$

$$
v(x \vee y) = S[v(x), v(y)].
\tag{15.4}
$$

We now consider complicating matters slightly. The idea is that we will come up with an expression that can be written two different ways. The result of each of the two relations must be the same, and we will use this to obtain our constraint equation. Consider the join of three elements $(x \vee y) \vee z$. Our relation (15.4) can then be written as

$$v((x \vee y) \vee z) = S[v(x \vee y), v(z)]. \tag{15.5}$$

We can now apply our relation again to the first argument of S on the right-hand side

$$v((x \vee y) \vee z) = S[S[v(x), v(y)], v(z)]. \tag{15.6}$$

However, we can also use the property of associativity of the join to write our multiple join as $x \vee (y \vee z)$. This results in the relation being written as

$$v(x \vee (y \vee z)) = S[v(x), v(y \vee z)], \tag{15.7}$$

which subsequently can be expanded into

$$v(x \vee (y \vee z)) = S[v(x), S[v(y), v(z)]]. \tag{15.8}$$

Now we have two expressions (15.6) and (15.8) that compute the same quantity and thus must be equal

$$S[S[v(x), v(y)], v(z)] = S[v(x), S[v(y), v(z)]]. \tag{15.9}$$

It is a little difficult to see what is going on here, so we clean things up a bit by introducing three quantities

$$u = v(x)$$
$$v = v(y)$$
$$w = v(z)$$

The result is a functional equation

$$S[S[u, v], w] = S[u, S[v, w]]. \tag{15.10}$$

This is known as the Associativity Equation, and the general solution for the function S [17] is

$$f(S[v(x), v(y)]) = f(v(x)) + f(v(y)), \tag{15.11}$$

which is

$$f(v(x \vee y)) = f(v(x)) + f(v(y)). \tag{15.12}$$

This suggests that there exists a more convenient representation for this valuation that is given by $f(v(\cdot))$. We can adopt this convenient valuation and call it by another name. For the purposes of simplicity, I shall just continue to call it $v(\cdot)$.

For the special case when $x \wedge y = \bot$, we now have a constraint equation that arises because of associativity

$$v(x \vee y) = v(x) + v(y). \tag{15.13}$$

These are all very basic considerations, and this relation must hold if one wishes to conform to associativity of the join. Any deviation from this constraint equation will lead to the serious consequence of not conforming to the algebraic properties of the set of elements, or equivalently a violation of the ordering relation among elements.

The last remaining step is to derive the constraint in the more general cases where $x \wedge y \neq \bot$. This can be done by considering the Möbius function for the distributive lattice, which enables us to avoid double-counting. The result is that

$$v(x \vee y) = v(x) + v(y) - v(x \wedge y), \tag{15.14}$$

which is equivalent to the well-known valuation equation

$$v(x) + v(y) = v(x \vee y) + v(x \wedge y). \tag{15.15}$$

Many readers may be familiar with Cox's derivation of the sum and product rules of Bayesian probability theory from Boolean algebra, and may also be equally familiar with many of the disagreements. However, it is important to point out that in the derivation above, we have not mentioned probability, logical statements, belief or any such concepts—only valuations which map a single lattice element to a real number. The result we have obtained here is seen to be a necessary constraint equation imposed by associativity of the distributive lattice.

A consequence of this result is that this constraint equation is fundamental and thus must appears in a wide variety of theories. Indeed this is the case as one may recognize this as Gian-Carlo Rota's famous *inclusion-exclusion principle* [18]. The inclusion-exclusion principle appears over and over in mathematics: for example, in probability theory

$$p(x \vee y|t) = p(x|t) + p(y|t) - p(x \wedge y|t),$$

where $x \wedge y$ represents the logical AND of two statements, and $x \vee y$ represents the logical OR, and t represents some statement assumed to be true. It appears in information theory

$$MI(x, y) = H(x) + H(y) - H(x, y), \tag{15.16}$$

where MI is the mutual information and H is the entropy. It is even relevant to integer relations as Rota [5] highlighted from Pólya and Szegö's work [19]

$$\mathtt{max}(x, y) = x + y - \mathtt{min}(x, y).$$

The Euler characteristic is an excellent example in geometry. For regular polytopes, the Euler characteristic is found by the number of faces minus the number of edges plus the number of vertices

$$\chi = F - E + V.$$

Despite the fact that the Euler characteristic has a value of two for all three-dimensional regular polytopes, it is an example of inclusion-exclusion as the faces of polytopes are comprised of edges, which themselves are comprised of vertices. The inclusion-exclusion relation also appears in quantum mechanics as the rule for summing quantum amplitudes and is the basis for the Feynman Path Integrals [12]. Again, this is a direct result of the associativity of the underlying lattice structure [9, 10]. This is a fundamental relation indeed, and it originates from associativity.

15.2.3 Co-valuations and bi-valuations

It may be that v is a decreasing function so that as

$$x \geq y \Rightarrow v(x) \leq v(y). \tag{15.17}$$

In this case, v is called a *co-valuation*. It is important to note that the derivation of the previous section does not hold for co-valuations, since it was explicitly assumed that $v(\perp) = 0$ and that $v(x) \geq v(\perp)$ for all elements x. Instead, with co-valuations, $v(\perp)$ is maximal, and $v(x) \leq v(\perp)$ for all elements x. In the case of a co-valuation, the constraint of associativity leads to a different constraint equation.

The special requirements of the co-valuation can be handled by first re-graduating the co-valuation so that $v(\perp) = 1$ and $v(\top) = 0$. Once this has been performed, one can define a new co-valuation by taking the logarithm $k(x) = \log(v(x))$. This mapping gives us a co-valuation where $k(\perp) = 0$ and $k(\top) = -\infty$. The constraint of associativity results in a sum-rule constraint equation in the log space, which gives a product-rule constraint for the original co-valuation. The details of this analysis will be published elsewhere.

The concept of valuations can be extended to include multiple lattice elements as arguments, such as the *bi-valuation*, which takes two lattice elements to a real number. The zeta function and Möbius functions above are two such examples. In general, we will refer to all such functions as valuations, with the exception when the fact that the function is a bi-valuation requires emphasis.

Since a bi-valuation has two arguments, it is not as easy to characterize the function as a valuation or a co-valuation. Consider the bi-valuation $w(a, b)$ defined on the lattice $\mathcal{L}$. If for every $b \in \mathcal{L}$, $x \leq y \rightarrow w(x, b) \leq w(y, b)$, then we can say that w is a valuation in its first argument. Similarly, if for every $b \in \mathcal{L}$, $x \geq y \rightarrow w(x, b) \leq w(y, b)$, then we can say that w is a co-valuation in its first argument. Similar definitions enable us to differentiate bi-valuations that are valuations or co-valuations in their second argument. In the remainder, we will focus on bi-valuations that are valuations in their first argument.

15.2.4 Distributivity

We now focus on distributive lattices, which satisfy the following distributive properties in addition to the previous properties P1-P3 and L1-L4 we have Distributivity of $\wedge$ over $\vee$ (D1), and Distributivity of $\vee$ over $\wedge$ (D2):

$$D1.\ x \wedge (y \vee z) = (x \wedge y) \vee (x \wedge z)$$
$$D2.\ x \vee (y \wedge z) = (x \vee y) \wedge (x \vee z)$$

It should be noted that a useful property of distributive lattices is that every distributive lattice can be expressed as a lattice of sets ordered by set inclusion.

Boolean lattices are distributive lattices $\mathcal{D}$ with an additional property of *complementation*, where for every element $x \in \mathcal{D}$, there exists a unique element $\sim x \in \mathcal{D}$, such that

$$C1.\ x \vee \sim x = \top$$
$$C2.\ x \wedge \sim x = \bot \qquad (Complementation)$$

Boolean lattices belong to the class of complemented distributive lattices. Since the number of elements of a Boolean lattice goes as 2^N, lattices where $N > 3$ are impractical to display. Figure 15.1 shows a Boolean lattice formed from three atomic elements (join-irreducible elements, which cover $\bot$). In this example the set of elements are logical statements ordered by logical

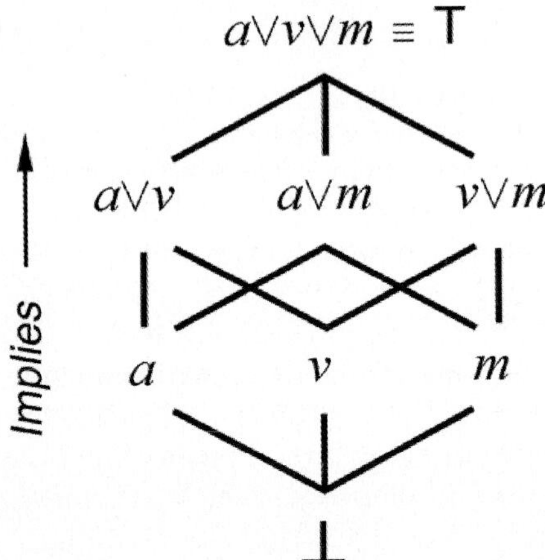

Fig. 15.1. A Boolean lattice of logical statements. The three atomic statements a, v, m, stand for animal, vegetable, and mineral. The elements are ordered by logical implication, and the top element, the truism is the statement "It is an animal, vegetable, or a mineral!" The bottom statement is the absurdity formed by a logical AND of any pair of atomic statements, such as "It is an animal and a mineral!"

implication so that y includes x, $x \leq y$, represents $x \to y$ with the atomic elements being the statements:[2]

$$a = \text{"It is an animal!"}$$
$$v = \text{"It is a vegetable!"}$$
$$m = \text{"It is a mineral!"}$$

The Boolean lattice consists of the set of these three atomic logical statements and all of the possible joins one could construct from them ordered by logical implication. This lattice is isomorphic to the lattice of the powerset of three elements ordered by set inclusion.

15.2.5 Generalizing the zeta function

We now consider a straightforward generalization of the zeta function (15.1). The motivation for this will become apparent.

First we define its dual $\zeta^{\partial}(x, y)$ [9]

$$\zeta^{\partial}(x, y) = \begin{cases} 1 \text{ if } x \geq y \\ 0 \text{ otherwise} \end{cases} \tag{15.18}$$

which is the original zeta function, but with the inclusion condition turned around. This function simply indicates whether the lattice element x includes the element y. We now introduce the generalization by defining the function $z(\cdot, \cdot)$, which expands on the "otherwise" case

$$z(x, y) = \begin{cases} 1 \text{ if } x \geq y \\ 0 \text{ if } x \wedge y = \bot \\ z \in (0, 1) \text{ otherwise} \end{cases} \tag{15.19}$$

Inclusion on the lattice has now been generalized to *degrees of inclusion*, which are represented by a real number in the interval $[0, 1]$. We could have defined the value of z to be over any finite real interval, but we have chosen the unit interval here both for convenience and to highlight the relationship to probability. In this case, this function is defined so that if we are certain that x includes y, then it returns a value of 1; however, if x does not include y this function can encode the *degree* to which x includes y. This extends the incidence algebra to an inclusion calculus. We have not yet stated the particular values that the function $z(\cdot, \cdot)$ takes in a given case. This will be discussed in some detail in the next section. It should be noted that such generalizations were independently introduced into the fuzzy lattice literature [20, 21], where the zeta function is called a 'crisp lattice inclusion relation' and the generalized zeta function is called a 'fuzzy lattice inclusion relation'.

---

[2] Here we use the notation introduced by Robert Fry where a logical statement is indicated with an exclamation mark.

This bi-valuation is required to follow Rota's inclusion-exclusion principle, which is a consequence of associativity of $\vee$ of the lattice. Enforcing this consistency requirement for the first argument of $z(\cdot,\cdot)$ defined on a distributive lattice gives [9, 10]

$$z(x_1 \vee x_2 \vee \cdots \vee x_n, t) = \sum_i z(x_i, t) - \sum_{i<j} z(x_i \wedge x_j, t)$$
$$+ \sum_{i<j<k} z(x_i \wedge x_j \wedge x_k, t) - \cdots . \quad (15.20)$$

Given $x, y, t \in \mathcal{D}$, we would like to be able to compute the degree to which t includes the meet $x \wedge y$, written $z(x \wedge y, t)$. The sum rule (15.20) imposes a relevant constraint

$$z(x \wedge y, t) = z(x, t) + z(y, t) - z(x \vee y, t). \quad (15.21)$$

However, another form can be found by requiring consistency with distributivity D1. Following Cox [7, 8], and relying on the consistency arguments given by Jaynes [13], Tribus [14], and Smith and Erickson [15], this degree can be written two ways as a function P a pair of lattice elements

$$z(x \wedge y, t) = P(z(x, t), z(y, x \wedge t)) = P(z(y, t), z(x, y \wedge t)). \quad (15.22)$$

The two expressions on the right are a consequence of commutativity, which we will address later.

The goal here is to use distributivity to impose a constraint on the possible forms for the function P. We focus on the first expression of P, and consider five elements $a, b, r, s, t \in \mathcal{D}$ where $a \wedge b = \perp$ and $r \wedge s = \perp$. Distributivity D1 of the meet over the join enables us to write $a \wedge (r \vee s)$ two ways

$$a \wedge (r \vee s) = (a \wedge r) \vee (a \wedge s). \quad (15.23)$$

Using the sum rule (15.20) and the form of P (15.22), distributivity requires that

$$P(z(a, t), z(r \vee s, a \wedge t)) = z(a \wedge r, t) + z(a \wedge s, t), \quad (15.24)$$

which simplifies to

$$P(z(a, t), z(r, a \wedge t) + z(s, a \wedge t)) =$$
$$P(z(a, t), z(r, a \wedge t)) + P(z(a, t), z(s, a \wedge t)). \quad (15.25)$$

This can be simplified by defining $u = z(a, t)$, $v = z(r, a \wedge t)$, and $w = z(s, a \wedge t)$, so that

$$P(u, v + w) = P(u, v) + P(u, w). \quad (15.26)$$

This functional equation for P captures the essence of distributivity D1.

Following Caticha [12], we now show that $P(u, v+w)$ is linear in its second argument. Defining $k = w + v$, and writing (15.26) as

$$P(u, k) = P(u, v) + P(u, w), \tag{15.27}$$

we can compute the second derivative with respect to x. Using the chain rule for differentiation, we find that

$$\frac{\partial}{\partial v} = \frac{\partial k}{\partial v}\frac{\partial}{\partial k} = \frac{\partial}{\partial k} = \frac{\partial k}{\partial w}\frac{\partial}{\partial k} = \frac{\partial}{\partial w}, \tag{15.28}$$

so that the second derivative with respect to k can be written as

$$\frac{\partial^2}{\partial k^2} = \frac{\partial}{\partial v}\frac{\partial}{\partial w}. \tag{15.29}$$

The second derivative of $P(u, k)$ with respect to k is then found to be

$$\frac{\partial^2}{\partial k^2}P(u, k) = 0, \tag{15.30}$$

which implies that P is linear in its second argument

$$P(u, v) = A(u)v + B(u), \tag{15.31}$$

where A and B are functions to be determined. Substitution of (15.31) into (15.26) gives $B(u) = 0$.

Now we consider $(a \vee b) \wedge r$, which using D1 can be written as

$$(a \vee b) \wedge r = (a \wedge r) \vee (b \wedge r). \tag{15.32}$$

This gives a similar functional equation

$$P(v + w, u) = P(v, u) + P(w, u), \tag{15.33}$$

where $u = z(r, t)$, $v = z(a, r \wedge t)$, $w = z(b, r \wedge t)$. By taking the appropriate derivatives, it is easy to show that P is also linear in its first argument

$$P(u, v) = A(v)u. \tag{15.34}$$

Together with (15.31), the general solution is

$$P(u, v) = Cuv, \tag{15.35}$$

where C is an arbitrary constant. Thus we have derived the *product rule*

$$z(x \wedge y, t) = Cz(x, t)z(y, x \wedge t), \tag{15.36}$$

as a constraint equation that enforces distributivity. This rule enables us to assign the degree to which t includes the meet $x \wedge y$. The constant C acts as a normalization factor, and is crucial when these valuations are normalized to values other than unity.

Last, commutativity of the meet further constrains the product rule resulting in a constraint equation that is analogous to Bayes' Theorem

$$z(y, x \wedge t) = \frac{z(y, t)z(x, y \wedge t)}{z(x, t)}. \tag{15.37}$$

15.3 Bayesian Inference From Valuations

The dual of the zeta function (15.18) indicates implication in a Boolean lattice of logical statements. The generalization to the z-function allows us to quantify degrees of implication. This generalization is quite useful. Even though the truism does not imply the statement a = "It is an animal!", we can use the defined bi-valuation to compute the degree to which the truism implies that "It is an animal!". In practical situations, this generalization is useful indeed.

The fact that the z-function quantifies degrees of implication, and is manipulated using the familiar sum rule, product rule and Bayes' Theorem, suggests that we have obtained an order-theoretic derivation of the notion of probability. With a simple change in notation, by defining $p(x|y) = z(x, y)$, one can see that we have derived a valuation that follows all of the expectations of Bayesian probability theory, where [9]

$$p(y|x) = \begin{cases} 1 \text{ if } x \rightarrow y \\ 0 \text{ if } x \wedge y = \bot \qquad (\textit{Probability}) \\ p \in (0, 1) \text{ otherwise} \end{cases} \qquad (15.38)$$

It is vital to understand that the area of discourse is a lattice of logical statements ordered by implication. This lattice represents a hypothesis space, and its structure, previously hidden by the algebra, is now made explicit by the perspective of lattice theory. The sum and product rules are also seen from a new viewpoint where they are merely constraint equations necessary to ensure that the valuation obeys the structure of the lattice. Any other rules for manipulating such degrees of implication would violate the properties P1-P3, L1-L4, D1 and D2.

The view of probability as degrees of belief represented by real numbers goes back to 1946 with Richard Cox who derived the sum and product rules by requiring that these degrees of belief be consistent with the rules of Boolean algebra [7, 8, 13]. However, here we do not rely on controversial quantities such as degrees of belief, but instead we merely generalize the zeta function, which quantifies inclusion on the lattice, to degrees of inclusion. For this reason, this development is more general in that it applies to valuations on all distributive lattices, and is not restricted to the arena of probability theory. What is meant by a degree of inclusion depends entirely on the meaning of inclusion on the partially ordered set.

Furthermore, Cox relied on complementation (which is valid for only Boolean lattices) to obtain the sum rule and associativity to obtain the product rule. This present work follows Caticha [12] and relies on the constraint of associativity to obtain the sum rule and the constraint of distributivity to obtain the product rule.

We are, of course, now left with the dilemma of probability assignments, which we find to be outside the realm of this discourse. The constraints of associativity and distributivity are clearly not sufficient to uniquely define the valuation on the lattice. Those who choose to view this as a limitation of the theory, may be comforted by the fact that there is a theorem by Rota that states:

Theorem 15.1 (Assigning Valuations [5]) *A valuation in a finite distributive lattice is uniquely determined by the values it takes on the set of join-irreducibles of $\mathcal{L}$, and these values can be arbitrarily assigned.*

This theorem holds when the value of the bottom element is zero, which in our probability theory example is $p(\perp|\top) = 0$. This theorem can be applied by considering $p(x|\top) \equiv p(x)$ as a valuation $p(\cdot)$ where x is any one of the join-irreducible assertions in the space, which comprise the N mutually exclusive atomic statements. The valuation $p(\cdot)$ can be seen to be equivalent to the Bayesian notion of a prior probability. The sum rule can be used to obtain $p(y|\top)$ for all $y \in \mathcal{L}$. One can then show that the product rule defines unique values for $p(y|z)$ for all $y, z \in \mathcal{L}$.

In short, the constraints imposed by the lattice structure, or equivalently the algebra, do not influence the possible values that a valuation assigns to the join-irreducible elements of the lattice. With reference to our example in Fig. 15.1, the probability that the object considered is an animal given the fact that we know that the object must be either an animal, a vegetable, or a mineral, $p(a|\top)$, can be assigned any value within the limits of the defined interval, and the calculus will suffer no contradictions. How then do we assign such values? We must take into account other consistency principles relevant to the problem, such as symmetry and constraints. Some of this groundwork has already been laid by Jaynes who introduced the *Principle of Group Invariance* [22] and the *Principle of Maximum Entropy* [23].

The Principle of Group Invariance is particularly interesting. It is useful when two problems are related via a symmetry. If particular valuation assignments are chosen for the join-irreducible elements of the lattice in one of the two problems, then consistency with respect to the symmetry between the two problems constrains the assignment of the valuations of the join-irreducible elements in the second problem. However, this principle does not tell you how to make the valuation assignments to the first problem. It merely asserts that the valuation assignments to the two problems must be commensurate with the symmetry relating the problems. This is the nature of a constraint.

15.4 Conclusions

In this paper we define a generalization of the zeta function on a lattice that enables us to encode the degree to which one element of the lattice includes another. This generalization does not tell us how to define all the values of the function. However, the lattice structure places strong constraints on the values that are assigned. We show that the constraint of associativity of the join results in a *sum rule*, which is equivalent to Rota's inclusion-exclusion principle. As expected, the generality of this result suggests that this rule should be ubiquitous in mathematics, and indeed we find that this is the case. Furthermore, the constraint of distributivity results in a rule that depends on the product of two degrees of inclusion. This result holds for *all* distributive lattices, which means that a valuation that generalizes set inclusion to degrees of inclusion will also follow a sum rule and a product rule.

The constraint of commutativity on a distributive lattice results in a constraint equation that is analogous to Bayes' Theorem. It is significant that this derivation makes no reference to a particular lattice. Instead the results hold equally for all distributive lattices. When it comes to the Boolean lattice of logical statements, we get a clearer picture of probability theory as arising from the generalization (or fuzzification) of lattice inclusion. This explicit generalization avoids the conceptual and linguistic pitfalls carried by terminology such as 'plausibility', 'degrees of belief' and 'frequencies of occurrences of events'. Instead the meaning of the zeta function implies that probability is more clearly though of as a degree of implication, and the rules of probability are not so much rules of manipulation as they are constraint equations limiting the values that the function can take.

However, these lattice constraints do not completely define the valuation. There are values of the valuation that can be arbitrarily assigned without violating the algebra, or equivalently, the lattice structure. Does this imply that the entire procedure is useless? Not at all. A valuation calculus that is completely constrained has no room to be applied to a specific problem. Instead, each problem has symmetries and constraints of its own, which now further constraint the valuation assignments. This perspective of constraints highlights some of the difficulties with the purely subjective Bayesian philosophy with which many fuzzy logic researchers disagree. Instead, it suggests that objective Bayesianism might be best viewed in terms of constraints imposed by symmetries and constraints imposed by the application itself.

It is exciting that a similar generalization of lattice inclusion was independently introduced by Vassilis Kaburlasos and colleagues [20, 21] from the perspective of fuzzification of the lattice. At long last, it is possible that fuzzy logic and probability theory have found common ground by elucidating the topic of discourse, which has remained implicit in both fields since their respective inceptions.

Acknowledgment

The author would like to thank John Skilling, Philip Goyal, Steve Gull, Ariel Caticha, Carlos Rodríguez, Janos Aczél, and Vassilis Kaburlasos for inspiring discussions, invaluable remarks and comments, and much encouragement. This work was supported in part by the College of Arts and Sciences and the College of Computing and Information of the University at Albany (SUNY), and the NASA SISM IS Program.

References

1. Knuth KH (2006) Valuations on lattices and their application to information theory (invited paper). In: Proc World Congress Computational Intelligence (WCCI) FUZZ-IEEE Program pp 813–820
2. Davey BA, Priestly HA (2002) Introduction to Lattices and Order. Cambridge Univ Press, Cambridge, UK
3. Birkhoff GD (1967) Lattice Theory. American Mathematical Society
4. Rota GC (1964) On the foundations of combinatorial theory I — theory of Möbius functions. Zeitschrift für Wahrscheinlichkeitstheorie und Verwandte Gebiete 2:340–368
5. Rota GC (1971) On the combinatorics of the Euler characteristic. In: Studies in Pure Mathematics (Presented to Richard Rado) pp 221–233. Academic Press
6. Barnabei M, Pezzoli E (1995) Gian-Carlo Rota on combinatorics. In: Kung JPS (ed) Möbius Functions pp 83–104. Birkhauser, Boston
7. Cox RT (1946) Probability, frequency, and reasonable expectation. Am J Physics 14:1–13
8. Cox RT (1961) The Algebra of Probable Inference. Johns Hopkins Press, Baltimore
9. Knuth KH (2005) Lattice duality: the origin of probability and entropy. Neurocomputing 67:245–274
10. Knuth KH (2004) Deriving laws from ordering relations. In: Erickson GJ, Zhai Y (eds) Bayesian Inference and Maximum Entropy Methods in Science and Engineering 707:204–235
11. Simpson PK (1992) Fuzzy min-max neural networks — part 1: classification. IEEE Trans Neural Networks 3(5):776–786
12. Caticha A (1998) Consistency, amplitudes and probabilities in quantum theory. Phys Rev A 57:1572–1582
13. Jaynes ET (2003) Probability Theory: The Logic of Science. Cambridge Univ Press, Cambridge, UK
14. Tribus M (1969) Rational Descriptions, Decisions and Designs. Pergamon Press, New York
15. Smith CR, Erickson GJ (1990) Probability theory and the associativity equation. In: Fougere P (ed) Maximum Entropy and Bayesian Methods pp 17–30. Kluwer
16. Skilling J (1988) The axioms of maximum entropy. In: Erickson GJ, Smith CR (eds) Maximum Entropy and Bayesian Methods in Science and Engineering pp 173–187. Kluwer
17. Aczél J (1966) Lectures on Functional Equations and Their Applications. Academic Press, NY
18. Klain DA, Rota GC (1997) Introduction to Geometric Probability. Cambridge Univ Press, Cambridge, UK
19. Pölya G, Szegö G (1964) Aufgaben und Lehrstze aus der Analysis 2:121. Springer, Berlin
20. Kaburlasos VG, Petridis V (2000) Fuzzy lattice neurocomputing (FLN) models. Neural Networks 13(10):1145–1170

21. Kaburlasos VG, Athanasiadis IN, Mitkas PA (2007) Fuzzy lattice reasoning (FLR) classifier and its application for ambient ozone estimation. Intl J Approximate Reasoning 45(1):152–188
22. Jaynes ET (1983) Prior probabilities. In: Rosenkrantz RD (ed) E. T. Jaynes: Papers on Probability, Statistics and Statistical Physics 158:114–130. D Reidel Publishing Company, Boston
23. Jaynes ET (1979) Where do we stand on maximum entropy?. In: Levine RD, Tribus M (eds) The Maximum Entropy Formalism pp 15–118. MIT Press

L-fuzzy Sets and Intuitionistic Fuzzy Sets

Anestis G. Hatzimichailidis and Basil K. Papadopoulos

Democritus University of Thrace, Dept. of Civil Engineering
Xanthi 67100, Greece
hatz_ane@hol.gr, papadob@civil.duth.gr

Summary. In this article we firstly summarize some notions on $L-$fuzzy sets, where L denotes a complete lattice. We then study a special case of $L-$fuzzy sets, namely the "intuitionistic fuzzy sets". The importance of these sets comes from the fact that the negation is being defined independently from the fuzzy membership function. The latter implies both flexibility and effectiveness in fuzzy inference applications. We additionally show several practical applications on intuitionistic fuzzy sets, in the context of computational intelligence.

16.1 Introduction

This work presents a novel fuzzy implication. Then, its extends it to intuitionistic fuzzy sets [14, 16]. Finally, it proposes useful geometric interpretations regarding intuitionistic fuzzy sets [15].

This chapter is organized as follows. Sections 16.2 and 16.3 summarize basic notations and definitions including useful geometrical interpretations. Sect. 16.4 presents a fuzzy implication. Sect. 16.5 presents an extension of the aforemenioned implication to intuitionistic fuzzy sets. Finally, Sect. 16.6 summarizes the contribution of this work.

16.2 $L-$Fuzzy Sets

The definition and notations can be found in [12, 19].

Definition 16.1 *Let X be a nonempty, ordinary set and let L be a complete lattice. An $L-$fuzzy subset, on X, is a mapping:*

$$A : X \rightarrow L$$

That is, the family of all $L-$fuzzy subsets, of X, is just L^X consisting of all mappings from X to L. Here, L^X is called an $L-$*fuzzy space*, X is called

A.G. Hatzimichailidis and B.K. Papadopoulos: *L-fuzzy Sets and Intuitionistic Fuzzy Sets*, Studies in Computational Intelligence (SCI) **67**, 325–339 (2007)
www.springerlink.com   © Springer-Verlag Berlin Heidelberg 2007

the *carrier domain*, of each $L-$fuzzy subset of it, and L is called the *value domain*, of each $L-$fuzzy subset of X (see [12, 19]).

Many authors study fuzziness in the case $L = [0, 1]$, and use the word "fuzzy", and not "$L-$fuzzy", to describe it. So, the word "fuzzy" possesses two levels of meaning. In the first case it means "$[0, 1]-$fuzzy" and in the other describes all fuzzy $L = [0, 1]$ cases including both $L-$fuzzy and $[0, 1]$ cases.

In the special case, in which $L = [0, 1] \times [0, 1]$ is equipped with the order $\leq$, where $(x_1, y_1) \leq (x_2, y_2)$, if and only if $x_1 \leq x_2$ and $y_1 \geq y_2$, we consider the *intuitionistic fuzzy set*.

16.3 Intuitionistic Fuzzy Sets

The definitions and notations, that we are going to use in this Section, can be found in [1, 2, 3, 4, 5, 6, 13, 15].

16.3.1 Some basic notions

Atanassov (see [1]) suggested, in 1983, a generalization of a classical fuzzy set, that was named *intuitionistic fuzzy set* [1]. We define such a set as follows.

Definition 16.2 *An* intuitionistic fuzzy set *(IFS) A, in X, is an object of the following form*

$$A = \{\langle x, \mu_A(x), \nu_A(x)\rangle : x \in X\},$$

where the functions

$$\mu_A : X \to [0, 1]$$

and

$$\nu_A : X \to [0, 1]$$

define the degree of membership and the degree of non-membership, of an element $x \in X$ and, evenmore, for each $x \in X$

$$0 \leq \mu_A(x) + \nu_A(x) \leq 1 .$$

Now, if $\pi_A(x) = 1 - \mu_A(x) - \nu_A(x)$, then $\pi_A(x)$ is called the *degree of non-determinance*, of an element $x \in X$, to the set A, where $\pi_A(x) \in [0, 1]$, $\forall x \in X$.

It can be easily verified that each fuzzy set is a particular case of the intuitionistic fuzzy set. Moreover, if A is a fuzzy set, then $\pi_A(x) = 0$, $\forall x \in X$.

We are particularly interested in the study of fuzzy sets for the case that $L - [0, 1]$, because many practical problems can be solved within this framework. In the same manner, a great interest is growing for the intuitionistic fuzzy sets, with $L = [0, 1] \times [0, 1]$.

---

[1] The term "intuitionistic" is controversial in the area of Fuzzy Logic [7, 10].

16.3.2 Geometric interpretations of IFSs

In the following we give two geometrical interpretations of intuitionistic fuzzy sets, which introduced in the literature (see [1, 15]).

(i) Let X denote the universe set. Let us also consider the Euclidean plane, with the Cartesion coordinates (see Fig. 16.1), and let F be defined as follows:

$$F = \{P/(p = \langle a, b \rangle) : a, b \geq 0,\ a + b \leq 1\}$$

Let $A \subseteq X$ be a fixed set. Then, we can construct a function $f_A : X \to F$, such that if $x \in X$, then

$$f_A(x) = p = \langle a, b \rangle \in F,$$

where

$$0 \leq a + b \leq 1 .$$

Note that the coordinates have been fixed such that $a = \mu_A(x)$ and $b = \nu_A(x)$.

(ii) The geometrical interpretation given below ([15]) could also be regarded as a generalization of the corresponding one given in [1].

Let T be an arbitrary triangle. It is known, from the Euclidean geometry, that for an arbitrary internal point P, of T, the following relation holds:

$$\frac{d_\mu}{v_\mu} + \frac{d_\nu}{v_\nu} + \frac{d_\pi}{v_\pi} = 1 ,$$

where d_μ, d_ν, d_π denote the distances between P and the sides μ, ν and π, of the respective hights (see Fig. 16.2).

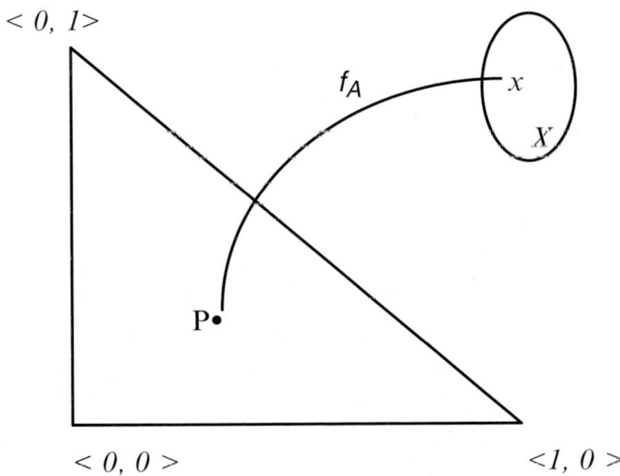

Fig. 16.1. Atanassov's geometrical interpretation of intuitionistic fuzzy sets

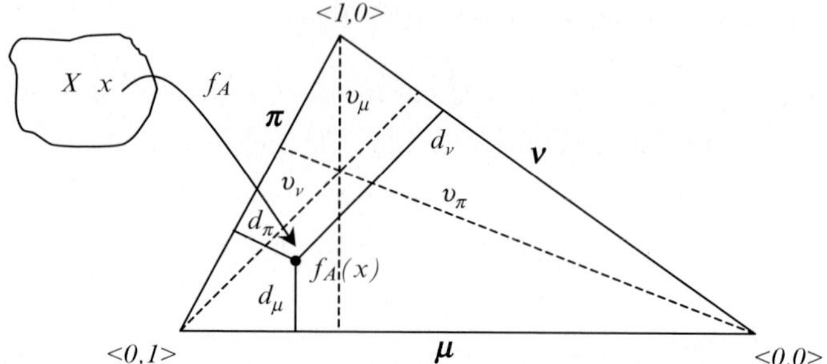

Fig. 16.2. Geometrical interpretation of some concepts in the intuitionistic fuzzy logics

We now assume that X is the universe set, and $A \subseteq X$ is a fixed set. We then define a function $f_A : X \to T$, as follows:

$$f_A(x) = p = \langle a, b \rangle \in T, \ \forall \, x \in X,$$

where

$$a, b \in [0, 1]$$

and

$$0 \leq a + b \leq 1 \,.$$

The coordinates a and b are defined by the relations:

$$a = \mu_A(x) = \frac{d_\mu}{v_\mu},$$

$$b = \nu_A(x) = \frac{d_\nu}{v_\nu}$$

and

$$1 - \pi_A(x) = \frac{d_\pi}{v_\pi} \,.$$

16.3.3 Operations on IFS

For any two IFSs, A and B, several relations and operations have been introduced (see references [1]-[6]). Here we shall introduce only those which are closely related to this article, namely the properties:

1. $A \subseteq B \Leftrightarrow \mu_A(x) \leq \mu_B(x)$ and $\nu_A(x) \geq \nu_B(x), \ \forall \, x \in X$
2. $A \supseteq B \Leftrightarrow B \subseteq A$
3. $A = B \Leftrightarrow \mu_A(x) = \mu_B(x)$ and $\nu_A(x) = \nu_B(x), \ \forall x \in X$
4. $\bar{A} = \{\langle x, \nu_A(x), \mu_A(x) \rangle \ : \ x \in X\}$
5. $A \cap B = \{\langle x, \min(\mu_A(x), \mu_B(x)), \max(\nu_A(x), \nu_B(x)) \rangle \ : \ x \in X\}$

6. $A \cup B = \{\langle x, \max\left(\mu_A(x), \mu_B(x)\right), \min\left(\nu_A(x), \nu_B(x)\right)\rangle : x \in X\}$

7. $A + B = \{\langle x, (\mu_A(x) + \mu_B(x) - \mu_A(x) \cdot \mu_B(x)), (\nu_A(x) \cdot \nu_B(x))\rangle : x \in X\}$

8. $A \cdot B = \{\langle x, (\mu_A(x) \cdot \mu_B(x)), (\nu_A(x) + \nu_B(x) - \nu_A(x) \cdot \nu_B(x))\rangle : x \in X\}$

9. $A @ B = \left\{\left\langle x, \frac{(\mu_A(x) + \mu_B(x))}{2}, \frac{(\nu_A(x) + \nu_B(x))}{2}\right\rangle : x \in X\right\}$

10. $A \$ B = \left\{\left\langle x, \sqrt{\mu_A(x) \cdot \mu_B(x)}, \sqrt{\nu_A(x) \cdot \nu_B(x)}\right\rangle : x \in X\right\}$

11. $A * B = \left\{\left\langle x, \frac{\mu_A(x) + \mu_B(x)}{2 \cdot (\mu_A(x) \cdot \mu_B(x) + 1)}, \frac{\nu_A(x) + \nu_B(x)}{2 \cdot (\nu_A(x) \cdot \nu_B(x) + 1)}\right\rangle : x \in X\right\}$

12. $A \triangleright \triangleleft B = \left\{\left\langle x, 2\frac{\mu_A(x) \cdot \mu_B(x)}{\mu_A(x) + \mu_B(x)}, 2\frac{\nu_A(x) \cdot \nu_B(x)}{\nu_A(x) + \nu_B(x)}\right\rangle : x \in X\right\}$,

where if $\mu_A(x) = \mu_B(x) = 0$, then $\frac{\mu_A(x) \cdot \mu_B(x)}{\mu_A(x) + \mu_B(x)} = 0$ and if $\nu_A(x) = \nu_A(x) = 0$, then $\frac{\nu_A(x) \cdot \nu_B(x)}{\nu_A(x) + \nu_B(x)} = 0$.

We introduce four operators, over IFS, with the following definitions.

Definition 16.3 *The* necessity *operator is defined as:*

$$\Box A = \{< x, \mu_A(x), 1 - \mu_A(x) > : x \in X\}$$

and the possibility *operator is defined as:*

$$\Diamond A = \{< x, 1 - \nu_A(x), \nu_A(x) > : x \in X\}$$

The above operators are similar to the operators of necessity and possibility that are defined in some modal logic, and they have no counterparts in the ordinary fuzzy set theory.

The operator below is an extension of the operators $\Box$ and $\Diamond$, but it can be extended even further.

Definition 16.4 *Let $a \in [0, 1]$ be a fixed number. Given an IFS A, the operator D_a is defined as follows:*

$$D_a(A) = \{< x, \mu_A(x) + a \cdot \pi_A(x), \nu_A(x) + (1 - a) \cdot \pi_A(x) > : x \in X\}.$$

Definition 16.5 *Let $\alpha, \beta \in [0, 1]$ and $\alpha + \beta \leq 1$. The operator $F_{\alpha,\beta}$, for the IFS, A is defined as:*

$$F_{\alpha,\beta}(A) = \{< x, \mu_A(x) + \alpha \cdot \pi_A(x), \nu_A(x) + \beta \cdot \pi_A(x) > : x \in X\}.$$

In Figs. 16.3 and 16.4, we give the geometric interpretations for the operators D_a and $F_{\alpha,\beta}$ respectively.

16.4 Implications on Fuzzy Sets

The definitions and notations can be found in [9, 11, 18, 20].

Let X denote a universe of discourse. Then, a fuzzy set A, in X, is defined as a set of ordered pairs $A = \{< x, \mu_A(x) > : x \in X\}$, where the function $\mu_A : X \to [0, 1]$ defines the degree of membership, of the element $x \in X$ [22].

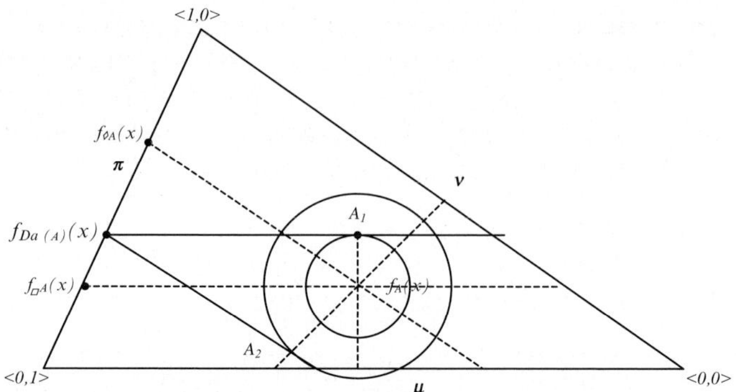

Fig. 16.3. The point $f_{D\alpha}(x)$ belongs to the segment defined by the points $f_{\square A}(x)$ and $f_{\diamond A}(x)$, and its exact position depends on the value of the arguments $\alpha \in [0,1]$

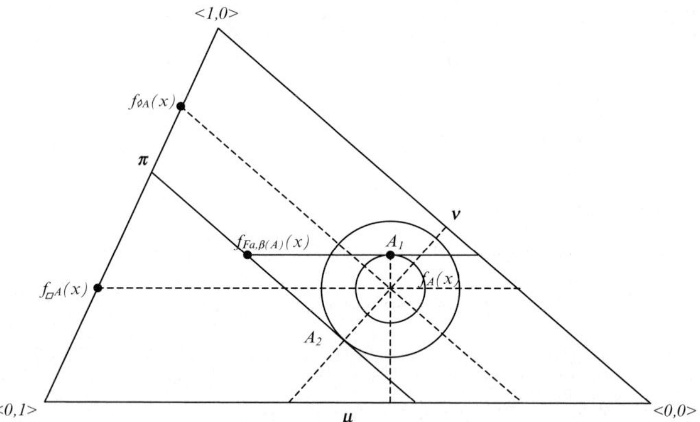

Fig. 16.4. The point $f_{F_{\alpha,\beta(A)}}(x)$ is an internal point of the triangle with vertices $f_A(x)$, $f_{\square A}(x)$ and $f_{\diamond A}(x)$. The point depends on the value of the argumets $\alpha, \beta \in [0,1]$, for which $\alpha + \beta \leq 1$

Definition 16.6 *A* *binary operation i, on the unit interval, (i.e. a $[0,1] \times [0,1] \to [0,1]$ mapping) is called a* fuzzy intersection, *if it is an extension of the classical Boolean intersection $i(a,b) \in [0,1]$, for every $a, b \in [0,1]$ where $i(0,0) = i(0,1) = i(1,0) = 0$ and $i(1,1) = 1$.*

A canonical model of fuzzy intersections is the family of triangular norms (briefly the $t-$norms).

Definition 16.7 *A $t-$norm is a function of the form:*

$$i : [0,1] \times [0,1] \to [0,1],$$

which is commutative, associative, non-decreasing, and $i(a,1) = a$, $a \in [0,1]$.

A $t-$norm is called *Archimedean*, if it is continuous, and for $a \in (0,1)$, $i(a,a) < a$ it is *nilpotent*, if it is continuous and $\forall a \in (0,1)$, there exists a $\nu \in \mathbb{N}$, such that $i\overbrace{(a,\ldots,a)}^{\nu} = 0$. So, the Archimedean norms appear in two forms, in the nilpotent and in the non-nilpotent ones. Moreover, those which are not nilpotent are called *strict*.

Definition 16.8 *A function* $n : [0,1] \to [0,1]$ *is called a* negation, *if it is non-increasing, i.e.* $n(a) \leq n(b)$, *if* $a \geq b$, *and* $n(0) = 1$, $n(1) = 0$.

A negation, n, is called *strict*, if and only if n is continuous and strictly decreasing $(n(a) < n(b)$, if $a > b$, for all $a, b \in [0,1])$. A strict negation, n, is called *strong*, iff it is self-inverse, i.e. $n(n(a)) = a$, for all $a \in [0,1]$. The most important, and widely used strong negation, is the standard negation $n_s : [0,1] \to [0,1]$, given by $n_S = 1 - a$.

A function $u : [0,1] \times [0,1] \to [0,1]$, satisfying the properties:

1. $u(a,0) = a$, for all $a \in [0,1]$,
2. $u(a,b) \leq u(c,d)$, if $a \leq c$ and $b \leq d$,
3. $u(a,b) = u(b,a)$, for all $a, b \in [0,1]$,
4. $u(u(a,b),c) = u(a,u(b,c))$, for all $a,b,c \in [0,1]$
   is called a *triangular conorm* (or $t-$conorm).

A *fuzzy implication*, g, is a function of the form:

$$g : [0,1] \times [0,1] \to [0,1],$$

which defines (for any possible truth values a, b, of some given fuzzy propositions p, q respectively) the truth value $I(a,b)$, of the conditional proposition "if p then q".

Function g should be an extension of the classical implication, from the domain $\{0,1\}$ to the domain $[0,1]$, of truth values in fuzzy logic.

Definition 16.9 *The* implication operator, *of the classical logic, it a mapping:*

$$m : \{0,1\} \times \{0,1\} \to \{0,1\},$$

which satisfies the conditions:

$$m(0,0) = m(0,1) = m(1,1) = 1 \; and \; m(1,0) = 0$$

These conditions are the least ones that we can demand, from a fuzzy implication operator. In other words, the fuzzy implications collapse to the classical implication, when the truth values are restricted to 0 and 1:

$$g(0,0) = g(0,1) = g(1,1) = 1 \; and \; g(1,0) = 0$$

One way of defining g in classical logic is to use the logic-formula:

$$g(a, b) = \bar{a} \vee b,$$

for all $a, b \in \{0, 1\}$.

Another way is to employ the formula:

$$g(a, b) = \max\{x \in \{0, 1\} : a \wedge x \leq b\},$$

for all $a, b \in \{0, 1\}$.

The extensions of these equations in fuzzy logic are, respectively:

$$g(a, b) = u(n(a), b), \tag{16.1}$$

$$g(a, b) = \sup\{x \in [0, 1] : i(a, x) \leq b\}. \tag{16.2}$$

for all $a, b \in [a, b]$, where u, i and n denote a fuzzy union, a continuous fuzzy intersection and a fuzzy negation, respectively. Furthermore, if u and i are dual with respect to n, we say that a $t-$conorm i and a $t-$conorm u are dual with respect to a fuzzy negation n, iff $n(i(a, b)) = u(n(a), n(b))$ and $n(u(a, b)) = i(n(a), n(b))$, for all $a, b \in [0, 1]$.

The fuzzy implications, that are obtained from (16.1), are usually referred to the literature as $S-$implications (S is often used for denoting $t-$conorms), and the fuzzy implications that are obtained from (16.2) are called $R-$implications.

Moreover, the formula $g(a, b) = \bar{a} \vee b$ may also be rewritten (due to the law of absorption of negation in classical logic), as either:

$$g(a, b) = \bar{a} \vee (a \wedge b)$$

or

$$g(a, b) = (\bar{a} \wedge \bar{b}) \vee b$$

The extensions of these equations, in fuzzy logic, are, respectively:

$$g(a, b) = u(n(a), i(a, b)), \tag{16.3}$$

$$g(a, b) = u(i(n(a), n(b)), b), \tag{16.4}$$

where u, i and n are required to satisfy the De Morgan laws.

The fuzzy implications that are obtained from (16.3) are called $QL-$implications, since they were originally employed in quantum logic.

Identifying various properties of the classical implication, and generalizing them appropriately, leads to the following properties, which may be viewed as reasonable axioms of fuzzy implications (see [18]):

(A1) $a \le b$ implies that $g(a, x) \ge g(b, x)$ (*monotonicity in first argument*). This means that the truth value of fuzzy implications increases, as the truth value of the antecedent increases.

(A2) $a \le b$ implies that $g(x, a) \le g(x, b)$ (*monotonicity in second argument*). This means that the truth value of of fuzzy implications increases, as the truth value of the consequent increases.

(A3) $g(a, g(b, x)) = g(b, g(a, x))$ (*exchange propery*). This is a generalization of the equivalence of $a \Rightarrow (b \Rightarrow x)$ and $b \Rightarrow (a \Rightarrow x)$, which holds for the classical implication.

(A4) $g(a, b) = g(n(b), n(a))$, for a fuzzy complement n (*contraposition*)

(A5) $g(1, b) = b$ (*neutrality of truth*)

(A6) $g(0, a) = 1$ (*dominance of falsity*)

(A7) $g(a, a) = 1$ (*identity*)

(A8) $g(a, b) = 1$, if and only if $a \le b$ (*boundary condition*)

(A9) g is a continuous function (*continuity*)

One can easily prove that all the $S-$implications fulfil axioms A1, A2, A3, A5, A6 and, when the negation is strong, A4. In addition, all the $R-$implications fulfil the axioms A1, A2, A5, A6 and A7.

Let now G be the fuzzy implications that are obtained from (16.4), when $a \le b$, i.e. $G(a, b) = u(i(n(a), n(b)), b)$, when $a \le b$. We also defined $G(a, b) = 0$, when $a > b$. Thus, we have a class of fuzzy implications, which we define as follows (see also [16]):

Definition 16.10 *Consider the following function* $G : [0, 1] \times [0, 1] \to [0, 1]$

$$G(a, b) = [1 - sg(a - b)] \cdot u(i(n(a), n(b)), b), \tag{16.5}$$

for all $a, b \in [0, 1]$ *where* u, i *and* n *are required to satisfy De Morgan laws, n is a strong negation and* $sg(x) = \begin{cases} 1, \text{ if } x > 0 \\ 0, \text{ otherwise} \end{cases}$. *Then* $G(a, b)$ *is a class of fuzzy implications, which collapse to the classic implication when the truth values are restricted to* 0 *and* 1, *i.e.* $G(0, 0) = G(0, 1) = G(1, 1) = 1$ *and* $G(1, 0) = 0$.

Choosing in (16.5) as a fuzzy union the standard one, $u(a, b) = \mathtt{max}(a, b) = a \vee b$, as a fuzzy intersection the standard $i(a, b) - \mathtt{min}(a, b) - a \wedge b$, and as a negation the standard fuzzy negation $n_S = 1 - a$, we will have the following fuzzy implication, which we introduce in the proposition below.

Proposition 16.1 *Let* g *be a function defined as follows for all* $a, b \in [0, 1]$

$$g_h(a, b) = [1 - sg(a - b)] \cdot [(1 - b) \vee b],$$

where $sg(x)$ *is given in the previous definition. Then* $g_h(a, b)$ *is a fuzzy implication, which collapses to the classical implication when the truth values are restricted between* 0 *and* 1, *i.e* $g_h(0, 0) = g_h(0, 1) = g_h(1, 1) = 1$ *and* $g_h(1, 0) = 0$.

A graphical interpretation of fuzzy implication $g_h(a, b)$ is shown in Fig. 16.5.

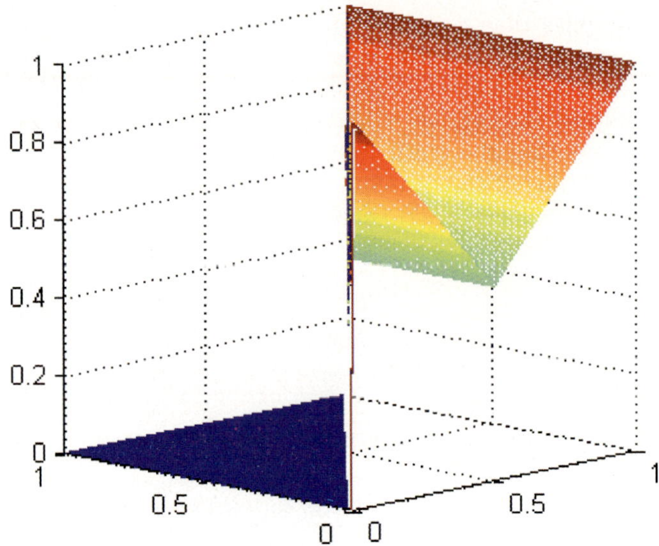

Fig. 16.5. The graphical representation of the fuzzy implication g_h

16.5 Implications on Intuitionistic Fuzzy Sets (IFSs)

Let S be a set of propositions p. Let also a valuation function, V, be defined over S, in the following way:

$$V(p) = \; <\mu(p), \nu(p)>,$$

where $\mu(p) + \nu(p) \leq 1$.

The function $V : S \to [0,1] \times [0,1]$ gives the truth and falsity degres of all propositions in S. The valuation function V assigns to the logical truth T, $V(T) =< 1, 0 >$ and to the logical falsity F, $V(F) =< 0, 1 >$. Also,

$$V(\dashv p) =< \nu(p), \mu(p) >,$$

$$V(p \wedge q) =< \min(\mu(p), \mu(q)), \max(\nu(p), \nu(q)) >,$$

$$V(p \vee q) =< \max(\mu(p), \mu(q)), \min(\nu(p), \nu(q)) > .$$

Now, considering the propositions p and q, the two most acceptable definitions of implications, in IFS, are the following:

1. The (max-min)-implication:

$$V(p \Rightarrow q) =< \max(\nu(p), \mu(q)), \min(\mu(p), \nu(q)) >$$

2. The $sg-$implication:

$$V(p \Rightarrow q) = \langle 1 - (1 - \mu(q)) \cdot sg(\mu(p) - \mu(q)), \nu(q) \cdot sg(\mu(p)$$
$$-\mu(q)) \cdot sg(\nu(q) - \nu(p))\rangle,$$

where $sg(x) = \begin{cases} 1, \text{ if } x > 0 \\ 0, \text{ otherwise} \end{cases}$

Implications in IFS collapse to the classical implication, when the truth values are restricted between 0 and 1, i.e.

$V(p)$	$V(q)$	$V(p \Rightarrow q)$
$< 0, 1 >$	$< 0, 1 >$	$< 1, 0 >$
$< 0, 1 >$	$< 1, 0 >$	$< 1, 0 >$
$< 1, 0 >$	$< 0, 1 >$	$< 0, 1 >$
$< 1, 0 >$	$< 1, 0 >$	$< 1, 0 >$

The following proposition introduces an implication in IFS, which we call "$\rightarrow$ implication" (see [14]).

Proposition 16.2 *Let A and B be two IFSs, and denote $A \rightarrow B = V(A \Rightarrow B)$. Then,*

$$A \rightarrow B = \begin{cases} \{< x, max(\mu_B(x), \nu_B(x), min(\mu_B(x), \nu_B(x)) >: x \in X\}, \\ \text{if } \mu_A(x) \leq \mu_B(x) \text{ and } \nu_A(x) \geq \nu_B(x) \\ \\ \{< x, 0, 1 >: x \in X\}, \text{otherwise} \end{cases}$$

The "$\rightarrow$ implication" collapses to the classical one, when the truth values are restricted between 0 and 1, i.e.

$V(A)$	$V(B)$	$V(A \rightarrow B)$
$< 0, 1 >$	$< 0, 1 >$	$< 1, 0 >$
$< 0, 1 >$	$< 1, 0 >$	$< 1, 0 >$
$< 1, 0 >$	$< 0, 1 >$	$< 0, 1 >$
$< 1, 0 >$	$< 1, 0 >$	$< 1, 0 >$

Also, the "$\rightarrow$ implication" will be an extension, in IFS, of the fuzzy implication g_h (see [16]) (where g_h is the fuzzy implication, which was introduced in Proposition 1).

Definition 16.11 *An IFS, A, is called an* Intuitionistic Fuzzy Tautological Set *(IFTS), iff $\mu_A(x) \geq \nu_A(x)$, $\forall x \in X$. (see [1]).*

In the following theorems (1-4) we introduce the relations, which the "$\rightarrow$ implication" satisfies. These theorems are easily proved (see [14]).

Theorem 16.1 *Let A, B and C be IFSs. Then the following sets are IFTS:*

1. $A \to A$
2. $A \cap B \to A$
3. $A \cap B \to B$
4. $\bar{\bar{A}} \to A$
5. $A \to (A \cup B)$
6. $B \to (A \cup B)$
7. $A \to (B \to A)$, when $B \subseteq A$
8. $A \to B$, if $A \subseteq B$
9. $(\bar{A} \to \bar{B}) \to [(\bar{A} \to B) \to A]$, when $max(\mu_B(x), \nu_B(x)) \leq \mu_A(x)$ and $min(\mu_B(x), \nu_B(x)) \geq \nu_A(x)$
10. $[A \cap (A \to B)] \to B$
11. $[(A \to B) \cap \bar{B}] \to A$
12. $[A \to (B \to C)] \to [(A \to B) \to (A \to C)]$ when $A \subseteq C$ and $max(\mu_B(x), \nu_B(x)) \leq max(\mu_C(x), \nu_C(x))$, $min(\mu_B(x), \nu_B(x)) \geq min(\mu_C(x), \nu_C(x))$
13. $[(A \to B) \cap (B \to C)] \to (A \to C)$, when $A \subseteq C$
14. $(A \to C) \to [(B \to C) \to ((A \cup B) \to C))]$.

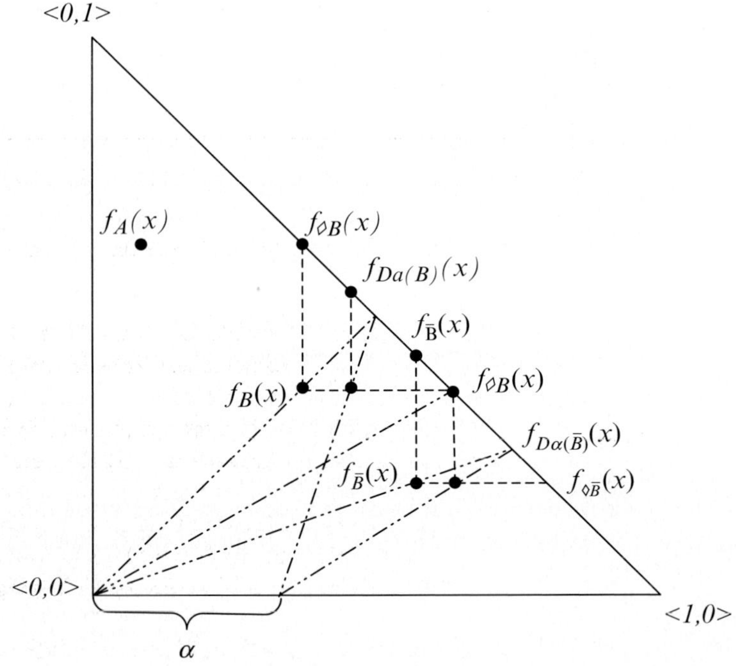

Fig. 16.6. The geometric interpretation of Theorem 16.4, statements 1. and 2

Theorem 16.2 *Let $A_1, A_2, \ldots, A_\nu$ be IFSs. If $max(\mu_{A_{\nu-1}}(x), \nu_{A_{\nu-1}}(x)) \leq \mu_{A_\nu}(x)$ and $min(\mu_{A_{\nu-1}}(x), \nu_{A_{\nu-1}}(x)) \geq \nu_{A_\nu}(x)$, then $(((A_1 \to A_2) \to \ldots) \to A_{\nu-1}) \to A_\nu$ is an IFTS.*

Theorem 16.3 *Let A and B be two IFSs. Then,*

1. $\Box(A \to B) \subseteq (\Box A) \to (\Box B)$ and
2. $\Diamond(A \to B) \supseteq (\Diamond A) \to (\Diamond B)$ when $\mu_A(x) \leq \mu_B(x)$.

Theorem 16.4 *Let A and B be two IFSs. Then, the following statements are true:*

1. If $A \subseteq B$, then $D_a(A \to B) \supseteq D_a(B)$.
2. If $A \not\subseteq B$, then $D_a(A \to B) \subseteq D_a(B)$.
3. If $A \subseteq B$, then $F_{\alpha,\beta}(A \to B) \supseteq F_{\alpha,\beta}(B)$.
4. If $A \not\subseteq B$, then $F_{\alpha,\beta}(A \to B) \subseteq F_{\alpha,\beta}(B)$.

The geometric interpretation of Theorem 16.4 is given in Fig. 16.6 (statements 1. and 2.) and in Fig. 16.7 (statements 3. and 4.).

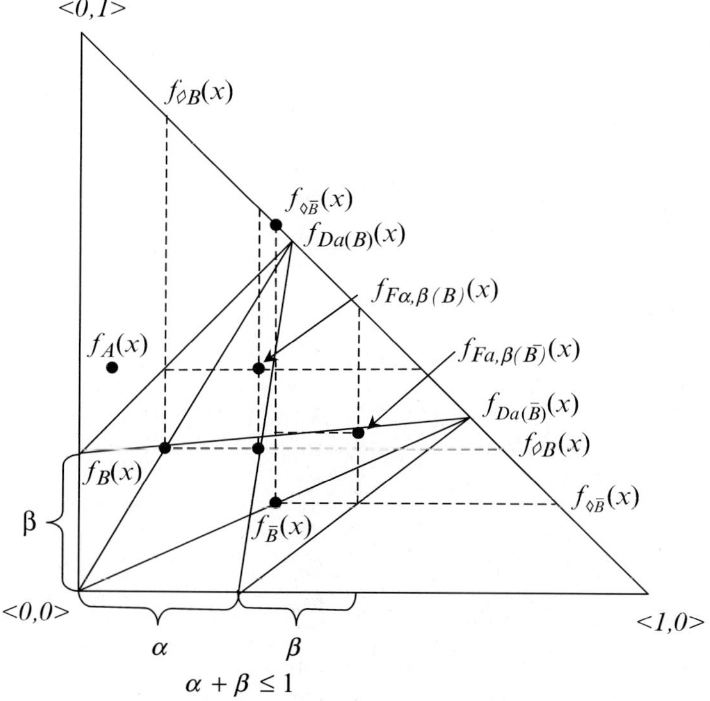

Fig. 16.7. The geometric interpretation of Theorem 16.4, statements 3. and 4.

16.6 Conclusion

This chapter has presented a fuzzy implication as well as its extension in intuitionistic fuzzy sets (IFSs). We pointed out that there is a strong connection between intuitionistic fuzzy sets and L-fuzzy sets. Therefore, the aforementioned implication might be expressed equivalently using either intuitionistic fuzzy sets or lattice fuzzy (L-fuzzy) sets.

References

1. Atanassov K (1998) Intuitionistic Fuzzy Sets: Theory and Applications. Springer-Verlag, Berlin, Germany
2. Atanassov K (1986) Intuitionistic fuzzy sets. Fuzzy Sets Systems 20(1):87–96
3. Atanassov K (1989) More on intuitionistic fuzzy sets. Fuzzy Sets and Systems 33(1):37–46
4. Atanassov K (1994) New operations defined over the intuitionistic fuzzy sets. Fuzzy Sets and Systems 61(2):137–142
5. Atanassov K, Gargov G (1989) Interval-valued intuitionistic fuzzy set. Fuzzy Sets and Systems 31(3):343–349
6. Atanassov K, Gargov G (1998) Elements of intuitionistic fuzzy logic. part I. Fuzzy Sets and Systems 95(1):39–52
7. Atanassov K (2005) Answer to D. Dubois, S. Gottwald, P. Hajek, J. Kacprzyk and H. Prade's paper "Terminological difficulties in fuzzy set theory – the case of intuitionistic fuzzy sets". Fuzzy Sets and Systems 156(3):496–499
8. Cornelis C, Deschrijver G, Kerre EE (2004) Implication in intuitionistic fuzzy and interval-valued fuzzy set theory: construction, classification, application. Intl J Approximate Reasoning 35(1):55–95
9. Dubois D, Prade H (1980) Fuzzy Sets and Systems: Theory and Applications. Academic Press, New York, USA
10. Dubois D, Gottwald S, Hajek P, Kacprzyk J, Prade H (2005) Terminological difficulties in fuzzy set theory – the case of "intuitionistic fuzzy sets". Fuzzy Sets and Systems 156(3):485–491
11. Fodor J, Roubens M (1994) Fuzzy preference modelling and multicriteria decision support. In: Theory and Decision Library. Kluwer Academic, Dordrecht
12. Goguen J (1967) L-fuzzy sets. J Math Analysis and Applications 18(1):145–174
13. Grzegorzewski P (2004) Distances between intuitionistic fuzzy sets and/or interval-valued fuzzy sets based on the Hausdorff metric. Fuzzy Sets and Systems 148(2):319–328
14. Hatzimichailidis AG, Papadopoulos BK (2002) A new implication in the intuitionistic fuzzy sets. NIFS 8(4):79–104
15. Hatzimichailidis AG, Papadopoulos BK (2005) A new geometrical interpretation of some concepts in the intuitionistic fuzzy logics. NIFS 11(2)
16. Hatzimichailidis AG, Kaburlasos VG, Papadopoulos BK (2006) An implication in fuzzy sets. In: Proc World Congress Computational Intelligence (WCCI) FUZZ-IEEE Program pp 203–208
17. Kaburlasos VG (2006) Towards a Unified Modeling and Knowledge-Representation Based on Lattice Theory — Computational Intelligence and Soft Computing Applications, ser Studies in Computational Intelligence 27. Springer, Heidelberg, Germany

18. Klir G, Yuan B (1995) Fuzzy Sets and Fuzzy Logic Theory and Applications. Prentice Hall, Upper Saddle River, New Jersey, USA
19. Liu Ying-Ming, Luo Mao-Kang (1997) Fuzzy Topology, Advances in Fuzzy Systems-Applications and Theory 9. Word Scientific
20. Nguyen HT, Walker EA (1997) A First Course in Fuzzy Logic. CRC Press, USA
21. Wang Guo-Jun, He Y (2000) Intuitionistic fuzzy sets and L-fuzzy sets. Fuzzy Sets and Systems 110(2):271–274
22. Zadeh LA (1965) Fuzzy sets. Information and Control 8(3):338–353

A Family of Multi-valued t-norms and t-conorms

Athanasios Kehagias

Faculty of Engineering, Aristotle University of Thessaloniki
`kehagiat@auth.gr`

Summary. Given a lattice $(X, \leq, \wedge, \vee)$ we define a *multi-valued* operation $\wedge^Q$ which is analogous to a t-norm (i.e. it is commutative, associative, has one as a neutral element and is monotone). The operation is parametrized by the set Q, hence we actually obtain an entire family of such *multi-valued* t-norms. Similarly we define a family of multi-valued t-conorms $\vee^P$. We show that, when P, Q are chosen appropriately, $\wedge^Q, \vee^P$ (along with a standard negation) form a de Morgan pair. Furthermore $\wedge^Q, \vee^P$ are order generating and $(X, \leq, \wedge^Q, \vee^P)$ is a *superlattice*, i.e. a multi-valued analog of a lattice.

17.1 Introduction

In this paper we generalize the concepts of t-norm and t-conorm. While many variants of t-norm and t-conorm have appeared in the literature, they always are *single-valued* functions. The current work is substantially different because it considers a *multi-valued* generalization.

As is well known, a t-norm is an extension of the classical logical operator AND to the realm of fuzzy logic. While in the classical case AND is a function from $\{0, 1\} \times \{0, 1\}$ to $\{0, 1\}$ (or, equivalently, from $\{\text{True,False}\} \times \{\text{True,False}\}$ to $\{\text{True,False}\}$), a t-norm is a function from $[0, 1] \times [0, 1]$ to $[0, 1]$. The interval $[0, 1]$ can be an interval of reals or, more generally, a lattice with bottom element 0 and top element 1. Recently considerable attention has been paid to t-norms for particular types of lattices, for instance the lattice of real intervals [2], the lattice of type-2 fuzzy sets [23] etc. All of these generalizations are concerned with the domain and range of the t-norm. Similar remarks hold for t-conorms, which extend from $\{0, 1\}$ to $[0, 1]$ the logical OR operator[1].

Many t-norms and t-conorms have been presented in the fuzzy literature but the above remarks apply to practically all of them. This also holds true for t-norms and t-conorms which apply to interval-valued fuzzy sets: in this

---

[1] Under another interpretation, a t-norm generalizes set intersection and a t-conorm generalizes set union.

A. Kehagias: *A Family of Multi-valued t-norms and t-conorms*, Studies in Computational Intelligence (SCI) **67**, 341–360 (2007)
`www.springerlink.com`

case a t-norm takes as input pairs of intervals and produces as output a single interval. The case with which we deal in the current work is significantly different. We are interested in *multi-valued* t-norms and t-conorms; a simple example would be a t-norm which takes as input two reals and produces as output an *interval* of reals. In this paper we discuss the generalization of this example.

Our motivation for studying multi-valued logical connectives is the following: fuzzy logic is a calculus of uncertain reasoning; but it seems that all the uncertainty is concentrated in the truth values of the logical propositions; the case where uncertainty is associated with the actual operation of the logical connectives has not been studied in the past. The introduction of multi-valued logical connectives is a reasonable way to introduce this type of uncertainty. On the other hand, much of the work presented concerns *intervals* (either of reals or, more generally, of lattice elements) which has been a common theme in the fuzzy and computationbal intelligenece literature; see for instance the use of "hyperboxes" (lattice intervals) in [11, 12, 17] for theoretical aspects and [1, 11, 18] for applications. We present our work in the general framework of lattice theory, which is quite popular for the general analysis of fuzzy systems [6, 8, 19].

There is a considerable literature on algebras equipped with multi-valued operations; see the book [5]. In this area the multi-valued operations are called *hyperoperations* and they yield *hyperalgebras* such as hypergroups [3, 4], hyperrings [15, 22], hyperlattices [14, 16, 20, 21] etc. Using similar terminology we will speak of the *hyper-t-norm*, the *hyper-t-conorm* etc. It is not stressed often enough, but a basic hyperoperation which finds frequent application in Computational Intelligence is *clustering* [17] which takes as input two (or a few) *points* to produce as output a *region*, i.e. a *set of points*.

Among all uni-valued t-norms, the min t-norm $\wedge$ has a special place. For example, it is the only idempotent t-norm. Also, in a certain sense, it is the simplest extension of AND. Finally, and perhaps more importantly, it is the only t-norm which *generates an order*, i.e. the double implication

$$aTb = a \Leftrightarrow a \leq b$$

only holds for $T = \wedge$. Similar remarks hold for the max t-conorm $\vee$. In the current work we present a family of multi-valued logical connectives which make essential use of $\wedge$ and $\vee$; our construction can be applied to other t-norms and t-conorms but the results are not as satisfactory. Let us mention however that the current generalization is only one of several ways to define multi-valued t-norms and t-conorms; we have presented other possibilities in [10, 20, 21].

The plan of this paper is as follows. In Section 17.2 we present some preliminary concepts. The rest of the paper assumes a basic de Morgan algebra $(X, \leq, \wedge, \vee, ')$ and constructs on it various algebraic structures. In Section 17.3 we introduce the basic objects of our study, namely the multi-valued

$\wedge^Q$ t-norm and $\vee^P$ t-conorm. In Section 17.4 we study the multi-valued *S-implication* obtained from $\vee^P$. In Section 17.5 we show that $\wedge^Q$ and $\vee^P$ generate a structure analogous to a lattice, the so-called (P, Q)-*superlattice*. In Sect. 17.6 we obtain additional results for the case when $(X, \leq, \wedge, \vee,')$ is a *Boolean* algebra. Finally, we summarize our results in Section 17.7.

17.2 Preliminaries

In this section we review some fundamental concepts which will be used in the main part of the paper; we also present some well known propositions (we omit their proofs, which can be found in standard texts [7]).

Given a set X, the *power set* of X will be denoted by $\mathbf{P}(X)$ and will be defined to be the set of all (crisp) subsets of X. We can equip X with an order relationship $\leq$ and thus obtain a *partially ordered set* $(X, \leq)$. If for every pair $x, y \in X$ the $\inf(x, y)$ and $\sup(x, y)$ exist, we say that $(X, \leq)$ is a *lattice*. We denote $\inf(x, y)$ by $x \wedge y$ and $\sup(x, y)$ by $x \vee y$; then $\wedge, \vee$ are binary operations on X and we sometimes say that $(X, \leq, \wedge, \vee)$ or $(X, \wedge, \vee)$ is a lattice. We also use the notation $x_1 \wedge x_2 \wedge ... \wedge x_N$ to denote the infimum of $x_1, x_2, ..., x_N$ (the $\wedge$ operation is associative) and $\wedge P$ to denote $\wedge_{x \in P} x$, the infimum of a set $P \subseteq X$ (if such an infimum exists); similarly we use $x_1 \vee x_2 \vee ... \vee x_N$ and $\vee P$. A lattice is called *distributive* iff $\forall a, b, c \in X$ it is

$$a \wedge (b \vee c) = (a \wedge b) \vee (a \wedge c), \qquad a \vee (b \wedge c) = (a \vee b) \wedge (a \vee c).$$

We will use the following proposition repeatedly in what follows.

Proposition 17.1 *In a distributive lattice $(X, \leq, \vee, \wedge)$ the following properties hold for all $a, b, x, y \in X$ such that $x \leq y, a \leq b$:*

$$a \vee [x, y] = [a \vee x, a \vee y]; \ [a, b] \vee [x, y] = [a \vee x, b \vee y];$$
$$a \wedge [x, y] = [a \wedge x, a \wedge y]; \ [a, b] \wedge [x, y] = [a \wedge x, b \wedge y].$$

We now introduce several "substructures" within the lattice $(X, \leq, \wedge, \vee)$.

1. A *sublattice* of a lattice is a set $Y \subseteq X$ such that: $x, y \in Y \Rightarrow x \wedge y \in Y, x \vee y \in Y$.
2. A *convex sublattice* is a set $Y \subseteq X$ such that: if $x, y \in Y$ then $[x \wedge y, x \vee y] \subseteq Y$. We denote the set of all convex sublattices of X by $\mathbf{C}(X)$.
3. A *filter* is a set $Y \subseteq X$ which satisfies the following two conditions: (a) if $x \in Y, y \in X$ and $x \leq y$, then $y \in Y$ (b) if $x, y \in Y$ then $x \wedge y \in Y$.
4. An *interval* in the lattice $(X, \leq, \wedge, \vee)$ is a set of the form $\{x : a \leq x \leq b\}$, where $a, b \in X$ are the *endpoints* of the interval; we denote this interval by $[a, b]$. We denote the set of all intervals of X by $\mathbf{I}(X)$.

Every filter is a convex sublattice; in a lattice with bottom element 0 and top element 1, every interval of the form $[q, 1]$ is a filter.

Given a poset (or lattice) $(X, \leq)$ we can define two "order-like" relations on $\mathbf{P}(X)$ (i.e. they hold between subsets of X).

1. $A \leq_1 B$ means that: $\begin{cases} \exists a \in A \text{ such that } \forall b \in B \text{ we have } a \leq b \\ \exists b \in B \text{ such that } \forall a \in A \text{ we have } a \leq b \end{cases}$.
2. $A \leq_2 B$ means that: $\forall a \in A,\, b \in B$ we have $a \vee b \in B$ and $a \wedge b \in A$.

These relations are not independent; also (under appropriate restrictions) they are orders.

Proposition 17.2 *For every $A, B \in \mathbf{C}(X)$, we have $A \leq_1 B \Rightarrow A \leq_2 B$.*

Proposition 17.3 *The relations $\leq_i (i = 1, 2)$ are orders on $\mathbf{C}(X)$ (and, a fortiori, on $\mathbf{I}(X)$).*

In the rest of the paper we will use a *generalized de Morgan lattice* defined as follows.

Definition 17.1 *A generalized deMorgan lattice is a structure $(X, \leq, \vee, \wedge, ')$, where $(X, \leq, \vee, \wedge)$ is a complete distributive lattice with minimum element 0 and maximum element 1; the symbol $'$ denotes a unary operation ("negation"); and the following properties are satisfied.*

1. *For all $x \in X$, $Y \subseteq X$ we have $x \wedge (\vee_{y \in Y} y) = \vee_{y \in Y}(x \wedge y)$, $x \vee (\wedge_{y \in Y} y) = \wedge_{y \in Y}(x \vee y)$ (Complete distributivity).*
2. *For all $x \in X$ we have: $(x')' = x$ (Negation is involutory).*
3. *For all $x, y \in X$ we have: $x \leq y \Rightarrow y' \leq x'$ (Negation is order reversing).*
4. *For all $Y \subseteq X$ we have $(\vee_{y \in Y} y)' = \wedge_{y \in Y} y'$, $(\wedge_{y \in Y} y)' = \vee_{y \in Y} y'$ (Complete deMorgan laws).*

In Section 17.6 we will make a stronger assumption, that $(X, \leq, \vee, \wedge, ')$ is a *generalized Boolean lattice*, i.e. it satisfies the following.

Definition 17.2 *A generalized Boolean lattice is a generalized deMorgan lattice $(X, \leq, \vee, \wedge, ')$ in which every $x \in X$ satisfies: $x \vee x' = 1$, $x \wedge x' = 0$.*

Notation 17.1 *Given a set $A \subseteq X$ we will denote the set of all negated elements of A by $A' = \{x : x' \in A\}$.*

Proposition 17.4 *Let $A = [a_1, a_2]$; then $A' = [a_2', a_1']$.*

We now turn to t-norms, t-conorms and their multi-valued extensions. Recall that a t-norm on a general lattice $(X, \leq)$ with 0 and 1 is a function $T : X \times X \to X$ which is commutative, associative and satisfies for all $a, b, c \in X$ the following

$$\text{absorption} : aT1 = a, \tag{A1}$$

$$\text{monotonicity} : a \leq b \Rightarrow aTc \leq bTc; \tag{A2}$$

similarly a t-conorm is a function $S : X \times X \to X$ which is commutative, associative and satisfies for all $a, b, c \in X$ the following

$$\text{absorption} : aS0 = a, \tag{B1}$$

$$\text{monotonicity} : a \leq b \Rightarrow aSc \leq bSc. \tag{B2}$$

Now, t-norms and t-conorms are examples of *operations*, i.e. *uni-valued* mappings from $X \times X$ to X; they map pairs of elements to elements. The following notations are standard in the hyperoperations literature:

1. if $*$ is an operation, for any $a \in X$ and $A, B \subseteq X$ we define $a * B = \cup_{b \in B} \{a * b\}$, $A * B = \cup_{a,b \in B} \{a * b\}$.
2. if $*$ is a hyperoperation, for any $a \in X$ and $A, B \subseteq X$ we define $a * B = \cup_{b \in B} a * b$, $A * B = \cup_{a,b \in B} a * b$ (remember that $a * b$ is a *set*).

Now we are ready to define hyper-t-norms and hyper-t-conorms by generalizing properties A1, A2 and B1, B2 to their multi-valued analogs.

1. A *hyper-t-norm* is a multi-valued map $\mathbf{T} : X \times X \to \mathbf{P}(X)$ which is commutative, associative and satisfies for all $a, b, c \in X$ the following

$$\text{absorption} : a\mathbf{T}1 \ni a, \tag{C1}$$

$$\text{monotonicity} : a \leq b \Rightarrow a\mathbf{T}c \leq b\mathbf{T}c. \tag{C2}$$

   (where $\leq$ is an order relationship on the range of $\mathbf{T}$).
2. A *hyper-t-conorm* is a multi-valued map $\mathbf{S} : X \times X \to \mathbf{P}(X)$ which is commutative, associative and satisfies for all $a, b, c \in X$ the following $a \leq b \Rightarrow a\mathbf{T}c \leq b\mathbf{T}c.$

$$\text{absorption} : a\mathbf{S}0 \ni a, \tag{D1}$$

$$\text{monotonicity} : a \leq b \Rightarrow a\mathbf{S}c \leq b\mathbf{S}c. \tag{D2}$$

Conditions C1 and D1 are straightforward generalizations of A1 and B1, respectively. Conditions C2 and D2 are more subtle. Note that in both C2 and D2 $\leq$ *is an order relationship between sets* (such as the above defined $\leq_1, \leq_2$ etc.). To complete the definitions of hyper-t-norm and hyper t-conorm, we must specify which order relationship we are using. There are several options for $\leq$ and, furthermore, it is only required that $\leq$ is an order *on the range of* $\mathbf{T} / \mathbf{S}$. As will be seen in Section 17.3, this will turn out to be significant when we examine specific hyperoperations with a restricted range.

17.3 The $\wedge^Q$ hyper-t-norm and the $\vee^P$ hyper-t-conorm

17.3.1 The $\wedge^Q$ hyper-t-norm

We define a hyperoperation which, under suitable restrictions, has the properties of a multivalued t-norm.

Definition 17.3 *Let Q be some subset of X and define the hyperoperation $\wedge^Q$ as follows:*

$$\forall a, b \in X : a \wedge^Q b \doteq \{a \wedge b \wedge q : q \in Q\}.$$

Remark 17.1. Hence we could write $a \wedge^Q b = a \wedge b \wedge Q$.

We will now show that under suitable conditions $\wedge^Q$ is a hyper-t-norm, i.e. it is commutative, associative and satisfies C1, C2. We will proceed in several steps. First we show that $\wedge^Q$ is *always* commutative and associative.

Proposition 17.5 *Take any set $Q \subseteq X$. Then, for all a, b, c we have*

$$a \wedge^Q b = b \wedge^Q a; \quad \left(a \wedge^Q b\right) \wedge^Q c = a \wedge^Q \left(b \wedge^Q c\right).$$

Proof. The first part is obvious. For the second, take any $x \in \left(a \wedge^Q b\right) \wedge^Q c$; then exists some y such that

$$y \in a \wedge^Q b \ \text{ and } \ x \in y \wedge^Q c;$$

also, $y \in a \wedge^Q b$ means that there exists $q_1 \in Q$ such that $y = a \wedge b \wedge q_1$; and $x \in y \wedge^Q c$ means that exists $q_2 \in Q$ such that $x = y \wedge c \wedge q_2$. Hence

$$x = (a \wedge b \wedge q_1) \wedge c \wedge q_2 = a \wedge (b \wedge c \wedge q_1) \wedge q_2.$$

But the last part of the above equality shows that $x \in a \wedge^Q \left(b \wedge^Q c\right)$. Hence $\left(a \wedge^Q b\right) \wedge^Q c \subseteq a \wedge^Q \left(b \wedge^Q c\right)$; similarly we can show that $a \wedge^Q \left(b \wedge^Q c\right) \subseteq \left(a \wedge^Q b\right) \wedge^Q c$; and so $a \wedge^Q \left(b \wedge^Q c\right) = \left(a \wedge^Q b\right) \wedge^Q c$. $\square$

Next we give a necessary and sufficient condition on Q for C1 to hold.

Proposition 17.6 *We have:* $\left(\forall a : a \in a \wedge^Q 1\right) \Leftrightarrow (1 \in Q)$.

Proof. If $1 \in Q$, then $a = a \wedge 1 \wedge 1 \in a \wedge^Q 1$. On the other hand, if $\forall a : a \in a \wedge^Q 1$, then $1 \in 1 \wedge^Q 1$, i.e. exists $q_1 \in Q$ such that $1 = 1 \wedge 1 \wedge q_1 = q_1$. $\square$

Finally we examine the question of monotonicity (i.e. condition C2). To this end we need an auxiliary proposition, regarding the nature of $a \wedge^Q b$.

Proposition 17.7 *Take any $Q \subseteq X$ such that $1 \in Q$. Then we have the following*

1. *(Q is a convex sublattice)* $\Leftrightarrow$ *$\left(\forall a, b \in X : a \wedge^Q b \text{ is a convex sublattice}\right)$;*
2. *(Q is a filter)* $\Rightarrow$ *$\left(\forall a, b \in X : a \wedge^Q b \text{ is a convex sublattice}\right)$;*
3. *(Q is an interval)* $\Leftrightarrow$ *$\left(\forall a, b \in X : a \wedge^Q b \text{ is an interval}\right)$.*

Proof. For 1, suppose Q is a convex sublattice. Take any $a, b \in X$, then take any $x, y \in a \wedge^Q b$; i.e. $x = a \wedge b \wedge q_1, q_1 \in Q$; and $y = a \wedge b \wedge q_2, q_2 \in Q$. Now consider $[x \wedge y, x \vee y]$; we have

$$[x \wedge y, x \vee y] = [a \wedge b \wedge (q_1 \wedge q_2), a \wedge b \wedge (q_1 \vee q_2)]$$
$$= a \wedge b \wedge [q_1 \wedge q_2, q_1 \vee q_2] \subseteq a \wedge b \wedge Q = a \wedge^Q b.$$

(since Q is a convex sublattice, $[q_1 \wedge q_2, q_1 \vee q_2] \subseteq Q$). Hence $a \wedge^Q b$ is a convex sublattice. Conversely, if $\forall a, b \in X : a \wedge^Q b$ is a convex sublattice, then $1 \wedge^Q 1 = 1 \wedge 1 \wedge Q = Q$ is a convex sublattice. For 2, simply note that every filter is a convex sublattice. For 3, suppose $Q = [q_1, q_2]$. Then

$$a \wedge^Q b = a \wedge b \wedge [q_1, q_2] = [a \wedge b \wedge q_1, a \wedge b \wedge q_2]$$

is an interval; and conversely, if $\forall a, b \in X : a \wedge^Q b$ is an interval, then $1 \wedge^Q 1 = 1 \wedge 1 \wedge Q = Q$ is an interval. $\square$

Remark 17.2. In fact, the third part of Proposition 17.7 can be strengthened. As will be seen in the sequel, we are mainly interested in the Q's which contain 1; if such a Q is also an interval, then $Q = [q, 1]$. Then we have the following.

Proposition 17.8 *If* $Q = [q, 1]$ *then* $a \wedge^Q b = [a \wedge b \wedge q, a \wedge b]$ *(* $\forall a, b \in X$ *).*

Proof. $a \wedge^Q b = a \wedge b \wedge [q, 1] = [a \wedge b \wedge q, a \wedge b]$. $\square$

Under certain conditions $\leq_1, \leq_2$ are orders *on the collection* $\{a \wedge^Q b\}_{a,b \in X}$.

Proposition 17.9 *Take any* $Q \subseteq X$ *such that* $1 \in Q$.

1. *If* Q *is a convex sublattice, then* $\leq_2$ *is an order on* $\{a \wedge^Q b\}_{a,b \in X}$.
2. *If* Q *is an interval, then* $\leq_1, \leq_2$ *are orders on* $\{a \wedge^Q b\}_{a,b \in X}$.

Proof. First we prove 1. Suppose Q is a convex sublattice.

i. Pick any $a, b \in X$ and any $x, y \in a \wedge^Q b$. Then $a \wedge^Q b$ is a convex sublattice (Prop. 17.7) and so $x \wedge y \in a \wedge^Q b$ and $x \vee y \in a \wedge^Q b$ and so $a \wedge^Q b \leq_2 a \wedge^Q b$.

ii. Pick any $a, b, c, d \in X$ such that $a \wedge^Q b \leq_2 c \wedge^Q d$ and $c \wedge^Q d \leq_2 a \wedge^Q b$. Now choose any $x \in a \wedge^Q b$ and any $y \in c \wedge^Q d$. Then $x \wedge y \in a \wedge^Q b$ and $x \vee y \in a \wedge^Q b$. So $y \in [x \wedge y, x \vee y] \subseteq a \wedge^Q b$ which implies that $c \wedge^Q d \subseteq a \wedge^Q b$. Similarly we can show that $a \wedge^Q b \subseteq c \wedge^Q d$ and so $c \wedge^Q d = a \wedge^Q b$.

iii. Pick any $a, b, c, d, e, f \in X$ such that $a \wedge^Q b \leq_2 c \wedge^Q d$ and $c \wedge^Q d \leq_2 e \wedge^Q f$. Now choose any $x \in a \wedge^Q b$, any $y \in c \wedge^Q d$ and any $z \in e \wedge^Q f$. Then $x \vee y \in c \wedge^Q d$ and so $x \vee y \vee z \in e \wedge^Q f$. So $x \vee z \in [z, x \vee y \vee z] \subseteq e \wedge^Q f$. Similarly we can show $x \wedge z \in a \wedge^Q b$. Hence $a \wedge^Q b \leq_2 e \wedge^Q f$.

2 follows from 1: if Q is an interval, it is also a convex sublattice. $\square$

Now we can prove a monotonicity property of $a \wedge^Q b$.

Proposition 17.10 *Take any $Q \subseteq X$ with $1 \in Q$.*

1. If Q is a filter, then we have: $a \leq b \Leftrightarrow \left(\forall c \in X : a \wedge^Q c \leq_2 b \wedge^Q c \right)$.
2. If Q is an interval, then (for $i = 1, 2$) we have:

$$ a \leq b \Leftrightarrow \left(\forall c \in X : a \wedge^Q c \leq_i b \wedge^Q c \right). $$

Proof. 1. Let Q be a filter. Suppose that $a \leq b$ and take $x \in a \wedge^Q c$, $y \in b \wedge^Q c$. That is, $x = a \wedge c \wedge q_1$ and $y = b \wedge c \wedge q_2$. Then

$$ x \wedge y = a \wedge b \wedge c \wedge (q_1 \wedge q_2) = a \wedge c \wedge (q_1 \wedge q_2) ; $$

since $q_1, q_2 \in Q$ and Q is a filter, then $q_1 \wedge q_2 \in Q$ and $x \wedge y \in a \wedge^Q c$. Also

$$
\begin{aligned}
x \vee y &= (a \wedge c \wedge q_1) \vee (b \wedge c \wedge q_2) = ((a \wedge q_1) \vee (b \wedge q_1)) \wedge c \\
&= ((a \vee b) \wedge (q_1 \vee b) \wedge (a \vee q_2) \wedge (q_1 \vee q_2)) \wedge c \\
&= (b \wedge (q_1 \vee b) \wedge (a \vee q_2) \wedge (q_1 \vee q_2)) \wedge c \\
&= (b \wedge (a \vee q_2) \wedge (q_1 \vee q_2)) \wedge c.
\end{aligned}
$$

Now, $a \vee q_2 \geq q_2 \in Q$ and so $a \vee q_2 \in Q$ (Q is a filter); also $q_1 \vee q_2 \in Q$ (Q is a sublattice) hence $q = (a \vee q_2) \wedge (q_1 \vee q_2) \in Q$. This means

$$ x \vee y = b \wedge c \wedge q \in b \wedge^Q c. $$

Since $x \wedge y \in a \wedge^Q c$ and $x \vee y \in b \wedge^Q c$, it follows that $a \wedge^Q c \leq_2 b \wedge^Q c$.
2. If Q is an interval, then $Q = [q, 1]$. Hence $a \leq b \Rightarrow a \wedge c \leq b \wedge c \Rightarrow a \wedge c \wedge q \leq b \wedge c \wedge q$. Hence

$$ a \wedge^Q c = [a \wedge c \wedge q, a \wedge c] \leq_1 [b \wedge c \wedge q, b \wedge c] = b \wedge^Q c $$

On the other hand, if $\left(\forall c \in X : a \wedge^Q c \leq_1 b \wedge^Q c \right)$, then taking $c = 1$ we get

$$ [a \wedge q, a] \leq_1 [b \wedge q, b] \Rightarrow a \leq b. $$

The same result follows for $\leq_2$, since this is equivalent to $\leq_1$ over a class of intervals. $\square$

Remark 17.3. Note the double implications in the above proposition. They show that $\wedge^Q$ has a stronger-than-monotonicity property (when $1 \in Q$).

From the above propositions we can state the following proposition which give conditions for $\wedge^Q$ to be a hyper-t-norm.

Proposition 17.11 *Let $Q \subseteq \mathbf{P}(X)$ with $1 \in Q$.*

1. If Q is a filter then $\wedge^Q$ is a hyper-t-norm (with respect to the order $\leq_2$).
2. If Q is an interval, then $\wedge^Q$ is a hyper-t-norm (with respect to the orders $\leq_1, \leq_2$).

Proof. This is a simple consequence of Propositions 17.7, 17.9, 17.10. $\square$

In the rest of the paper we will always assume that Q is (at least) a filter and that $1 \in Q$. Here are some additional properties of $\wedge^Q$ regarding order.

Proposition 17.12 *For all $a, b \in X$ we have:* $\boldsymbol{max}(a \wedge^Q b) = a \wedge b$.

Proof. Indeed, $x \in a \wedge^Q b \Rightarrow x = a \wedge b \wedge q \leq a \wedge b \wedge 1 = a \wedge b$. And $a \wedge b = a \wedge b \wedge 1 \in a \wedge^Q b$. $\square$

Proposition 17.13 *For all $a, b \in X$ we have:* $a \wedge^Q b = a \wedge^Q a \Leftrightarrow a \leq b$.

Proof. Assume $a \leq b$. Then $a\wedge^Q b = \cup_{q \in Q} a \wedge b \wedge q = \cup_{q \in Q} a \wedge q = \cup_{q \in Q} a \wedge a \wedge q = a \wedge^Q a$. Conversely, if $a \wedge^Q b = a \wedge^Q a$, then by Proposition 17.12

$$a \wedge b = \mathtt{max}\left(a \wedge^Q b\right) = \mathtt{max}\left(a \wedge^Q a\right) = a \wedge a = a \Rightarrow a \leq b. \square$$

17.3.2 The $\vee^P$ hyper-t-conorm

In completely analogous manner to that of the previous section, we define a hyperoperation which, under suitable restrictions, has the properties of a multivalued t-conorm.

Definition 17.4 *Let P be some subset of X and define the hyperoperation $\vee^P$ as follows:*

$$\forall a, b \in X : a \vee^P b \doteq \{a \vee b \vee p' : p \in P\}.$$

Remark 17.4. We could also write $a \vee^P b = a \vee b \vee P'$.

Proposition 17.14 *Take any set $P \subseteq X$. Then, for all a, b, c we have*

$$a \vee^P b = b \vee^P a; \quad \left(a \vee^P b\right) \vee^P c = a \vee^P \left(b \vee^P c\right).$$

Proposition 17.15 *We have:* $\left(\forall a : a \in a \vee^P 0\right) \Leftrightarrow (1 \in P)$.

Proposition 17.16 *Take any $P \subseteq X$ such that $1 \in P$. Then we have the following*

1. $(P$ is a convex sublattice$) \Leftrightarrow \left(\forall a, b \in X : a \vee^P b$ is a convex sublattice$\right)$;
2. $(P$ is a filter$) \Rightarrow \left(\forall a, b \in X : a \vee^P b$ is a convex sublattice$\right)$;
3. $(P$ is an interval$) \Leftrightarrow \left(\forall a, b \in X : a \vee^P b$ is an interval$\right)$.

Proposition 17.17 *If $P = [p, 1]$ then $a \vee^P b = [a \vee b, a \vee b \vee p']$ $(\forall a, b \in X)$.*

Proposition 17.18 *Take any $P \subseteq X$ such that $1 \in P$.*

1. If P is a convex sublattice, then $\leq_2$ is an order on $\left\{a \vee^P b\right\}_{a, b \in X}$.
2. If P is an interval, then $\leq_1, \leq_2$ are orders on $\left\{a \vee^P b\right\}_{a, b \in X}$.

Proposition 17.19 *Take any $P \subseteq X$ with $1 \in P$.*

1. *If P is a filter, then we have: $a \le b \Leftrightarrow \left(\forall c \in X : a \vee^P c \le_2 b \vee^P c \right)$.*
2. *If P is an interval, then (for $i = 1, 2$) we have:*

$$a \le b \Leftrightarrow \left(\forall c \in X : a \vee^P c \le_i b \vee^P c \right).$$

Proposition 17.20 *Let $P \subseteq \mathbf{P}(X)$ with $1 \in P$.*

1. *If P is a filter then $\vee^P$ is a hyper-t-conorm (with respect to the order $\le_2$).*
2. *If P is an interval, then $\vee^P$ is a hyper-t-conorm (with respect to the orders $\le_1, \le_2$).*

Proposition 17.21 *For all $a, b \in X$ we have: $min(a \vee^P b) = a \vee b$.*

Proposition 17.22 *For all $a, b \in X$ we have: $a \vee^P b = a \vee^P a \Leftrightarrow a \le b$.*

17.3.3 Further Properties of $\wedge^Q$ and $\vee^P$

The following proposition establishes that $\wedge^Q$ and $\vee^P$ have a weak form of distributivity.

Proposition 17.23 *For all $a, b, c \in X$ we have*

$$\left(a \wedge^Q \left(b \vee^P c \right) \right) \cap \left(\left(a \wedge^Q b \right) \vee^P \left(a \wedge^Q c \right) \right) \ne \emptyset, \tag{17.1}$$
$$\left(a \vee^P \left(b \wedge^Q c \right) \right) \cap \left(\left(a \vee^P b \right) \wedge^Q \left(a \vee^P c \right) \right) \ne \emptyset. \tag{17.2}$$

Proof. Since $1 \in Q$ and $1 \in P$, then

$$a \wedge (b \vee c) = a \wedge (b \vee c \vee 1') \wedge 1 \in a \wedge^Q \left(b \vee^P c \right). \tag{17.3}$$

Also

$$(a \wedge b) \vee (a \wedge c) = (a \wedge b \wedge 1) \vee (a \wedge c \wedge 1) \vee 1' \in \left(a \wedge^Q b \right) \vee^P \left(a \wedge^Q c \right). \tag{17.4}$$

Finally,

$$a \wedge (b \vee c) = (a \wedge b) \vee (a \vee c). \tag{17.5}$$

From (17.3), (17.4) and (17.5) we obtain (17.1); (17.2) is proved dually. $\square$

Here are some more properties related to order.

Proposition 17.24 *For all $a, b \in L$ we have: $x \in a \vee^P b, \; y \in a \wedge^Q b \Rightarrow y \le x$.*

Proof. True, since $y \le \max \left(a \wedge^Q b \right) = a \wedge b \le a \vee b = \min \left(a \vee^P b \right) \le x$. $\square$

Proposition 17.25 *For all $a, b \in L$ we have: $\left(a \vee^P b \right) \cap \left(a \wedge^Q b \right) \ne \emptyset \Rightarrow a = b$.*

Proof. Take some $x \in \left(a \vee^P b\right) \cap \left(a \wedge^Q b\right)$; then for some $p \in P, q \in Q$ we have

$$a \wedge b \geq a \wedge b \wedge q = x = a \vee b \vee p' \geq a \vee b \Rightarrow a = a \wedge b = a \vee b = b. \square$$

We also have a sort of "reduction property".

Proposition 17.26 *For all $a, x, y \in L$ we have:*

$$\left(a \vee^P x = a \vee^P y \text{ and } a \wedge^Q x = a \wedge^Q y\right) \Rightarrow x = y.$$

Proof.

$$a \vee^P x = a \vee^P y \Rightarrow \max\left(a \vee^P x\right) = \max\left(a \vee^P y\right) \Rightarrow a \vee x = a \vee y \quad (17.6)$$
$$a \wedge^Q x = a \wedge^Q y \Rightarrow \max\left(a \wedge^Q x\right) = \max\left(a \wedge^Q y\right) \Rightarrow a \wedge x = a \wedge y \quad (17.7)$$

From (17.6), (17.7) and a standard property of distributive lattices we get $x = y$. $\square$

Proposition 17.27 *For every $'$ and Q, we obtain the de Morgan triple* $\left(\wedge^Q, \vee^Q, '\right)$. *In other words, $\vee^{Q'}$ is a hyper-t-conorm, $\wedge^Q$ is a hyper-t-norm and the following analogs of de Morgan's laws hold:*

$$\left(a \wedge^Q b\right)' = a' \vee^Q b' \text{ and } \left(a \vee^Q b\right)' = a' \wedge^Q b'$$

Proof. Straightforward. $\square$

So far we have assumed that P, Q are filters. Let us now strengthen this assumption; for the rest of this section we will asume that P, Q are intervals of the form $P = [p, 1]$, $Q = [q, 1]$ and we prove several additional results. It will be convenient to introduce a new notation.

Notation 17.2 *When $Q = [q, 1]$ we will denote $\wedge^Q$ by $\wedge_q$; when $P = [p, 1]$ (and $P' = [0, p]$) we will denote $\vee^P$ by $\vee_p$.*

The following proposition establishes that $\wedge_q$ and $\vee_p$ have a weak form of distributivity.

Proposition 17.28 *For all $p, q \in X$ and for all $a, b, c \in X$ we have*

$$a \wedge_q (b \vee_p c) \subseteq (a \wedge_q b) \vee_p (a \wedge_q c), \qquad a \vee_p (b \wedge^q c) \subseteq (a \vee_p b) \wedge_q (a \vee^P c).$$

Proof. Ít is straightforward to show that

$$a \wedge_q (b \vee_p c) = [a \wedge (b \vee c) \wedge q, a \wedge (b \vee c \vee p')] \quad (17.8)$$
$$(a \wedge_q b) \vee_p (a \wedge_q c) = [(a \wedge b \wedge q) \vee (a \wedge c \wedge q), (a \wedge b) \vee (a \wedge c) \vee p']. \quad (17.9)$$

Also

$$a \wedge (b \vee c) \wedge q = (a \wedge b \wedge q) \vee (a \wedge c \wedge q) \quad (17.10)$$
$$a \wedge (b \vee c \vee p') = (a \wedge b) \vee (a \wedge c) \vee (a \wedge p') \leq (a \wedge b) \vee (a \wedge c) \vee p' \quad (17.11)$$

and (17.8) – (17.11) show that the first part of the proposition holds; the second part is proved dually. $\square$

Proposition 17.29 *For all $p_1, p_2, q_1, q_2 \in X$ and $a, b, c \in X$ the following properties hold.*

$$(a \wedge_{q_1} b) \wedge_{q_2} c = (a \wedge_{q_1} b) \wedge_{q_2} c = a \wedge_{q_1 \wedge q_2} b \wedge_{q_1 \wedge q_2} c. \tag{17.12}$$

$$(a \vee_{p_1} b) \vee_{p_2} c = (a \vee_{p_1} b) \vee_{p_2} c = a \vee_{p_1 \vee p_2} b \vee_{p_1 \vee p_2} c. \tag{17.13}$$

Proof. Straightforward. $\square$

17.4 The $\to^P$ S-hyper-implication

We can also introduce an additional hyperoperation, the *hyper-implication*. This is a straightforward generalization of Boolean and fuzzy implications. Recall that in Boolean logic we can define the implication $\to$ as follows

$$\forall a, b : a \to b \doteq a' \vee b. \tag{17.14}$$

Several other equivalent definitions can be used. In fuzzy logic we generalize (17.14) by introducing the class of S-implications: given a t-conorm S (and a negation $'$) we define

$$\forall a, b : a \to b \doteq a' S b. \tag{17.15}$$

We can define a hyper-implication (i.e. a multi-valued implication) by using $\vee^P$ in place of S in (17.15). In other words, we introduce the following hyper-implication, denoted by $\to^P$ and defined as follows

$$a \to^P b \doteq a' \vee^P b.$$

This hyper-implication has several interesting properties (analogous to these of the classical implication).

Proposition 17.30 *Given a filter $P \subseteq X$ such that $1 \in P$, we have for every a, b, c the following.*

1. $a \le b \Rightarrow \left((a \to^P c) \ge_2 (b \to^P c) \text{ and } (c \to^P a) \le_2 (c \to^P b) \right),$

2. $1 \in \left(0 \to^P a \right)$ *and* $a \in \left(1 \to^P a \right),$

3. $\left(a \to^P b \right) = \left(b' \to^{P'} a' \right).$

Proof. This is fairly straightforward.

1. Assume $a \le b$. Then $a' \ge b'$ and, by monotonicity of $\vee^P$ we have $\left(a \to^P c \right) = a' \vee^P c \ge_2 b' \vee^P c = \left(b \to^P c \right)$. Similarly, if $a \le b$ then $\left(c \to^P a \right) = c' \vee^P a \le_2 c' \vee^P b = \left(c \to^P b \right)$.

2. $\min \left(0 \to^P a \right) = \min \left(0' \vee^P a \right) = 1 \vee a = 1$; hence $1 \in \left(0 \to^P a \right)$. Also $1 \to^P a = 1' \vee^P a = \cup_{p \in P} [0 \vee a, 0 \vee a \vee p']$. Since $0 \in P'$, $a = 0 \vee a \vee 0 \in \left(1 \to^P a \right)$.

3. $\left(a \to^P b \right) = a' \vee^P b = (b')' \vee^P a' = \left(b' \to^{P'} a' \right)$. $\square$

The above properties are similar to the ones established in [9] and other places about (uni-valued) implications.

17.5 The (P, Q)-superlattice

As we have already mentioned in Section 17.1, $\wedge$ has several special properties, as compared to other t-norms. Among all these properties, we believe the most special is that $\wedge$ is both monotone and "order-generating". In other words, while by definition every t-norm T satisfies

$$\text{monotonicity:} \qquad \forall a, b, c : a \leq b \Rightarrow aTc \leq bTc \qquad (17.16)$$

the only t-norm that satisfies

$$\text{order generation:} \qquad \forall a, b : a \leq b \Leftrightarrow aTb = a. \qquad (17.17)$$

is $T = \wedge$. The order generated by $\wedge$ can then be defined as follows

$$a \leq b \text{ iff } a \wedge b = a. \qquad (17.18)$$

Similar things hold for the t-conorm $\vee$. The algebra $(X, \leq, \wedge, \vee)$ is a *lattice*; now we will show that the *hyper*algebra $(X, \leq, \wedge^Q, \vee^P)$ is a *superlattice*, i.e. the multivalued analog of a lattice. First consider the following general definition.

Definition 17.5 *Given a poset* $(U, \sqsubseteq)$ *and hyperoperations* $\sqcap$ *and* $\sqcup$ *mapping* $U \times U$ *to* $\mathbf{P}(U)$, *the structure* $(U, \sqsubseteq, \sqcap, \sqcup)$ *is called a* superlattice *iff the following conditions hold for all* $a, b, c \in U$.

G1. $a \in (a \sqcup a)$, $a \in (a \sqcap a)$.
G2. $a \sqcup b = b \sqcup a$, $a \sqcap b = b \sqcap a$.
G3. $(a \sqcup b) \sqcup c = a \sqcup (b \sqcup c)$, $(a \sqcap b) \sqcap c = a \sqcap (b \sqcap c)$.
G4. $a \in (a \sqcup b) \sqcap a$, $a \in (a \sqcap b) \sqcup a$.
G5. $a \sqsubseteq b \Rightarrow b \in a \sqcup b$, $a \in a \sqcap b$.
G6. $b \in a \sqcup b \Leftrightarrow a \in a \sqcap b \Leftrightarrow a \sqsubseteq b$.

The similarity ("order generation") between $\sqcap$ and $\wedge$, as well as between $\sqcup$ and $\vee$ is especially seen in properties G5, G6.

We now show that the hyperalgebra $(X, \wedge^Q, \vee^P)$ is a superlattice, in the sense of Definition 17.5. To this end we must prove the following auxiliary proposition.

Proposition 17.31 *Take any* $P, Q \subset X$ *such that* $1 \in Q$ *and* $1 \in P$. *Then* $(X, \leq, \wedge^Q, \vee^P)$ *is a superlattice, i.e. it satifies*

S1. $a \in (a \vee^P a)$, $a \in (a \wedge^Q a)$.
S2. $a \vee^P b = b \vee^P a$, $a \wedge^Q b = b \wedge^Q a$.
S3. $(a \vee^P b) \vee^P c = a \vee^P (b \vee^P c)$, $(a \wedge^Q b) \wedge^Q c = a \wedge^Q (b \wedge^Q c)$.
S4. $a \in (a \vee^P b) \wedge^Q a$, $a \in (a \wedge^Q b) \vee^P a$.
S5. $a \leq b \Rightarrow b \in a \vee^P b$, $a \in a \wedge^Q b$.
S6. $b \in a \vee^P b \Leftrightarrow a \in a \wedge^Q b \Leftrightarrow a \leq b$.

Proof. S1 is straightforward. S2 and S3 were proved in Propositions 17.5, 17.14. For S4, we know that

$$a \in [(a \vee b) \wedge a \wedge 1, (a \vee b \vee 0) \wedge a] \subseteq [a \vee b, a \vee b \vee 0] \wedge^Q a \subseteq \left(a \vee^P b\right) \wedge^Q a;$$

and $a \in \left(a \wedge^Q b\right) \vee^P a$ is proved similarly. For S5, $a \le b \Rightarrow a \vee b = b$ and so $[b, b] = [a \vee b, a \vee b \vee 0] \subseteq a \vee^P b$. Conversely, if $b \in a \vee^P b$, then $a \vee b \le b \le a \vee b \vee 0$ and so $b = a \vee b$, $a \le b$. Hence we have proved $a \le b \Leftrightarrow \left(b \in a \vee^P b\right)$. Similarly we can prove $a \le b \Leftrightarrow \left(a \in a \wedge^Q b\right)$. For S6, if $b \in a \vee^P b$, then (for some $p \in P$) we have $b = a \vee b \vee p' \ge a$; similarly for $a \in a \wedge^Q b$. $\square$

The superlattice $\left(X, \le, \wedge^Q, \vee^P\right)$ has been studied in [20, 21], under the name of "(P, Q)-superlattice". A case of special interest is the (Q', Q)-superlattice (i.e. using $P = Q$).

17.6 The Boolean Case

We now turn to a special case which yields additional results. Up to this point we have assumed that the lattice $(X, \le, \wedge, \vee, ')$ is de Morgan; now we make the stronger assumption that it is *Boolean*. In other words we assume that $\forall a \in X : a \wedge a' = 0$ and $a \vee a' = 1$. Furthermore in this section (and in the rest of the paper) we assume that $P = [p, 1]$ ($P' = [0, p']$), $Q = [q, 1]$. We will also use again the notation $\wedge_q$, $\vee_p$ and $\to_p$ for the implication. Under the interval assumption we can obtain additional properties of $\wedge_q$ and $\vee_p$.

The $(X, \wedge_q, \vee_p)$ hyperalgebra can be characterized as a *weak* Boolean superlattice. By this we mean that it is a *weakly distributive* and *hypercomplemented* superlattice, i.e. in addition to S1-S6 the following properties hold

$$\text{weak distributivity} : a \wedge_q (b \vee_p c) \subseteq (a \wedge_q b) \vee_p (a \wedge_q c),$$
$$a \vee_p (b \wedge_q c) \subseteq (a \vee_p b) \wedge_q (a \vee_p c),$$
$$\text{hypercomplementation} : 1 \in (a \vee_p a'), \quad 0 \in (a \wedge_q a').$$

These properties hold in the general de Morgan case; now we see additional properties for the Boolean case.

Proposition 17.32 *For every $a, b, c \in X$, $i = 1, 2$ and $p \in X$ we have the following:*

1. $a \le b \Rightarrow (a \to_p c) \ge_i (b \to_p c)$ *and* $a \le b \Rightarrow (c \to_p a) \le_i (c \to_p b)$,
2. $1 \in (0 \to_p a)$ *and* $a \in (1 \to_p a)$,
3. $(a \to_p b) = (b' \to_p a')$,
4. $1 \in (a \to_p a)$ *and* $a \in (1 \to_p a)$.
5. $\left(a \to^P (b \to_p c)\right) = (b \to_p (a \to_p c))$ *(strong exchange property)*;

Proof. Parts 1 and 2 are proved exactly as in Proposition 17.30, except that now we use the order $\leq_1$. 3 is obvious. For 4 $(a \to_p a)= [a' \vee a, a' \vee a \vee p']= [1,1]$; for the second part we have $(1 \to_p a)= [1' \vee a, 1' \vee a \vee p']= [a, a \vee p']$; the second part is proved similarly. For 5 (exchange property) we have

$$(a \to_p (b \to_p c)) = a' \vee_p [b' \vee c, b' \vee c \vee p'] = [a' \vee b' \vee c, a' \vee b' \vee c \vee p']$$
$$(b \to_p (a \to_p c)) = b' \vee_p [a' \vee c, a' \vee c \vee p'] = [a' \vee b' \vee c, a' \vee b' \vee c \vee p']$$

which are obviously equal. $\square$

Furthermore in the Boolean case the implication $\to_p$ has Modus Ponens, Modus Tollens and syllogistic reasoning properties.

Proposition 17.33 *For every $a, b, c \in X$ and $p, q \in X$ we have the following*

$$Modus\ Ponens : a \wedge_q b \subseteq a \wedge_q (a \to_p b) \tag{17.19}$$
$$Modus\ Tollens : a' \wedge_q b' \subseteq b' \wedge^p (a \to_p b) \tag{17.20}$$
$$Syllogistic\ Reasoning : (a \to_p b) \wedge_q (b \to_p c) \leq_2 (a \to_p c) \tag{17.21}$$

Proof. For (17.19) we note that

$$a \wedge_q b = [a \wedge b \wedge q, a \wedge b]$$
$$a \wedge_q (a \to_p b) = a \wedge_q [a' \vee b, a' \vee b \vee p'] = [a \wedge (a' \vee b) \wedge q, a \wedge (a' \vee b \vee p')] .$$

Now, $a \wedge (a' \vee b) \wedge q= (a \wedge a' \wedge q) \vee (a \wedge b \wedge q)= a \wedge b \wedge q$ and $a \wedge (a' \vee b \vee p')= (a \wedge a') \vee (a \wedge b) \vee (a \wedge p')= a \wedge (b \vee p') \geq a \wedge b$ which complete the proof of (17.19). We omit the proof of (17.20), which is similar. Regarding (17.21), it follows $(a \to_p b)= [a' \vee b, a' \vee b \vee p']$, $(b \to_p c)= [b' \vee c, b' \vee c \vee p']$, $(a \to_p c)= [a' \vee c, a' \vee c \vee p']$. Also, with $f = (a' \vee b) \wedge (b' \vee c) \wedge q$ and $g = (a' \vee c \vee p') \wedge (b' \vee c \vee p')$, we have

$$[a' \vee b, a' \vee b \vee p'] \wedge_q [b' \vee c, b' \vee c \vee p'] = [f, g] .$$

Now

$$(a' \vee b) \wedge (b' \vee c) = (a' \wedge b') \vee (b \wedge b') \vee (a' \wedge c) \vee (b \wedge c) \leq a' \vee c;$$

since the first term in the middle expression above is less than a', the second is 0, the third term is less than a' and the last less than c. Also

$$(a' \vee b \vee p') \wedge (b' \vee c \vee p') = x \vee y \vee z$$

where

$$x = (a' \wedge b') \vee (b \wedge b') = (a' \wedge b') \leq a'$$
$$y = (a' \wedge c) \vee (b \wedge c) \vee (p' \wedge c) \leq c$$
$$z = (p' \wedge b') \vee (a' \wedge p') \vee (b \wedge p') \vee (p' \wedge p') \leq p'$$

hence $(a' \vee b \vee p') \wedge (b' \vee c \vee p') = a' \vee c \vee p'$ from which follows (17.21). $\square$

The $\to_p$ implication also induces an "order like" relationship.

Proposition 17.34 *For all $a, b, p \in X$ we have:* $1 \in (a \to_p b) \Leftrightarrow a \vee p' \leq b \vee p'$

Proof. On the one hand, $1 \in (a \to_p b) = [a' \vee b, a' \vee b \vee p']$ implies that $1 = a' \vee b \vee p'$. Then

$$a = (a' \vee b \vee p') \wedge a = (b \vee p') \wedge a \Rightarrow a \leq b \vee p' \Rightarrow a \vee p' \leq b \vee p'.$$

Conversely,

$$a \vee p' \leq b \vee p' \Rightarrow a' \vee a \vee p' \leq a' \vee b \vee p' \Rightarrow 1 \leq a' \vee b \vee p' \Rightarrow 1 \in (a \to_p b). \,\Box$$

This motivates us to define the relations $\leq_p, =_p$.

Definition 17.6 *For every $p \in X$ we define $\leq_p$ and $=_p$ as follows:*

$$a \leq_p b \Leftrightarrow a \vee p' \leq b \vee p', \qquad a =_p b \Leftrightarrow a \vee p' = b \vee p'.$$

Proposition 17.35 *For all $a, b, p \in X$ we have*

$$a \leq_p b \Leftrightarrow a \wedge p \leq b \wedge p, \qquad a =_p b \Leftrightarrow a \wedge p = b \wedge p. \tag{17.22}$$

Proof. In the one direction

$$a \leq_p b \Rightarrow a \vee p' \leq b \vee p' \Rightarrow (a \vee p') \wedge p \leq (b \vee p') \wedge p \Rightarrow$$
$$(a \wedge p) \vee (p' \wedge p) \leq (b \wedge p) \vee (p' \wedge p) \Rightarrow a \wedge p \leq b \wedge p.$$

Conversely,

$$a \wedge p \leq b \wedge p \Rightarrow (a \wedge p) \vee p' \leq (b \wedge p) \vee p' \Rightarrow$$
$$(a \vee p') \wedge (p \vee p) \leq (b \vee p') \wedge (p \vee p) \Rightarrow a \vee p' \leq b \vee p' \Rightarrow a \leq_p b.$$

This proves the first part of (17.22); the second part is proved similarly. $\Box$

Proposition 17.36 *The relation $\leq_p$ is a* preorder *(i.e. it is reflexive and transitive) and $=_p$ is the natural equivalence indiced from $\leq_p$.*

Proof. Straightforward (also see [7]). $\Box$

Let us now establish some properties of the equivalence relation $=_p$.

Notation 17.3 *For every $p \in X$ we denote by $\overline{a}^p$ the class of a under $=_p$, i.e. $\overline{a}^p = \{x : a \vee p' = x \vee p'\}$.*

Proposition 17.37 *For every $a, p \in X$, $\overline{a}^p$ is an interval. More specifically $\overline{a}^p = [a \wedge p, a \vee p']$.*

Proof. If $x \in \overline{a}^p$ then clearly $x \leq x \vee p' = a \vee p'$. Also

$$x \vee p' = a \vee p' \Rightarrow (x \vee p') \wedge p = (a \vee p') \wedge p \Rightarrow x \wedge p = a \wedge p \Rightarrow x \geq a \wedge p.$$

Hence

$$\overline{a}^p \subseteq [a \wedge p, a \vee p']. \tag{17.23}$$

On the other hand, let us show that $\overline{a}^p$ is a convex sublattice. Indeed, $\overline{a}^p$ is a sublattice: take any $x, y \in \overline{a}^p$ then

$$\left. \begin{array}{l} a \vee p' = x \vee p' \\ a \vee p' = y \vee p' \end{array} \right\} \Rightarrow \left\{ \begin{array}{l} a \vee p' = (x \vee p') \vee (y \vee p') = (x \vee y) \vee p' \\ a \vee p' = (x \vee p') \wedge (y \vee p') = (x \wedge y) \vee p' \end{array} \right.$$

and so $x \vee y, x \wedge y \in \overline{a}^p$. Further, take any $z \in [x \wedge y, x \vee y]$, then

$$(x \wedge y) \vee p' \leq z \vee p' \leq x \vee y \vee p' \Rightarrow$$
$$a \vee p' = (x \vee p') \wedge (y \vee p') \leq z \vee p' \leq (x \vee p') \vee (y \vee p') = a \vee p' \Rightarrow$$
$$z \vee p' = a \vee p'$$

and $(x \wedge y) \wedge p \leq z \wedge p \leq (x \vee y) \wedge p \Rightarrow$

$$a \wedge p = (x \wedge p) \wedge (y \wedge p) \leq z \wedge p \leq (x \wedge p) \vee (y \wedge p) = a \wedge p \Rightarrow z \wedge p = a \wedge p \tag{17.24}$$

So $z \in \overline{a}^p$ and $\overline{a}^p$ is a convex sublattice. And $a \wedge p$ and $a \vee p'$ belong to $\overline{a}^p$, hence

$$[a \wedge p, a \vee p'] \subseteq \overline{a}^p. \tag{17.25}$$

Combining (17.24) with (17.25) we get the desired result. $\square$

Proposition 17.38 *For all $p, a, b, c \in X$ we have:*

$$\overline{a}^p = \overline{b}^p \Rightarrow \left\{ \begin{array}{l} \overline{a \vee c}^p = \overline{b \vee c}^p \\ \overline{a \wedge c}^p = \overline{b \wedge c}^p \end{array} \right.$$

Proof. We only prove the first part (the second is proved similarly). We have

$$\overline{a}^p = \overline{b}^p \Rightarrow [a \wedge p, a \vee p'] = [b \wedge p, b \vee p'] \Rightarrow \left\{ \begin{array}{l} a \wedge p = b \wedge p \\ a \vee p' = b \vee p' \end{array} \right.$$

Now

$$a \vee p' = b \vee p' \Rightarrow a \vee c \vee p' = b \vee c \vee p' \tag{17.26}$$

and

$$\left. \begin{array}{l} b \wedge p = a \wedge p \\ c \wedge p = c \wedge p \end{array} \right\} \Rightarrow (a \wedge p) \vee (c \wedge p) = (b \wedge p) \vee (c \wedge p) \Rightarrow (a \vee c) \wedge p = (b \vee c) \wedge p. \tag{17.27}$$

From (17.26) and (17.27) we get

$$[(a \vee c) \wedge p, (a \vee c) \vee p'] = [(b \vee c) \wedge p, (b \vee c) \vee p'] \Rightarrow \overline{a \vee c}^p = \overline{b \vee c}^p. \square$$

We will prove a Proposition similar to Proposition 17.38 for $\vee_p$ and $\wedge_p$, but first we need the following.

358   A. Kehagias

Definition 17.7 *The set of classes $\overline{A}^p$ is defined as follows $\overline{A}^p = \{\overline{x}^p : x \in A\}$.*

Proposition 17.39 *For all $p, a, b \in X$ we have:*

$$\overline{a \vee_p b}^p = \{[(a \vee b) \wedge p, (a \vee b) \vee p']\}, \qquad \overline{a \wedge_q b}^q = \{[(a \vee b) \wedge q, (a \vee b) \vee q']\} \tag{17.28}$$

i.e. each of $\overline{a \vee_p b}^p, \overline{a \wedge_q b}^q$ contains a single class.

Proof. We only prove the first part of (17.28):

$$\overline{a \vee_p b}^p = \{\overline{x}^p : a \vee b \leq x \leq a \vee b \vee p'\} = \{[x \wedge p, x \vee p'] : a \vee b \leq x \leq a \vee b \vee p'\}$$

Take any x such that $\overline{x}^p \in \overline{a \vee_p b}^p$. Then $a \vee b \leq x \leq a \vee b \vee p'$ and so $a \vee b \vee p' \leq x \vee p' \leq a \vee b \vee p' \Rightarrow a \vee b \vee p' = x \vee p'$. Hence

$$(a \vee b) \wedge p \leq x \wedge p \leq (a \vee b \vee p') \wedge p = ((a \vee b) \wedge p) \vee (p' \wedge p) = (a \vee b) \wedge p$$

which implies $(a \vee b) \wedge p = x \wedge p$. Hence

$$\overline{x}^p = [x \wedge p, x \vee p'] = [(a \vee b) \wedge p, (a \vee b) \vee p'] = \overline{a \vee b}^p$$

and the proof is complete. $\square$

Now we prove the analog of Proposition 17.38.

Proposition 17.40 *For all $p, a, b, c \in X$ we have:*

$$\overline{a}^p = \overline{b}^p \Rightarrow \begin{cases} \overline{a \vee_p c}^p = \overline{b \vee_p c}^p \\ \overline{a \wedge_p c}^p = \overline{b \wedge_p c}^p \end{cases}.$$

Proof. Take any $\overline{x}^p \in \overline{a \vee_p c}^p, \overline{y}^p \in \overline{b \vee_p c}^p$. Then, by the previous Proposition,

$$\overline{x}^p = [(a \vee c) \wedge p, (a \vee c) \vee p'], \qquad \overline{y}^p = [(b \vee c) \wedge p, (b \vee c) \vee p'] \tag{17.29}$$

also $\overline{a}^p = \overline{b}^p \Rightarrow [a \wedge p, a \vee p'] = [b \wedge p, b \vee p']$ and so

$$\begin{rcases} a \wedge p = b \wedge p \\ a \vee p' = b \vee p' \end{rcases} \Rightarrow \begin{rcases} (a \vee c) \wedge p = (b \vee c) \wedge p \\ (a \vee c) \vee p' = (b \vee c) \vee p' \end{rcases}. \tag{17.30}$$

From (17.29) and (17.30) we obtain the first part of the proposition; the second part is proved similarly. $\square$

Proposition 17.41 *For all $p, a, b, c \in X$ we have:*

$$\begin{rcases} \overline{a \vee c}^p = \overline{b \vee c}^p \\ \overline{a \wedge c}^p = \overline{b \wedge c}^p \end{rcases} \Rightarrow \overline{a}^p = \overline{b}^p \quad \text{and} \quad \begin{rcases} \overline{a \vee_p c}^p = \overline{b \vee_p c}^p \\ \overline{a \wedge_{p'} c}^p = \overline{b \wedge_{p'} c}^p \end{rcases} \Rightarrow \overline{a}^p = \overline{b}^p.$$

Proof. We have $\overline{a \vee c}^{\,p} = \overline{b \vee c}^{\,p} \Rightarrow$

$$\left\{ \begin{array}{l} (a \vee c) \wedge p = (b \vee c) \wedge p \\ (a \vee c) \vee p' = (b \vee c) \vee p' \end{array} \right\} \Rightarrow \left\{ \begin{array}{l} (a \wedge p) \vee (c \wedge p) = (b \wedge p) \vee (c \wedge p) \\ (a \vee p') \vee (c \vee p') = (b \vee p') \vee (c \vee p') \end{array} \right. .$$

$$(17.31)$$

Similarly, $\overline{a \wedge c}^{\,p} = \overline{b \wedge c}^{\,p} \Rightarrow$

$$\left\{ \begin{array}{l} (a \wedge c) \wedge p = (b \wedge c) \wedge p \\ (a \wedge c) \vee p' = (b \wedge c) \vee p' \end{array} \right\} \Rightarrow \left\{ \begin{array}{l} (a \wedge p) \wedge (c \wedge p) = (b \wedge p) \wedge (c \wedge p) \\ (a \vee p') \wedge (c \vee p') = (b \vee p') \wedge (c \vee p') \end{array} \right. .$$

$$(17.32)$$

From the first parts of (17.31), (17.32) we get $a \wedge p = b \wedge p$ and from the second parts $a \vee p' = b \vee p'$, which together imply $\overline{a}^{\,p} = \overline{b}^{\,p}$. $\square$

17.7 Conclusion

We have introduced two crisp *hyperoperations*, $\wedge^Q$ and $\vee^P$, which are natural multi-valued generalizations of the t-norm $\wedge$ and the t-conorm $\vee$. The hyperoperations depend on the sets Q and P. In the special case $Q = [q, 1]$, $P = [0, p]$, we have the hyperoperations denoted as $\wedge_q$ and $\vee_p$ Clearly, the new hyperoperations have a great potential for applications to computational intelligence, where they can extend the concepts and procedures of fuzzy reasoning.

We intend to further pursue our research especially in the following directions: a formulation of the orders $\leq_1, \leq_2$ in terms of the zeta function discussed in [13] and secondly an in-depth study of the multi-valued implication $\rightarrow^P$ along the lines of [9].

References

1. Athanasiadis IN (2007) The fuzzy lattice reasoning (FLR) classifier for mining environmental data. This volume, Chapter 9
2. Cornelis C, Deschrijver G, Kerre EE (2006) Advances and challenges in interval-valued fuzzy logic. Fuzzy Sets and Systems 157:622–627
3. Corsini P, Tofan I (1997) On fuzzy hypergroups. Pure Math Appl 8:29–37
4. Corsini P, Leoreanu V (2000) Hypergroups and binary relations. Algebra Universalis 43:321–330
5. Corsini P, Leoreanu V (2003) Applications of Hyperstructure Theory. Kluwer
6. Cripps A, Nguyen N (2007) Fuzzy lattice reasoning (FLR) classification using similarity measures. This volume, Chapter 13
7. Davey BA, Priestly HA (1990) Introduction to Lattices and Order. Cambridge University Press, Cambridge, UK
8. Graña M, Villaverde I, Moreno R, Albizuri FX (2007) Convex coordinates from lattice independent sets for visual pattern recognition. This volume, Chapter 6
9. Hatzimichailidis AG, Papadopoulos BK (2007) L-fuzzy sets and intuitionistic fuzzy sets. This volume, Chapter 16

360   A. Kehagias

10. Kehagias A (2003) L-fuzzy $\curlyvee$ and $\curlywedge$ hyperoperations and the associated L-fuzzy hyperalgebras. Rend Circ Mat Palermo 52:322–350
11. Kaburlasos VG (2004) FINs: Lattice theoretic tools for improving prediction of sugar production from populations of measurements. IEEE Trans Systems, Man and Cybernetics - B, Cybernetics 34(2):1017–1030
12. Kaburlasos VG, Petridis V (2000) Fuzzy lattice neurocomputing (FLN) models. Neural Networks 13(10):1145–1170
13. Knuth KH (2007) Valuations on lattices: fuzzification and its implications. This volume, Chapter 15
14. Konstantinidou M, Mittas J (1977) An introduction to the theory of hyperlattices. Math Balkanica 7:187–193
15. Massouros CG (1985) On the theory of hyperrings and hyperfields. Algebra i Logika 24:728–742
16. Mittas J, Konstantinidou M (1989) Sur une nouvelle generalisation de la notion de treillis: les supertreillis et certaines de leurs proprietes generales. Ann Sci Univ Clermont-Ferrand II Math 25:61–83
17. Petridis V, Kaburlasos VG (2003) FINkNN: a fuzzy interval number k-nearest neighbor classifier for prediction of sugar production. Journal of Machine Learning Research 4:17–37
18. Petridis V, Syrris V (2007) Machine learning techniques for environmental data estimation. This volume, Chapter 10
19. Ritter GX, Urcid G (2007) Learning in lattice neural networks that employ dendritic computing. This volume, Chapter 2
20. Serafimidis K, Kehagias A, Konstantinidou M (2004) The structure of the (P, Q)-superlattice and order related properties. Ital J Pure Appl Math 15:133–150
21. Serafimidis K, Kehagias A (2004) Some representation results for (P, Q)-superlattices. Ital J Pure Appl Math 15:151–164
22. Vougiouklis T (1991) The fundamental relation in hyperrings — the general hyperfield. Algebraic hyperstructures and applications 203–211. World Sci
23. Walker E, Walker C (2005) The algebra of fuzzy truth values. Fuzzy Sets and Systems 149:309–347

The Construction of Fuzzy-valued t-norms and t-conorms

Athanasios Kehagias

Faculty of Engineering, Aristotle University of Thessaloniki
kehagiat@auth.gr

Summary. In this paper we present a method to construct *fuzzy-valued* t-norms and t-conorms, i.e. operations which map pairs of lattice elements to fuzzy sets, and are commutative, associative and monotone. The fuzzy-valued t-norm and t-conorm are synthesized from their α-cuts which are obtained from families of *multi-valued* t-norms and t-conorms.

18.1 Introduction

We are interested in generalizations of the concepts of t-norm and t-conorm. In a companion chapter in this volume [14] we have presented a family of *hyper-t-norms* $\wedge_q$ and a family of *hyper-t-conorms* $\vee_p$. The prefix *hyper* is used to indicate *multi-valued* operations, also known as *hyperoperations* (see [5, 14]), i.e. operations which map pairs of elements to *sets* of elements. See [14] for the construction of hyperoperations which generalize t-norms and t-conorms. These hyperoperations are *crisp*, i.e. their output is a crisp set. A natural generalization is to consider *fuzzy* hyperoperations, i.e. operations which map elements to *fuzzy* sets.

Hence in this chapter we will present a procedure to construct fuzzy-valued t-norms and t-conorms.

Relatively little work in this direction has appeared in the literature. A pioneering paper is [4] which introduces fuzzy hypoperations which induce *fuzzy hypergroups*. A fuzzy hypergroup, different from the one used by Corsini, is [13] and a version of fuzzy min and fuzzy max operations appears in [11, 12][1].

As explained also in [14], our motivation to study multi-valued and fuzzy-valued connectives is that, while fuzzy logic is a calculus of uncertain reasoning, not much attention has been paid to the case where uncertainty is

---

[1] Let us also note that t-norms and t-conorms appropriate for the lattice of fuzzy-valued truth values are studied in [19, 20, 21]. But all of these works concern single- not multi-valued operations. (Also, t-norms and t-conorms for the lattice of real intervals are studied in [1, 2, 3]).

A. Kehagias: *The Construction of Fuzzy-valued t-norms and t-conorms*, Studies in Computational Intelligence (SCI) **67**, 361–370 (2007)
www.springerlink.com

associated with the actual operation of the logical connectives. This is exactly the situation we want to capture with the proposed fuzzy-valued t-norm and t-conorm.

There is some literature on the use of fuzzy *uni*-valued operations, in the context of type-2 fuzzy sets [9, 17] which is related to the present chapter.

The plan of this paper is as follows. In Sect. 18.2 we present some preliminary concepts. In Sect. 18.3 we construct the basic objects of our study, namely the fuzzy-valued t-norm λ and t-conorm Υ. In Sect. 18.4 we study the fuzzy-valued *S-implication* obtained from Υ. In Sect. 18.5 we summarize our results.

18.2 Preliminaries

In this section we review some fundamental concepts which will be used in the main part of the paper.

We will work with a *generalized deMorgan lattice* $(X, \leq, \vee, \wedge, ')$ (for the corresponding definition see [14] in this volume). We will study *lattice-valued fuzzy sets*, also termed *L-fuzzy sets* or, for the sake of brevity, simply *fuzzy sets*. We take these to be identical to their membership functions and we consider the special case where both the domain and range of the membership function is $(X, \leq, \vee, \wedge, ')$. In short, we define fuzzy sets as follows.

Definition 18.1 *An* L-fuzzy set *is a function* $M : X \to X$.

The collection of all L-fuzzy sets is denoted by $\mathbf{F}(X)$.

The α-*cut* of the fuzzy set M is denoted by M_α and defined by $M_\alpha \doteq \{x : M(x) \geq \alpha\}$. We will use the following properties of α-cuts (see [18]).

Proposition 18.1 *Given a fuzzy set* $M \in \mathbf{F}(X)$ *we have*

1. $M_0 = X$.
2. *For all* $p, q \in X$ *we have:* $p \leq q \Rightarrow M_q \subseteq M_p$.
3. *For all* $p, q \in X$ *we have:* $M_q \cap M_p = M_{p \vee q}$, *for all* $P \subseteq X$ *we have:* $\cap_{p \in P} M_p = M_{\vee P}$.

Proposition 18.2 *Suppose a family of sets* $\left\{ \widetilde{M_p} \right\}_{p \in X}$ *is given which satisfies*

1. $\widetilde{M_0} = X$.
2. *For all* $p, q \in X$ *we have:* $p \leq q \Rightarrow \widetilde{M_q} \subseteq \widetilde{M_p}$.
3. *For all* $p, q \in X$ *we have:* $\widetilde{M_q} \cap \widetilde{M_p} = \widetilde{M_{p \vee q}}$, *for all* $P \subseteq X$ *we have:* $\cap_{p \in P} \widetilde{M_p} = \widetilde{M_{\vee P}}$.

Then, defining for every $x \in X$

$$M(x) = \sup \left\{ p : x \in \widetilde{M_p} \right\},$$

we obtain the fuzzy set $M \in \mathbf{F}(X)$ *which, for every* $p \in X$, *satisfies* $M_p = \widetilde{M_p}$.

From the above propositions we see that a fuzzy set M is in a 1-to-1 correspondence with its α-cuts $\{M_p\}_{p \in X}$. A special class of fuzzy sets are the fuzzy intervals [10].

Definition 18.2 *A fuzzy set is called a* fuzzy interval *iff all its α-cuts are closed intervals. We denote the set of all fuzzy intervals of X by $\widetilde{\mathbf{I}}(X)$.*

In the rest of the paper we will deal with crisp and fuzzy *hyper*operations. Crisp hyperoperations map pairs of elements to crisp sets; fuzzy hyperoperations map pairs of elements to fuzzy sets. We will use the following.

Notation 18.1 *Let $*$ be a fuzzy operation and $a, b, x \in X$. We will denote the membership grade of x in the fuzzy set $a * b$ by $(a * b)(x)$.*

18.3 Fuzzy Valued t-norm and t-conorm

Our goal in the current section is to define *fuzzy-valued operations* which are analogous to t-norms and t-conorms. Recall the definition of the *hyperoperations* $\wedge_q$, $\vee_p$ (see [14] in this volume).

$$a \wedge_q b = [a \wedge b \wedge q, a \wedge b], \quad a \vee_p b = [a \vee b, a \vee b \vee p'].$$

The fuzzy-valued operations will be defined in terms of the *crisp hyperoperations* $\wedge_q$ and $\vee_p$. To this end, let us first prove the "α-cut properties" for $\wedge_q$ and $\vee_p$.

Proposition 18.3 *Take any $q_1, q_2 \in X$ and $R \subseteq X$. Then, for every $a, b, c \in X$ we have:*

1. $a \wedge_0 b = [0, a \wedge b]\,; a \wedge_1 b = [a \wedge b, a \wedge b]\,;$
2. $q_1 \leq q_2 \Rightarrow a \wedge_{q_2} b \subseteq a \wedge_{q_1} b;$
3. $(a \wedge_{q_1} b) \cap (a \wedge_{q_2} b) = a \wedge_{q_1 \vee q_2} b,\ \cap_{q \in R} (a \wedge_q b) = a \wedge_{\vee R} b.$

Proof. For 1   we have: $a \wedge_0 b = [a \wedge b \wedge 0, a \wedge b] = [0, a \wedge b]$, $a \wedge_1 b = [a \wedge b \wedge 1, a \wedge b] = [a \wedge b, a \wedge b]$.

For 2 we have $a \wedge_{q_2} b = [a \wedge b \wedge q_2, a \wedge b] \subseteq [a \wedge b \wedge q_1, a \wedge b] = a \wedge_{q_1} b$ since $a \wedge b \wedge q_1 \leq a \wedge b \wedge q_2$.

For the second (more general) part of 3:

$$\cap_{q \in R} (a \wedge_q b) = \vee_{q \in R} [a \wedge b \wedge q, a \wedge b] = [\vee_{q \in R} (a \wedge b \wedge q), a \wedge b]$$
$$= [a \wedge b \wedge (\vee_{q \in R} q), a \wedge b] = [a \wedge b \wedge (\vee R), a \wedge b] = a \wedge_{\vee R} b. \square$$

Proposition 18.4 *For $p_1, p_2 \in X$ and $\overline{P} \subseteq X$ and $a, b, c \in X$ we have:*

1. $a \vee_1 b = [0, a \vee b]\,; a \vee_1 b = [a \vee b, a \vee b]\,;$
2. $p_1 \leq p_2 \Rightarrow a \vee_{p_2} b \subseteq a \vee_{p_1} b;$
3. $(a \vee_{p_1} b) \cap (a \vee_{p_2} b) = a \vee_{p_1 \vee q_2} b;\ \cap_{p \in R} (a \vee_p b) = a \vee_{\vee R} b.$

Proof. The proof is similar to that of Proposition 18.3 and hence is omitted.
$\square$

We now construct the L-fuzzy hyperoperations $\curlyvee$ and $\curlywedge$. Following a standard approach, we will construct $\curlyvee$ and $\curlywedge$ through their α-cuts, which will be the $\vee_p$ and $\wedge_p$ families studied previously. First, for compatibility with the usual interpretation of α-cuts, we redefine for every a, b the symbols

$$a \wedge_0 b = [0, 1], \qquad a \vee_0 b = [0, 1].$$

Now, for every $a, b \in X$ we can define an L-fuzzy valued hyperoperation.

Definition 18.3 *For all $a, b \in X$*

1. *We define the L-fuzzy set $a \curlyvee b$ by defining for every $x \in X$: $(a \curlyvee b)(x) \doteq \vee \{q : x \in a \vee_q b\}$;*
2. *We define the L-fuzzy set $a \curlywedge b$ by defining for every $x \in X$: $(a \curlywedge b)(x) \doteq \vee \{q : x \in a \wedge_q b\}$;*

Proposition 18.5 *For all $a, b \in X$ and $p \in X$ we have $(a \curlyvee b)_p = a \vee_p b$, $(a \curlywedge b)_p = a \wedge_p b$.*

Proof. It follows from the construction of $a \curlyvee b$, $a \curlywedge b$ in Definition 18.3 (for details see [18]). $\square$

Proposition 18.6 *For all $a, b \in X$, the L-fuzzy sets $a \curlyvee b$ and $a \curlywedge b$ are L-fuzzy intervals.*

Proof. As already mentioned (Proposition 18.5), for any $p \in X$ the p-cut of $a \curlyvee b$ is $(a \curlyvee b)_p = a \vee_p b$ and by construction $a \vee_p b$ is an interval. The same is true for $a \curlywedge b$. $\square$

Before proceeding, we will need an auxiliary definition.

Definition 18.4 *Let $\circ : X \times X \to \mathbf{F}(X)$ be an L-fuzzy hyperoperation.*

1. *For all $a \in X$, $B \in \mathbf{F}(X)$ we define the L-fuzzy set $a \circ B$ by*

$$(a \circ B)(x) \doteq \vee \left([(a \circ b)(x)] \wedge B(b) \right).$$

2. *For all $A, B \in \mathbf{F}(X)$ we define the L-fuzzy set $A \circ B$ by*

$$(A \circ B)(x) \doteq \vee \left([(a \circ b)(x)] \wedge A(a) \wedge B(b) \right).$$

Remark 18.1. The above definition also applies to *crisp* operations if we take the view that a crisp operation gives as output not an element but an *indicator function*. An example should make this clear. Take the operation $\wedge$. For any $a, b \in X$ we can write that

$$(a \wedge b)(x) = \begin{cases} 1 \text{ iff } x = \inf(a,b) \\ 0 \text{ else.} \end{cases}$$

The same approach can be used for crisp *hyper*operations. In other words, we take crisp sets to be a special case of fuzzy sets and identify every set (crisp or fuzzy) with its membership function.

Remark 18.2. The above construction of the L-fuzzy valued hyperoperations is similar to the construction of fuzzy interval numbers (FINs) used in [7, 8].

Proposition 18.7 *For all $a, p \in X$, for all $A, B \in \mathbf{F}(X)$ we have*

1. $a \vee_p B_p \subseteq (a \curlyvee B)_p$; $A_p \vee_p B_p \subseteq (A \curlyvee B)_p$.
2. $a \wedge_p B_p \subseteq (a \curlywedge B)_p$; $A_p \wedge_p B_p \subseteq (A \curlywedge B)_p$.

Proof. We only prove the first part of 1, since the remaining items are proved similarly. Choose any $x \in a \vee_p B_p$. If $p = 0$, then, by definition, $(a \curlyvee B)_p = X$ and obviously $a \vee_0 B_0 \subseteq X$. If $p > 0$ then there is some $b \in B_p$ and so $B(b) \geq p$. Also $x \in a \vee_p b = (a \curlyvee b)_p$ implies that $(a \curlyvee b)(x) \geq p$. Hence

$$(a \curlyvee B)(x) = \vee_{u \in X}([(a \curlyvee u)(x)] \wedge [B(u)]) \geq [(a \curlyvee b)(x)] \wedge [B(b)] \geq p$$

which implies that $x \in (a \curlyvee B)_p$. We have thus $a \vee_p B_p \subseteq (a \curlyvee B)_p$. $\square$

Let us now prove some simple properties of $\curlyvee, \curlywedge$.

Proposition 18.8 *For all $a, b, c \in X$ the following hold.*

1. $(1 \curlyvee a)(1) = 1$, $(1 \curlywedge a)(a) = 1$, $(0 \curlyvee a)(a) = 1$, $(0 \curlywedge a)(0) = 1$.
2. $a \curlyvee b = b \curlyvee a$, $a \curlywedge b = b \curlywedge a$.
3. $a \vee_p b \vee_p c \subseteq (a \curlyvee (b \curlyvee c))_p \cap ((a \curlyvee b) \curlyvee c)_p$, $a \wedge_p b \wedge_p c \subseteq ((a \curlywedge b) \curlywedge c) \cap (a \curlywedge (b \curlywedge c))$.
4. $(a \curlyvee a)(a) = 1$, $(a \curlywedge a)(a) = 1$.
5. $(a \curlywedge b)(a \wedge b) = 1$, $(a \curlyvee b)(a \vee b) = 1$.
6. $[(a \curlywedge b) \vee a](a) = 1$, $[(a \curlyvee b) \wedge a](a) = 1$.
7. $((a \curlywedge b) \curlyvee a)(a) = 1$, $((a \curlyvee b) \curlywedge a)(a) = 1$.

Proof. 1. We have: $(1 \curlyvee a)(1) \doteq \vee\{q : 1 \in 1 \vee_q a\}$. Since $1 \in 1 \vee_1 a = [(1 \vee a), (1 \vee a) \vee 1']$, it follows that $1 \in \{q : 1 \in 1 \vee_q a\}$ and so $(1 \curlyvee a)(1) = 1$. The remaining parts of 1 are proved similarly.

2. This is obvious.

3. We apply Proposition 18.7.1 using $B = b \curlyvee c$; in this manner we show that $a \vee_p b \vee_p c = a \vee_p (b \vee_p c) = a \vee_p (b \curlyvee c)_p \subseteq (a \curlyvee (b \curlyvee c))_p$. Similarly $a \vee_p b \vee_p c \subseteq ((a \curlyvee b) \curlyvee c)_p$ and we are done.

4. Note that $a \in [a, a] = a \vee_1 a = (a \curlyvee a)_1$ and so $(a \curlyvee a)(a) \geq 1$. Similarly we show $(a \curlywedge a)(a) = 1$.

5. Note that $(a \curlywedge b)(a \wedge b) = \vee\{q : a \wedge b \in a \wedge_q b\} = 1$ (since $a \wedge b \in a \wedge_1 b$); the case $(a \curlyvee b)(a \vee b) = 1$ is proved similarly.

6. From 5 we have $(a \curlywedge b)(a \wedge b) = 1 > 0$. Also $[a \vee (a \wedge b)](a) = 1$. Now

$$[(a \curlywedge b) \vee a](a) = \vee_{z \in X} [(a \curlywedge b)(z)] \wedge [(z \vee a)(a)] \geq \tag{18.1}$$
$$[(a \curlywedge b)(a \wedge b)] \wedge [((a \wedge b) \vee a)(a)] = 1 \wedge 1 = 1 \tag{18.2}$$

Similarly we can prove $[(a \curlyvee b) \wedge a](a) = 1$.

7. We prove the first part as follows. We already have $(a \curlywedge b)(a \wedge b) = 1$ and

$$[((a \wedge b) \curlyvee a)(a)] = \vee_{z \in X} [(z \curlyvee a)(a)] \wedge [(a \wedge b)(z)] \geq \tag{18.3}$$
$$[((a \wedge b) \curlyvee a)(a)] \wedge [(a \wedge b)(a \wedge b)] = 1 \wedge 1 = 1; \tag{18.4}$$

hence

$$[(a \curlywedge b) \curlyvee a](a) = \vee [(a \curlywedge b)(z)] \wedge [(z \curlyvee a)(a)] \geq \tag{18.5}$$
$$[(a \curlywedge b)(a \wedge b)] \wedge [((a \wedge b) \curlyvee a)(a)] = 1 \wedge 1 = 1. \tag{18.6}$$

The second part is proved similarly. $\square$

Proposition 18.9 *For all $a, b, c, p \in X$ we have*

1. $a \vee_p (b \wedge_p c) \subseteq (a \curlyvee (b \curlywedge c))_p \cap ((a \curlyvee b) \curlywedge (a \curlyvee c))_p$.
2. $a \wedge_p (b \vee_p c) \subseteq (a \curlywedge (b \curlyvee c))_p \cap ((a \curlywedge b) \curlyvee (a \curlywedge c))_p$.

Proof. From Proposition 18.7.1 we have

$$a \vee_p (b \wedge_p c) \subseteq (a \curlyvee (b \curlywedge c))_p. \tag{18.7}$$

From Proposition 28 of [14], with $p = q$, we have

$$a\vee_p(b \wedge_p c) \subseteq (a \vee_p b)\wedge_p(a \vee_p c) = (a \curlyvee b)_p \wedge_p (a \curlyvee c)_p \subseteq ((a \curlyvee b) \curlywedge (a \curlyvee c))_p. \tag{18.8}$$

From (18.7) and (18.8) follows the first part of the proposition; the second part can be proved similarly. $\square$

Remark 18.3. Part 3 of Proposition 18.8 shows that the associativity of $\curlyvee, \curlywedge$ holds in a *limited* sense. Proposition 18.9 shows a *limited* form of distributivity. Both of these limitations can be traced to the fact that, in Proposition 18.7, we do not have equality of sets but set inclusion.

Proposition 18.10 *For all $a, b, c \in X$ we have*

$$\left. \begin{array}{l} a \curlyvee c = b \curlyvee c \\ a \curlywedge c = b \curlywedge c \end{array} \right\} \Rightarrow a = b.$$

Proof. We have

$$a \curlyvee c = b \curlyvee c \Rightarrow \left(\forall p \in X : (a \curlyvee c)_p = (b \curlyvee c)_p \right) \Rightarrow \tag{18.9}$$

$$(\forall p \in X : a \vee_p c = b \vee_p c) \Rightarrow a \vee_1 c = b \vee_1 c \Rightarrow a \vee c = b \vee c; \tag{18.10}$$

similarly $a \curlywedge c = b \curlywedge c \Rightarrow a \wedge c = b \wedge c$; finally, as is well known, in a distributive lattice we have

$$\left.\begin{array}{r} a \vee c = b \vee c \\ a \wedge c = b \wedge c \end{array}\right\} \Rightarrow a = b. \qquad \square$$

In [14] we have introduced an order $\leqslant_2$ on crisp intervals. We now extend this order to $\widetilde{\mathbf{I}}(X)$, the collection of all L-fuzzy intervals of X.

Definition 18.5 *For every $A, B \in \widetilde{\mathbf{I}}(X)$, we write $A \preceq B$ iff for all $p \in X$ we have $A_p \leqslant_2 B_p$.*

Proposition 18.11 $\preceq$ *is an order on $\widetilde{\mathbf{I}}(X)$ and $(\widetilde{\mathbf{I}}(X), \preceq)$ is a lattice.*

Proof. This follows from the fact that a fuzzy set is uniquely specified by its α-cuts. $\square$

The $\curlyvee$, $\curlywedge$ hyperoperations are monotone in the following sense.

Proposition 18.12 *For all $a, b \in X$ we have:* $a \leq b \Rightarrow \begin{cases} a \curlyvee c \preceq b \curlyvee c, \\ a \curlywedge c \preceq b \curlywedge c. \end{cases}$

Proof. Take any $p \in X$. Then

$$a \leq b \Rightarrow \left\{\begin{array}{c} a \vee c \leq b \vee c \\ (a \vee c) \vee p' \leq (b \vee c) \vee p' \end{array}\right\} \Rightarrow a \vee_p c \preceq b \vee_p c \Rightarrow (a \curlyvee c)_p \preceq (b \curlyvee c)_p.$$

Since the above is true for every $p \in X$, it follows that $a \curlyvee c \preceq b \curlyvee c$. Similarly we show that $a \curlywedge c \preceq b \curlywedge c$. $\square$

$\curlyvee$, $\curlywedge$ and $'$ are interrelated as seen by Proposition 18.14.

Definition 18.6 *For every $A \in \mathbf{F}(X)$ define A' by its α-cuts, i.e. A' is the (unique) fuzzy set which for every $p \in X$ satisfies*

$$(A')_p = (A_p)' = \{x'\}_{x \in A_p}.$$

Proposition 18.13 *If A is a fuzzy interval, then A' is also a fuzzy interval.*

Proof. Take any $p \in X$. Suppose that $A_p = [a_1, a_2]$. Then

$$(A')_p = (A_p)' = \{x' : a_1 \leq x \leq a_2\} = [a_2', a_1'] . \qquad \square$$

Proposition 18.14 *For every $a, b \in X$ we have:*

$$(a \curlyvee b)' = a' \curlywedge b', \qquad (a \curlywedge b)' = a' \curlyvee b'.$$

Proof. Choose any $p \in X$. Then

$$\left((a \curlyvee b)'\right)_p = \left((a \curlyvee b)_p\right)' = (a \vee_p b)' = a' \wedge_p b' = (a' \curlywedge b')_p .$$

Since for all $p \in X$ the fuzzy sets $(a \curlyvee b)'$ and $a' \curlywedge b'$ have the same cuts, we have $(a \curlyvee b)' = a' \curlywedge b'$. $\square$

In conclusion, we have constructed a *fuzzy hyperoperation* λ. Let us now see in what sense it can be called a fuzzy t-norm. To see the similarity consider the following Table 18.1, which compares the properties of any crisp t-norm T with those of λ; the properties of λ are obtained from Proposition 18.8.

Table 18.1. This list compares a *crisp t-norm* T properties with those of *fuzzy t-norm* λ

A crisp t-norm T	The fuzzy t-norm λ
$aTb = bTa$	$a \lambda b = b \lambda a$
$(aTb)Tc = aT(bTc)$	$(a \lambda b) \lambda c = a \lambda (b \lambda c)$
$a = aT1$	$(a \lambda 1)(a) = 1$
$a \leq b \Rightarrow aTc \leq bTc$	$a \leq b \Rightarrow a \lambda c \preceq b \lambda c$

A similar table can be used to compare Υ with some t-conorm S. The analogies between the "classical" operations T, S and the fuzzy-valued hyperoperations λ, Υ are obvious. Hence we can justifiably say that λ is a fuzzy t-norm and Υ is a fuzzy t-conorm.

18.4 Fuzzy-valued S-implication

We can also construct a fuzzy hyperoperation which behaves like an S-implication. This is done as follows.

Definition 18.7 *The fuzzy implication is denoted by $\rightsquigarrow$ and defined for every* $a, b \in X$ *by*

$$(a \rightsquigarrow b) = (a' \Upsilon b).$$

It is easy to prove the following.

Proposition 18.15 *For all* $a, b, p \in X$ *we have*

$$(a \rightsquigarrow b)_p = a \rightarrow_p b.$$

Proof. Indeed $(a \rightsquigarrow b)_p = (a' \Upsilon b)_p = a' \vee_p b = a \rightarrow_p b.$ $\square$

From Proposition 18.15 and the properties of $\vee_p$, described in Sect. 3 of [14] we can immediately prove the following proposition, which summarizes the properties of $\rightsquigarrow$; it can be easily seen that these are analogous to the classical implication.

Proposition 18.16 *We have for every* $a, b, c \in X$ *the following.*

1. $a \leq b \Rightarrow ((a \rightsquigarrow c) \succeq (b \rightsquigarrow c)$ *and* $(c \rightsquigarrow a) \preceq (c \rightsquigarrow b))$,
2. $(0 \rightsquigarrow a)(1) = 1$, $\left(1 \rightarrow^P a\right)(a) = 1$,
3. $(a \rightsquigarrow b) = (b' \rightsquigarrow a')$.

Proof. Straightforward. $\square$

Remark 18.4. The above are similar to the properties proved in [6] and other places about (crisp, uni-valued) implications.

18.5 Conclusion

We have introduced two fuzzy-valued hyperoperations, λ and $\curlyvee$, which are natural generalizations of the t-norm $\wedge$ and the t-conorm $\vee$. Clearly, the new (crisp and fuzzy) hyperoperations have a great potential for applications to computational intelligence, where they can extend the concepts and procedures of fuzzy reasoning.

In particular, the definitions and results of Sect. 18.4 constitute a first step in the study of fuzzy-valued implications; in the future we plan to work further in this direction and apply the resulting implications to the study of *fuzzy cognitive maps* [15, 16].

References

1. Deschrijver G, Kerre EE (2003) On the relationship between some extensions of fuzzy set theory. Fuzzy Sets and Systems 133:227–235
2. Deschrijver G, Kerre EE (2005) Implicators based on binary aggregation operators in interval-valued fuzzy set theory. Fuzzy Sets and Systems 153:229–248
3. Cornelis C, Deschrijver G, Kerre EE (2006) Advances and challenges in interval-valued fuzzy logic. Fuzzy Sets and Systems 157:622–627
4. Corsini P, Tofan I (1997) On fuzzy hypergroups. Pure Math Appl 8:29–37
5. Corsini P, Leoreanu V (2003) Applications of Hyperstructure Theory. Kluwer
6. Hatzimichailidis AG, Papadopoulos BK (2007) L-fuzzy sets and intuitionistic fuzzy sets. This volume, Chapter 16
7. Kaburlasos VG (2006) Towards a Unified Modeling and Knowledge-Representation Based on Lattice Theory — Computational Intelligence and Soft Computing Applications, ser Studies in Computational Intelligence 27, Springer, Heidelberg, Germany
8. Kaburlasos VG, Kehagias A (2006) Novel fuzzy inference system (FIS) analysis and design based on lattice theory — part I: working principles. Int J of General Systems 35:45–67
9. Karnik NN, Mendel JM (2001) Operations on type-2 fuzzy sets. Fuzzy Sets and Systems 122:327–348
10. Kehagias A (2002) The lattice of fuzzy intervals and sufficient conditions for its distributivity, Arxiv preprint cs.OH/0206025
11. Kehagias A (2003) Lattice-fuzzy meet and join hyperoperations. In: Proc Algebraic hyperstructures and applications pp 171–182
12. Kehagias A (2003) L-fuzzy $\curlyvee$ and λ hyperoperations and the associated L-fuzzy hyperalgebras. Rend Circ Mat Palermo 52:322–350
13. Kehagias A, Serafimidis K (2005) The L-fuzzy Nakano hypergroup. Inform Sci 169:305–327
14. Kehagias A (2007) A family of multi-valued t-norms and t-conorms. This volume, Chapter 17
15. Kim HS, Lee KC (1998) Fuzzy implications of fuzzy cognitive map with emphasis on fuzzy causal relationship and fuzzy partially causal relationship. Fuzzy Sets and Systems 97:303–313

16. Marchant T (1999) Cognitive maps and fuzzy implications. European Journal of Operational Research 114:626–637
17. Liang Q, Mendel JM (2000) Interval type-2 fuzzy logic systems: theory and design. IEEE Trans on Fuzzy Systems 8:535–550
18. Nguyen HT, Walker EA (1999) A First Course in Fuzzy Logic. Chapman & Hall/CRC
19. Walker E, Walker C (2003) Algebraic structures on fuzzy sets of type-2. In: Proc Int Conf on Fuzzy Information Processing: Theories and Applications pp 97–100
20. Walker E, Walker C (2003) t-norms on fuzzy sets of type-2. In: Proc 24th Linz Seminar on Fuzzy Set Theory: Triangular Norms and Related Operators in Many-Valued Logics
21. Walker E, Walker C (2005) The algebra of fuzzy truth values. Fuzzy Sets and Systems 149:309–347

Index

σ-FLNMAP, 226, 227
ε-insensitive zone, 201, 203

Active contours, 115
Aggregation operator, 289
Air quality assessment, 183
  daily vegetation index, 185
  ozone levels estimation, 187
Algorithms
  binary pattern recall using the
    max-memory M, 92
  binary pattern recall using the
    min-memory W, 91
  Multi-class SLLP training, 36
  real valued pattern recall using the
    max-memory M, 96
  real valued pattern recall using the
    min-memory W, 96
  SLLP training by elimination, 31
  SLLP training by merging, 33
ART, 233
ART over-training, 234
Artificial neural networks
  based on lattice theory, 25
Associative memory, 156
  fuzzy morphological, 163
  gray-scale morphological, 150
  heteroassociative morphological, 154
  implicative fuzzy, 150
  Lukasiewicz fuzzy morphological,
    164–166
  morphological, 149, 150, 153
Associativity Equation, 313

Autoassociative memory
  gray-scale morphological, 154, 159
Automatic recognition, 215, 219
AVHRR, 215, 217, 219

Batch history data, 197
Bayesian Inference, 320
Bi-valuation, 315
Biological neural networks
  dendrites, 25
  synapse, 25
Borel algebra, 291

Capacity control, 201, 203
Category, 60–69, 71, 72, 74, 75
  categorical model theory, 61, 65, 74
Category function, 265
Category proliferation problem, 234
Chain, 10
Chebyshev distance, 29, 32, 34, 38, 41
Ciao Prolog, 297
Cleveland heart, 279
Closed World Assumption, 288
CLP, 297
CLP($\mathcal{R}$), 297
Co-valuation, 315
Colimit, 64, 65, 68–72, 75
Colour vector ordering
  in the HSV colour model, 135
  in the Lab colour model, 137
  in the RGB colour model, 134
Complement coding, 17
Concept, 63–69, 71–75
Cone, 13, 15

Constraint logic programing, 297
Content based image retrieval (CBIR),
     102, 109, 124
   shape, 115
Convex optimization problem, 201
Correlation based features selection,
     221

Decision boundaries, 47, 54
Default value, 291
Degree of inclusion, 9, 317
Diagonal, 11
Diagram, 62–65, 68, 69, 71, 73, 75
   base, 64, 65, 68, 69, 71, 75
   commutative, 62, 63, 69, 71, 72, 75
   defining, 64, 69, 71
Differential evolution, 268, 282
Dilation, 130, 132
Distributivity, 343, 350, 351, 354, 366,
     367
Dual variables, 202

Ellipsoidal ARTMAP, 233
Endmember induction, 105
   convex cone analysis, 108, 114
   morphologically independent sets,
     107, 109, 114, 122
Endmember induction
   convex cone analysis, 108
ENVISAT, 215
Erosion, 130, 132
Evolution of ART Architectures, 235

Feature extraction, 101, 102, 112, 115,
     124
   convex coordinates, 103
   linear methods, 104, 117, 122
Filter, 343, 348–351
FIN, 13, 365
   representation, 16
Fit, 265
   first, 265, 271
   maximal, 265, 271
   ordered, 265, 271, 274
   selective, 265, 271
   tightest, 265, 271

FL-framework, 196, 197
FLN, 176, 197
FLNN, 225
FLR, 16, 226
FLR classifier, 179
   induction, 180
Functor, 72–75
Fuzzy lattice, 9, 198
Fuzzy lattice rule
   definition, 179
   degree of truth, 179
   engine, 179
   generalization, 181
   induction, 180
Fuzzy set, 362, 363, 365, 367
   interval valued, 341
   type-2, 362
Fuzzy-ART, 4
Fuzzy-ARTMAP, 6, 233, 239

GA category delete operator, 247
GA parameter selection process, 251
GA parameters for ART evolution, 248
GART: Genetically engineered ART,
     254, 255, 258, 261
Gaussian ARTMAP, 233
GEAM, 236, 248, 258, 261
Generalized intervals, 11
Genetic algorithm, 234, 267, 281
Genetic design of ART, 243
Genetic encoding of ART, 244
GFAM, 236, 248, 258, 261
GGAM, 236, 248, 258, 261
Granular computing, 3
Graphic expert system, 224, 227, 230
grSOM, 18

Hasse diagram, 8
Heteroassociative memory
   morphological, 160
Hydroelectric plant
   streamflow, 166
Hyperalgebra, 342, 353
Hyperbox, 10, 16, 48, 51, 239, 241, 270,
     272, 274, 277, 342
   volume maximization, 50
Hyperoperation, 342, 349, 359, 361, 363,
     364, 367–369

Hyperspectral image, 103, 124
  AVIRIS, 114
  CBIR, 109
    similarity measure, 110
  endmember induction, 105, 107
  supervised segmentation, 102, 112
  unsupervised segmentation, 102

Image analysis, 219, 221
Image denoising, 145
Image processing, 52
Image segmentation, 159
Implication
  $\rightarrow$, 335
  fuzzy, 331
    $g_h(a, b)$, 333
  fuzzy-valued, 362, 368
  hyper-implication, 352
  multi-valued, 343, 352
Incidence algebra, 310
Inclusion measure, 9, 197, 198
Information granule, 4
Interval, 341–343, 346–351, 354, 356
  fuzzy, 363, 364, 367
Interval-aggregation, 290
Intuitionistic fuzzy sets, 326
  geometric interpretations, 327
  implications, 334
  operations, 328
  tautological, 335

Kernel, 203
Knowledge coherence, 73, 75
Knowledge representation, 64, 65, 72–75

Lattice, 59–62, 65, 67, 74, 75, 197
  algebra-based definition, 7
  Boolean, 316, 344, 354
  complete, 149, 151, 152, 163, 197
  convex sublattice, 343, 346
  deMorgan, 342, 344, 354, 362
  distributive, 316, 343, 344, 351, 367
  hyperlattice, 342
  of intervals, 341, 361
  of type-2 fuzzy sets, 341
  operators, 46, 47
  order-based definition, 7
  product, 176
  sublattice, 343

Lattice associative memories
  activation function, 38, 39
  almost perfect recall, 91, 95
  binary masking examples, 88
  dendritic type, 37, 42, 43
  fixed points, 86
  gray-scale masking examples, 90
  hidden neurons, 38
  masking rules, 88
  matrix type, 81, 82, 85
  max-memory M, 85
  min-memory W, 85
  noise, 39, 87
  noise masking, 88, 90, 91, 98
  noise parameter, 38
  pattern associations, 37, 85
  perfect recall, 39, 86, 87, 91
  recall performance examples, 40, 93, 96
  recall response, 86
  robustness, 40, 87, 94, 97
  storage capacity, 37, 43, 82
Lattice independent, 101, 107, 117
Lattice interval, 342
Lattice neural networks, 25
  orthonormal basis, 49, 50
  standard basis, 48, 49
Lattice perceptrons
  activation function, 27
  dendritic computing, 26
  learning by elimination, 29, 31
  learning by merging, 29, 33
  multiple class learning, 36
  single class learning, 28
  single layer, 26, 28, 42
  single neuron computation, 26
  training examples, 32, 35
Limit, 64, 65, 68, 71, 72, 75
Linear space, 13
Local orientation, 50, 54
Logic, 59–61, 65, 66
  categorical, 61, 64
  fuzzy, 59
  geometric, 59–61, 74
Logical connective, 341
  multi-valued, 342

Möbius function, 310, 314
Matrix
  exponential, 51, 58
  rotation, 50
  skew symmetric, 52, 58
Measure
  cosine distance, 266, 271
  cosine similarity, 266
  delta, 270–273, 281, 282
  distance, 263–265
  inclusion, 264, 273
  metric, 266
  semimetric, 266
  sigma, 269, 282
  similarity, 263, 265
  weighted cosine, 271, 282
micro-ARTMAP, 254, 258, 261
Minimax algebra, 83
  fixed point, 83
  linear combination, 84
  linear span, 84
  matrix operations, 83
  matrix transform, 86
  max-product, 84
  min-product, 84
  outer product, 84
Minimum and maximum operators
  in the HSV colour model, 138
  in the Lab colour model, 139
Modelling of digital images, 129
Modus Ponens, 355
Modus Tollens, 355
Monotonicity, 344, 346–348, 352, 353,
    361, 367
Morphism, 62–69, 71–75
Morphology
  greyscale
    threshold approach, 132
  binary, 130
  colour
    component-based approach, 134
    vector-based approach, 142
  greyscale
    fuzzy approach, 132
  mathematical, 129, 149, 151, 164
  fuzzy, 163
Moving Average, 205
Multi-valued
  operation, 342

Naive Bayes, 222, 224, 226, 228
Natural transformation, 73, 75
NEFCLASS, 225, 227
NEFPROX, 225
Neural network, 59, 61, 65, 72–75
Neural networks
  fuzzy morphological, 164
  morphological, 149, 150
Normalized absolute error, 95
Normalized Hamming distance, 91
Normalized mean squared error, 95

Object, 62–64, 66, 69, 70, 72–75
  apical, 63, 69, 71, 75
Occam razor, 18
Ocean structures, 215, 217
Ontology, 64, 72, 75
Open World Assumption, 288
Operations between colours
  in the HSV colour model, 139, 141
  in the Lab colour model, 140, 141
  in the RGB colour model, 139, 140
Order, 341–345, 347, 349, 350, 353, 355,
    359
Order Theory, 309
Orthonormal basis, 49, 50

Particle swarm optimization, 268
Perception, 65
Persistence, 205
Prediction horizon, 197
Principle
  inclusion-exclusion, 314
  of Group Invariance, 321
  of Induction, 312
  of Maximum Entropy, 321
Product rule, 319
Prolog, 287

Remote sensing, 215
Robot visual localization, 102, 120, 125

SAR images, 145
Segmentation, 221
Semi-supervised ART, 254
SOM, 5
Sort, 66, 69
Spatial correlation, 195, 196
ssEAM, 248, 254, 258, 261

ssFAM, 248, 254, 258, 261
ssGAM, 248, 254, 258, 261
Superlattice, 341, 343, 353, 354
SV models, 204
SVR, 196, 201
    linear regression, 201
    nonlinear regression, 203
Syllogistic reasoning, 355

T-conorm, 331
T-norm, 330
    fuzzy-valued, 361–363, 368, 369
    hyper-t-norm, 341, 342, 345, 361, 369
    multi-valued, 342, 345, 359, 361
    multivalued, 342
Tabu search, 267, 282
Topological space, 61, 74
Tree Augmented Naive Bayes, 224

Uncertainty, 287
Union-aggregation, 290

Valuation, 197, 312, 321
    hyperplane, 271
    hypersphere, 269
    positive, 9, 264, 265
    positive hyperplane, 265
    unit, 265, 271
Valuation function, 177
    positive, 177
    sigmoid, 182
Vigilance parameter, 199, 242, 261

Wind prediction, 195

Zeta function, 310, 317

Printing: Krips bv, Meppel
Binding: Stürtz, Würzburg